CONCEPTUAL PHYSICS

Laboratory Manual
Teacher's Edition

Paul Robinson
San Mateo High School
San Mateo, California
Illustrated by Paul G. Hewitt

PAUL G. HEWITT

Needham, Massachusetts
Upper Saddle River, New Jersey
Glenview, Illinois

About the Author

Paul Robinson has been teaching physics since 1974 after getting his BA in Physics at the University of California, Santa Barbara. He completed his MA in Physics at Fresno State University in 1984. In 1983 Robinson was recruited to assist with the design and opening of Edison Computech—a magnet science, computer, and math school for grades 7–12 in Fresno. Encouraged by the administration to be innovative, he put a conceptual physics course at the *beginning* of the science sequence instead of at the *end*. This untraditional approach is now gaining nationwide acceptance. In 1987, Robinson received the Presidential Award for Excellence in Science Teaching. He has also received the Distinguished Service Citation from the Northern California/Nevada Section of the American Association of Physics Teachers and was the first high school teacher elected president of that association. Robinson now teaches physics and astronomy at San Mateo High School in San Mateo, California.

Contributors

Roy Unruh
University of Northern Iowa
Cedar Falls, Iowa

Tim Cooney
Price Laboratory School
Cedar Falls, Iowa

Clarence Bakken
Gunn High School
Palo Alto, California

Consultants

Kenneth Ford
Germantown Academy
Fort Washington, Pennsylvania

Jay Obernolte
University of California
Los Angeles, California

Cover photograph: Motor Press Agent/Superstock, Inc.

ISBN 0-13-054258-X
1 2 3 4 5 6 7 8 9 10 05 04 03 02 01

Contents of Teacher Section

The teacher section concludes with a Teacher Feedback and Evaluation Form. Your feedback will help to keep future editions in touch with teacher needs.

Using the Manual

For maximum benefit, please invest the time to read over this prefatory material before using this lab manual. It will save time and effort and will increase your effectiveness.

Activities and Experiments

This lab manual contains 61 activities and 38 experiments. Most activities are designed to provide hands-on experience that relates to a specific concept. Experiments are usually designed to give practice using a particular piece of apparatus. The goals for activities and experiments, while similar, are in some ways different.

The chief goal of activities is to acquaint students with a particular physical phenomenon that they may or may not already know something about. The emphasis during an activity is on observing relationships, identifying variables, and developing tentative explanations of phenomena in a qualitative fashion. In some cases, students are asked to design experiments or formulate models that lead to a deeper understanding.

Experiments are more quantitative in nature and generally involve acquiring data in a prescribed manner. Here, a greater emphasis is placed on learning how to use a particular piece of equipment, making measurements, identifying and estimating errors, organizing data, and interpreting data.

Learning Cycles

This manual is more than an assembly of clever or attractive activities and experiments. It utilizes a particular learning strategy called a *learning cycle.* The *Conceptual Physics* program puts learning the concept before any mathematical problem solving; the laboratory part of the *Conceptual Physics* program puts experiencing physical phenomena before attempting to quantify it. Students do activities that familiarize them with the concepts and relationships before they attempt to graph or otherwise analyze their data. This procedure avoids the frustration that students feel when they are placed in a laboratory situation where they are making measurements of things they don't understand.

The learning cycle includes three phases: exploratory, concept development, and application. Many of the labs in this manual have been adapted from PRISMS (Physics Resources and Instructional Strategies for Motivating Students). For more information about this excellent program, contact Roy Unruh, Director, PRISMS, University of Northern Iowa, Cedar Falls, Iowa 50614.

Exploratory. Exploratory activities encourage students to observe relationships, identify variables, and develop tentative explanations of phenomena. The laboratory should be concluded before the students' interest begins to subside—some may span only 10 to 20 minutes.

Concept Development. Concept development should be based on experiences acquired from exploratory activities. Students are more prepared to be receptive to a concept if they have engaged directly in a concrete experience that has raised an issue in their minds that creates a need for further understanding. The nonlaboratory instructional methods for the concept-development phase can include assigned reading, individualized or group instruction, audiovisuals, and problem sets. In concept-development labs, the emphasis is on verifying a particular concept or relationship, rather than on introducing it.

Application. The emphasis of an application laboratory is not to discover or verify, but rather to *use* the concept. Application labs are likely to be more structured with a more specific purpose. The ideal application lab would allow students to analyze a phenomenon using the simple models that were first formulated during an exploratory lab and clarified during concept development. Depending on the ability and interest of your students, and on whether time permits, you may wish to follow up concept development with an application lab.

Not all physics topics lend themselves to qualitative exploratory activities that can be followed by quantitative concept-development labs. But many—such as motion, heat, light, energy, vibrations and waves, and electricity—do. Keep this in mind as you plan your units.

Teacher Lab Planning Notes

A quick thumbnail sketch of the appropriateness of a particular lab for your circumstances is provided in the teacher lab planning notes, which are found in the margin of the first page of every lab.

Setup. This is an estimate of the time you will need to gather the equipment and supplies (provided you have them) and assemble them on a cart or other manner in your classroom or lab. A plain numeral 1 means that the setup takes approximately one period of about 55 minutes. The symbols <1 and >1 mean less than and greater than one class period, respectively. When the equipment requires some first-time assembly (such as in Activity 31, "Where's Your CG?"), the indicated prep time is always greater than one class period. Once the equipment has been assembled, of course, it should take considerably less time thereafter.

Lab Time. Most labs *always* take longer than expected due to unforeseen interruptions, equipment maladies, shortages, student absences, and so on. A plain numeral 1 indicates that the lab may be expected to take one period of about 55 minutes. The symbol <1, less than one period, means anywhere from 10 minutes to 40 minutes. The symbol >1, greater than one period, means anywhere from $1\frac{1}{2}$ lab periods to several. The actual time you allot will depend on you, your students, and your supply of equipment.

Learning Cycle. This indicates what phase of the learning cycle the lab is intended to serve: exploratory, concept development, or application.

Conceptual Level. This roughly identifies the level of difficulty of the concepts explored in the lab material. The conceptual level is rated as *easy, moderate,* or *difficult.* The ability to handle moderate and difficult concepts will vary a great deal from class to class as well as from student to student.

Mathematical Level. Conceptual physics does *not* mean physics without math—rather, it is *physics in which the comprehension of concepts precedes computation.* Lab experiments do not avoid computation. *Easy* usually means that the lab involves reasoning or computation with direct proportions or linear equations. *Moderate* might mean using the direct or inverse-square relationships. *Difficult* usually means reasoning using more than one variable.

Course Planning

When planning your course, you must choose what topics from the text you want to cover. *Nobody* is expected to do all or even most of the 99 labs in this manual. A reasonable average is somewhere between 25 and 40. Which labs you choose will depend on your preferences and background, your students' abilities, and availability of equipment and supplies. Because *all* students, no matter what level and ability, have a variety of conceptual misconceptions that impede comprehension, *it is strongly urged that in no case should exploratory activities be dropped or deleted in order to save time; they form the bedrock on which the foundation of the course is to be built.* As your familiarity with the material grows, your efficiency will increase, allowing you to do the exploratories prior to class discussion or lectures.

The Laboratory Planning Guide is a chart that can help you select what labs are appropriate for your students. It correlates each lab to a chapter in the *Conceptual Physics* text, lists what phase of the learning cycle it serves, and gives its conceptual and mathematical levels. For a brief summary of the *content* of the labs, see the table of contents in the *student* portion of the lab manual.

If you have open lab during lunch or after school, you might wish to permit students to do labs that you can't fit into your class schedule for extra credit.

Most labs should be done when studying the corresponding chapter in the text. There are two possible major exceptions. If you have a computer in the laboratory, "Computerized Gravity" (Experiment 37) makes an excellent concept-development experiment during the study of acceleration. The graphing and data reduction skills acquired by doing "Trial and Error" (Experiment 40) are powerful tools students will use all year long. It is recommended that this lab be done just as soon as students have had a chance to do some graphing by hand, so they will appreciate the ease, elegance, and power of the computer. For those anxious to introduce the use of the computer in their course, "Trial and Error" can be done right after "Making Hypotheses" (Activity 1) as an example of the scientific method in Chapter 1 of the text.

Lab Reports

Some activities are so short they require very little or no formal write-up. On the other hand, a formal write-up for some of the more complex experiments with great amounts of data is appropriate. How many reports you require your students to do is discretionary on your part, of course, but it will undoubtedly be necessary to spend time tutoring your students on how to write their reports. Do not take the joy out of a 15-minute exploratory lab by requiring a 45-minute report. In place of having your students write a lab report that does not substantially add to their experience, consider conducting a class discussion instead. You might make some reports optional or extra credit.

Laboratory Planning Guide

The following chart gives an overview of all the labs in the manual. The labs are grouped by the chapter in the accompanying text that correlates with them. The chart indicates whether the lab is an activity or an experiment, how it fits into the learning cycle, and its conceptual and mathematical levels. (For a full explanation of these terms, see pages T5–T6.) The chart uses the following abbreviations for these terms.

Activity vs. Experiment	*Learning cycle*	*Conceptual level and Mathematical level*	
Act = Activity	Explor = Exploratory	0 = none	2 = moderate
Exp = Experiment	Conc Dev = Concept development	1 = easy	3 = difficult
	Applic = Application		

	Learning Cycle	**Conceptual Level**	**Mathematical Level**
1 About Science			
Act 1 Making Hypotheses	Explor	1	0
Unit I Mechanics			
2 Linear Motion			
Act 2 The Physics 500	Explor	1	1
Act 3 The Domino Effect	Conc Dev	1	1
Exp 4 Merrily We Roll Along!	Conc Dev	3	2
Act 5 Conceptual Graphing	Conc Dev	2	0
Act 6 Race Track	Conc Dev	1	0
3 Projectile Motion			
Exp 7 Bull's Eye	Applic	3	3
4 Newton's First Law of Motion—Inertia			
Act 8 Going Nuts	Explor	1	0
Act 9 Buckle Up!	Applic	1	0
Act 10 24-Hour Towing Service	Explor	3	2
5 Newton's Second Law of Motion—Force and Acceleration			
Act 11 Getting Pushy	Explor	2	1
Exp 12 Constant Force and Changing Mass	Conc Dev	2	3
Exp 13 Constant Mass and Changing Force	Conc Dev	3	3
Exp 14 Impact Speed	Applic	3	3
Exp 15 Riding with the Wind	Conc Dev	3	2
6 Newton's Third Law of Motion—Action and Reaction			
Act16 Balloon Rockets	Explor	1	0
Exp 17 Tension	Conc Dev	2	0
Exp 18 Tug-of-War	Conc Dev	2	0
7 Momentum			
Act 19 Go Cart	Explor	2	2
Exp 20 Tailgated by a Dart	Applic	2	2
8 Energy			
Act 21 Making the Grade	Explor	1	1
Act 22 Muscle Up!	Explor	1	1
Act 23 Cut Short	Explor	1	0
Exp 24 Conserving Your Energy	Conc Dev	2	2
Exp 25 How Hot Are Your Hot Wheels®?	Applic	1	2
Exp 26 Wrap Your Energy in a Bow	Applic	2	2
Exp 27 On a Roll	Conc Dev	2	2
Exp 28 Releasing Your Potential	Explor	3	3
Exp 29 Slip-Stick	Conc Dev	2	2
9 Circular Motion			
Act 30 Going in Circles	Explor	3	0
10 Center of Gravity			
Act 31 Where's Your CG?	Explor	3	0
11 Rotational Mechanics			
Act 32 Torque Feeler	Explor	1	0

	Learning Cycle	Conceptual Level	Mathematical Level
Exp 33 Weighing an Elephant	Conc Dev	2	2
Exp 34 Keeping in Balance	Conc Dev	2	2
Act 35 Rotational Derby	Conc Dev	2	2
12 Universal Gravitation (none)			
13 Gravitational Interactions			
Exp 36 Acceleration of Free Fall	Explor	2	3
Exp 37 Computerized Gravity	Conc Dev	2	3
Act 38 Apparent Weightlessness	Explor	1	0
14 Satellite Motion			
Act 39 Getting Eccentric	Explor	1	0
Exp 40 Trial and Error	Conc Dev	3	3
15 Special Relativity—Space and Time (none)			
16 Special Relativity—Length, Momentum, and Energy (none)			
Unit II Properties of Matter			
17 The Atomic Nature of Matter			
Act 41 Flat as a Pancake	Explor	1	2
Exp 42 Extra Small	Applic	2	3
18 Solids			
Exp 43 Stretch	Conc Dev	2	2
Act 44 Geometric Physics	Explor	3	1
19 Liquids			
Act 45 Eureka!	Explor	2	2
Exp 46 Sink or Swim	Conc Dev	3	2
20 Gases			
Act 47 Weighty Stuff	Explor	2	2
Act 48 Inflation	Applic	1	2
Unit III Heat			
21 Temperature, Heat, and Expansion			
Act 49 Heat Mixes: Part I	Explor	1	1
Act 50 Heat Mixes: Part II	Explor	2	2
Exp 51 Antifreeze in the Summer?	Applic	3	3
Act 52 Gulf Stream in a Flask	Explor	1	0
Exp 53 The Bridge Connection	Applic	3	3
22 Heat Transfer			
Act 54 Cooling Off	Applic	3	3
Exp 55 Solar Equality	Applic	2	2
Act 56 Solar Energy	Applic	2	2
23 Change of Phase			
Act 57 Boiling Is a Cooling Process	Explor	1	0
Exp 58 Melting Away	Conc Dev	3	2
Exp 59 Getting Steamed Up	Conc Dev	2	2
Act 60 Changing Phase	Conc Dev	2	1
Act 61 Work for Your Ice Cream	Conc Dev	2	1
Act 62 The Drinking Bird	Applic	2	0
24 Thermodynamics			
Exp 63 The Uncommon Cold	Conc Dev	2	2
Unit IV Sound and Light			
25 Vibrations and Waves			
Act 64 Tick-Tock	Explor	2	1
Exp 65 Grandfather's Clock	Explor	2	2
Act 66 Catch a Wave	Explor	2	2
Act 67 Ripple While You Work	Conc Dev	2	1
26 Sound			
Act 68 Chalk Talk	Explor	1	1

	Learning Cycle	Conceptual Level	Mathematical Level
Exp 69 Mach One	Conc Dev	2	2
27 Light			
Act 70 Shady Business	Explor	1	0
Exp 71 Absolutely Relative	Conc Dev	2	3
Act 72 Shades	Explor	2	0
28 Color			
Exp 73 Flaming Out	Conc Dev	1	0
29 Reflection and Refraction			
Act 74 Satellite TV	Applic	1	0
Act 75 Images	Explor	2	0
Act 76 Pepper's Ghost	Conc Dev	2	0
Act 77 The Kaleidoscope	Applic	1	1
Exp 78 Funland	Conc Dev	2	2
30 Lenses			
Act 79 Camera Obscura	Explor	1	0
Act 80 Thin Lens	Explor	1	0
Act 81 Lensless Lens	Conc Dev	2	0
Exp 82 Bifocals	Conc Dev	2	2
Exp 83 Where's the Point?	Conc Dev	2	2
Act 84 Air Lens	Applic	1	0
31 Diffraction and Interference			
Act 85 Rainbows Without Rain	Explor	2	0
Unit V Electricity and Magnetism			
32 Electrostatics			
Act 86 Static Cling	Conc Dev	2	0
33 Electric Fields and Potential (none)			
34 Electric Current			
Act 87 Sparky, the Electrician	Explor	2	2
Act 88 Brown Out	Explor	2	0
Exp 89 Ohm Sweet Ohm	Conc Dev	2	0
35 Electric Circuits			
Act 90 Getting Wired	Explor	1	0
Exp 91 Cranking Up	Conc Dev	2	2
Act 92 3-Way Switch	Applic	2	0
36 Magnetism			
Act 93 3-D Magnetic Field	Explor	1	0
Act 94 You're Repulsive	Explor	2	0
37 Electromagnetic Induction			
Act 95 Jump Rope Generator	Applic	1	0
Unit VI Atomic and Nuclear Physics			
38 The Atom and the Quantum			
Act 96 Particular Waves	Conc Dev	2	0
39 The Atomic Nucleus and Radioactivity			
Act 97 Nuclear Marbles	Conc Dev	2	2
Act 98 Half-Life	Conc Dev	1	1
40 Nuclear Fission and Fusion			
Act 99 Chain Reaction	Conc Dev	1	0

Using the Computer in the Physics Laboratory

The popularity and the speed of Macintosh and IBM-PC compatible computers along with the advent of HyperCard, CD-ROM, interfacing probeware, graphing calculators, desktop publishing, laser discs, digital audio-video, telecommunications, and the Internet are forever changing our instructional delivery and strategy. As the car and television have changed society in the 20th century, our classrooms—and our students—will never be the same.

However, despite the fact that computer technology has grown and developed immensely since this lab manual was first published, the underlying philosophy that the computer is a tool has not changed. The computer serves a variety of functions—word processing, drill and practice, simulations, data reduction, and as a laboratory instrument. Although word processing has many practical rewards, data reduction and learning how to use the computer as a laboratory instrument are far more important scientifically.

Interfacing, Graphing, and Data Reduction

Courseware consisting of both probeware and simulations software was specifically developed for *Conceptual Physics* by Laserpoint* in 1986 when the Apple II platform was dominant in the classroom and laboratory by the author and student Jay Obernolte. The Apple II has the added advantage of being relatively easy and inexpensive to connect with light and temperature probes (i.e., interfacing) for accurate timing as well as light sensing and temperature monitoring. *LabKit*, which includes both the hardware and the interfacing software *LabTools* and the accompanying graphing program *Data Plotter*, increases precision and student productivity in the physics lab. In 1987 Laserpoint developed one of the first sonic ranging systems, *Super Sonic Plus*.

The Apple II Series computer is comparatively antiquated technology. Although they are still relatively plentiful and available at little or low cost, Apple II computers are being replaced by the Macintosh or IBM-PC–based platforms. This manual has been revised to reflect that evolution. For those teachers whose labs are equipped with Apple IIs using Laserpoint software, specific references will be restricted to the margin notes in the Teacher's Edition. Otherwise, the instructions in the student edition of the lab manual will be generic.

For teachers whose labs are equipped with Macintosh or IBM-PC–based machines, there is a wide variety of vendors from which to select both software and hardware for either platform. Two dominant major manufacturers have emerged: Vernier Software and PASCO Scientific.

Vernier Software, founded by former physics teacher David Vernier and Christine Vernier—whose service and support to an ever-growing product line is a hallmark in physics education—features simulation and interfacing systems for both the Macintosh and IBM-PC platforms. Their software and the Universal Lab Interface (ULI) are supported by Mac, DOS, and Windows, can be used with a wide variety of probes (light, temperature, sonic ranger, sound, force, voltage, and pH) that will perform nearly every probeware experiment conceivable. Another probeware system called the Multi-Purpose Lab Interface (MPLI) supports both DOS and Windows and is capable of very rapid data acquisition, much like a computerized oscilloscope. In addition to these probeware systems, Vernier has an excellent graphing program, *Graphical Analysis*, for Apple II, MS-DOS, Windows, or Macintosh. They also have a large selection of simulation software and interfacing hardware for the Apple II platform. Detailed information on these products and their availability can be obtained from their Web site at www.vernier.com or by contacting the Verniers at the address and phone numbers listed on page T14.

Another way to use the computer with interfacing is with devices known as MBL. A variety of experiments and data acquisition techniques are possible both in and outside of the classroom. Clarence Bakken (Gunn High School, Palo Alto, CA) uses MBL devices to measure the velocity and acceleration of roller coaster rides at Great America theme park. Vernier Software offers a complete line of interfacing devices and probes.

PASCO Scientific was founded by Paul A. Stokstad, whose support and service of quality physics laboratory equipment is unparalleled in the industry. The PASCO interface systems, which support both the Macintosh and Windows platforms, feature the *Science Workshop* software program. This single program is used for all experiments and operates the same for both Macintosh and Windows machines. It has powerful built-in data acquisition and graphing routines. Three interfaces, for different levels of data collection, are available, including one that is battery powered and may be used for remote data logging away from a computer and outside a classroom setting. Over 40 sensors are available for use with

*For more information on Laserpoint software and hardware for the Apple II platform, contact Paul Robinson at LaserPablo@aol.com or at Laserpoint, 424 Quartz Street, Redwood City, CA 94062.

Science Workshop. In addition, PASCO manufactures many pieces of apparatus (in mechanics, thermodynamics, optics, waves and sound, and atomic physics) that are specifically designed for computer data collection.

Detailed information on these products and their availability can be obtained from their Web site at www.pasco.com or by contacting PASCO at the address and phone numbers listed on page T14.

Simulations

Nine simulations with the same names as labs in the manual were developed by Laserpoint for the Apple II: *Race Track, Impact Speed, Bull's Eye, Tailgated by a Dart, Flat as a Pancake, Extra Small, Wrap Your Energy in a Bow, Solar Equality,* and *Nuclear Marbles.* Some, like *Impact Speed, Wrap Your Energy in a Bow,* and *Solar Equality,* are unique and are valuable additions to the physics lab a decade later. Others, like *Bull's Eye, Flat as a Pancake,* and *Extra Small,* verify student data or help them perform complex calculations. Detailed information on these products and their availability can be obtained by contacting Laserpoint at the address and phone numbers listed on page T13.

Another set of wonderful simulations is *Good Stuff* by Robert H. Good, California State University, Hayward, CA. These simulations are some of the best demonstration software ever written for the Apple II. They include *Radiating Dipole, Moving Charge, Free Particle, Gas, Maxwell's Demon, Doppler Effect, Thin Lens, Longitudinal Waves, Sum of Two Waves, Frequency Analyzer, Waves on a String, Chain Reaction, Music,* and *Life.* These wonderful programs are to computer software as the classic film *Frames of Reference* is to videos. Some of these simulations are available as *Physics Simulation Programs* published by Physics Academic Software in DOS format (see T14 for more information).

Wouldn't it be nice if your students could make a computer simulation of a set of toppling dominoes to complement the lab *The Domino Effect*? (See Lab 3, p. 5.) Well, here's a program that can do just that—*Interactive Physics*™ by Knowledge Revolution. *Interactive Physics* makes it easy to integrate modeling and simulation into your curriculum by drawing onscreen with a powerful and easy-to-use graphic interface. You can add objects like springs, dampers, ropes, and joints and measure physical quantities like velocity, acceleration, momentum, and energy. *Interactive Physics* comes with a large collection of prepared simulations. A collection of interactive physics simulations designed to accompany *Conceptual Physics* is available from Scott Foresman - Addison Wesley.

Have advanced students try equation-based modeling using *Modellus*™, which enables students to interact with models in real time to gain insight into the underlying mathematics. *Modellus* lets students manipulate mathematical variables during a simulation to study the effects of their changes on animations, graphs, and tables. Detailed information on these products and their availability can be obtained from their Web site at www.krev.com or by contacting The Knowledge Revolution at the address and phone numbers listed on page T13.

Another source of excellent simulations is Physics Academic Software. Titles include *Physics Simulation Programs* by Robert H. Good (Apple II/DOS), *Electric Field Hockey* (DOS/Windows/Mac), *EM Field* (DOS/Windows/Mac), *Chart of the Nuclides* (Mac), *Conceptual Kinematics* (Mac/DOS), *Excel Spreadsheet Tutorial with Workshop Physics Tools* (Mac/Windows), *Freebody* (DOS/Mac), *Graphs & Tracks* (DOS/Mac), *Lighting Up Circuits* (Mac), *Objects in Motion* (DOS/Mac), *PEARLS* (Mac/Windows), *Vectors* (Mac/Windows), *VideoGraph* (Mac), *Ray* (Mac), *RelLab* (Mac), *Spacetime* (Mac/DOS), as well as many other titles. Site licenses are available to high schools at 2.5 times the cost of the program. Detailed information on these products and their availability can be obtained from their Web site at www.aip.org/pas/ or by contacting Physics Academic Software at the address and phone numbers listed on page T14.

If you are interested in creating your own lab activity, try using *VideoPoint*—a video analysis software package (available from PASCO) that allows you to collect position and time data from digital video in the form of "video points." Data is collected by clicking on the objects of interest for each frame of a QuickTime™ movie. These points can be combined to form other calculations such as center-of-mass locations and distances between points. Additionally, reference frames can be used to analyze relative motion.

The World Wide Web is becoming an increasingly significant source of information as well as a source of simulation software. Because of its volatile and rapidly changing evolutionary nature, it's difficult to designate specific Web sites, since they may quickly become outdated. An up-to-date list of relevant sites will be posted at www.cpsurf.com, the Conceptual Physics Internet site.

Also available from the Learning Team is the *Physics Info-Mall,* the single largest collection of physics resources ever assembled. Included within this extraordinary CD-ROM (for either Mac or Windows) is the complete text from 19 highly regarded introductory textbooks, the text from 11 books that cover a variety of science-related topics, 3000 problems and solutions, and 3000 articles from journals such as *The Physics Teacher, Physics Today,* and other well known publications, as well as 1000 demonstrations and laboratory exercises.

The Learning Team also publishes an excellent new program, designed to teach students about renewable energy, called *The Sun's Joules*. It includes 1000 screens of text and videos that teach students about the environment with accompanying activities including *The School Energy Doctor*, as well as an excellent astronomy observatory program called *RedShift*.

Home Projects

A few labs, such as "Impact Speed" (Experiment 14), can be easily adapted for extra-credit home projects. Allowing students to do labs you don't have time for is an excellent way to familiarize yourself with the lab.

A good way to encourage independent study is to have students build a project at home. Many such projects are listed as "Activities" at the end of each chapter in the text. Another source, which gives students hands-on experience building things that will inexpensively increase your repertoire of computer-related accessories, is David Vernier's *How to Build a Better Mousetrap* (Vernier Software). This book/disk includes 14 science projects for students who want to learn about laboratory interfacing and control via the Apple II game port. Each chapter consists of a "core project" which involves assembling an electronic circuit. Background information and detailed step-by-step instructions are provided. Sample programs for use with the core project are included on the disk. The projects are aimed at high school students who have minimal experience with computers or electronics. All electrical connections to the computer are through the game port. All projects require an Apple II+, IIe, or IIGS with a disk drive.

Projects include

Reaction Timer
Photogate Timer
Frequency Measurements
Resistance/Capacitance Meter
IC Temperature Probe
Voltage Input Unit
Thermocouples
pH Meter
Stepper Motors
A Better Mousetrap

David Vernier's second book is called *Chaos in the Laboratory and 13 Other Science Projects for the Apple® II*. It includes projects that range from simple ones to those geared toward students with more experience in electronics. This book includes such projects as

Computer Vision
Colorimeter
Barometer
Magnetic Field Measurement
Speed of Sound
Sonic Ranger
Chaos in the Laboratory

Equipment and Software Vendors

The following is not meant to be a personal endorsement, but rather a general guide for the inexperienced teacher based on my own personal experience. Your own needs, time, ability and desire to make your own equipment, and budget restraints will influence your final decision whether to build your own equipment or buy what is commercially available.

AAPT
One Physics Ellipse
College Park, MD 20740-3845
(310) 209-3300
Fax (310) 209-0845
e-mail: memb-aapt@aapt.org

The American Assocation of Physics Teachers (AAPT) is a good resource for general information, publications, and programs geared for physics teachers.

Arbor Scientific Co.
P.O. Box 2750
Ann Arbor, MI 48106-2750
(800) 367-6695 or (313) 913-6200
Fax (313) 913-6201
e-mail: mail@arborsci.com

Supplier of ready-made equipment and supplies called for in this lab manual, including specialty items such as the bouncing dart, acceleration of gravity kit, "Releasing Your Potential" pendulum apparatus, as well as general lab equipment such as Genecons, nichrome wire apparatus, parallel bulb apparatus, parallel protractor, balances, digital stop watches, lasers, holograph supplies, and so forth; has a catalog for *Conceptual Physics* that cross-references all equipment in this manual by lab number.

Central Scientific Company
3300 Cenco Parkway
Franklin Park, IL 60131
(800) 262-3626
Fax (708) 451-0231
e-mail: cencophys@aol.com

Good general physics equipment, thermal expansion apparatus, glassware, Franklin's flasks.

Edmund Scientific Co.
101 E. Gloucester Pike
Barrington, NJ 08007-1380
(609) 573-6270
Fax (609) 573-6295

Inexpensive source of diffraction-grating replicas, lenses, mirrors, polarizing sheets, and solar cells.

HelpWare Educational Materials
8300 Kestrel Drive
Raleigh, NC 27615
(919) 848-0596
e-mail: Beichner@ncsu.edu

Excellent software that teaches good laboratory techniques, significant figures, uncertainties, accuracy, and precision.

Knowledge Revolution
66 Bovet Road, Suite 200
San Mateo, CA 94402
(800) 766-6615 or (650) 574-7777
Fax (650) 574-7541
e-mail: info@krev.com

A leader in modeling and simulation software. Publisher of *Interactive Physics* and *Modullus* software.

The Learning Team
84 Business Park Drive
Armonk, NY 10504
(800) 793-8326 or (914) 273-2226
Fax (914) 273-2227
e-mail: learningtm@aol.com

The Learning Team is a not-for-profit organization devoted to the improvement of math and science education. They publish *The Physics Info Mall, The Sun's Joules,* as well as other interesting titles.

Laserpoint
424 Quartz Street
Redwood City, CA 94062
Phone (650) 369-1813
e-mail: LaserPablo@aol.com

Developer and publisher of *Conceptual Physics* courseware including interfacing software and hardware, simulation software, sonic ranging systems, test generation software, and clip-art. *LabKit* uses a simple, reliable, computer-interfacing system with temperature and light-sensing probes along with *Data Plotter,* which graphs data quickly and simply. *Super Sonic Plus* is a durable sonic ranging system that enables students to investigate sophisticated phenomena. *Laboratory Simulations* and *Good Stuff* complement labs from the lab manual; like other components of this program, they stress conceptual comprehension so that students know what they are measuring; often times the simulations provide students with helpful hints on how to perform complicated computations like scientific notation.

PASCO Scientific Co.
10101 Foothills Blvd.
P.O. Box 619011
Roseville, CA 95661-9011
(800) 772-8700
Fax (916) 786-8905
e-mail: sales@pasco.com

State-of-the-art physics equipment: timing systems, dynamic cart and track systems, photogates, air tracks, modular electronic systems, electrostatic equipment, multimeters, nuclear counters and detectors, Atwood's pulleys, low-friction pulleys, CASTLE Kits.

Physics Academic Software
Box 8202
North Carolina State University
Raleigh, NC 27695-8202
(800) 955-8275 or (919) 515-7447
Fax (919) 515-2682
e-mail: pas@aip.org

PAS reviews, selects, and publishes high-quality software suitable for use in high school, undergraduate, and graduate education in physics.

Pinhole Resource
Start Route 15
P.O. Box 1355
San Lorenzo, NM 88041
(505) 536-9942
e-mail: http://www.yatcom.com/~pinhole

Publisher of the *Pinhole Journal* three times a year and maintains photographic archives; also markets three types of pinhole cameras as well as pinhole accessories, including laser-drilled pinholes.

PRISMS
Physics Department
University of Northern Iowa
Cedar Falls, IA 50614
(319) 273-2380
Fax (319) 273-5813
e-mail: Unruh@uni.edu

Highly recommended PRISMS guide for physics laboratory activities. Contact the Director, Roy Unruh, for more information on teacher inservices and training workshops.

Sargent-Welch Scientific Co.
911 Commerce Court
Buffalo Grove, IL 60089-2375
(800) 727-4368
Fax (800) 676-2540

Good quality general physics equipment; glassware, clamps, thermal expansion apparatus, chemicals.

Vernier Software
8565 S.W. Beaverton-Hillsdale Hwy.
Portland, OR 97225-2429
(503) 297-5317
Fax (503) 297-1760
e-mail: dvernier@vernier.com

Publisher of excellent physics software and interfacing systems for Apple, Macintosh, and IBM-PC compatibles. The service and support Christine and David Vernier provide in addition to their ever-growing product line is a hallmark in physics education.

Master List of Equipment and Supplies

The following list gives recommended quantities of equipment and supplies for a class of 24 using this laboratory manual. If your school is not already equipped with these quantities, the expensive items can generally be shared by more students, so you may not need to order the full quantity the first year.

Quantity	Item	Lab Number
6	accelerometers	30
4 pkg	alligator leads	88, 90
6	aluminum channels (6-ft)	4, 7, 25, 27
2 rolls	aluminum foil	55, 74, 79
6	aluminum "sails"	15
12	ammeters or milliammeters	88, 89, 91, 94
1 gal	antifreeze (ethylene glycol)	51, 52
24	balances, spring (assorted capacities)	11,12,17,18, 20, 21, 24, 29, 35, 46, 56, 59, 61, 64
12	balances, triple beam	20, 24, 33, 34, 35, 45, 46, 47, 56, 58, 59
6 bags	balloons	16
6	bathroom scales (preferably metric)	21, 79
12	batteries (dry cell)	87, 88, 89, 90, 91
6	battery jars	53
4	battery holders	88, 90
24	beakers, assorted	1, 45, 46, 49, 55, 60, 61
12	boards (for inclines)	5, 21, 29, 35
12	boats, toy	46
6	bows (archer's)	26
6 jars	brass fasteners, jars of 200	99
36	bulbs, flashlight	87, 92
12	bulbs, 200-watt	96
12	bulbs, 100-watt	56, 73, 90
24	bulbs, 7-watt	78, 82
24	bulbs (long)	88
12	bulb sockets, large, with power cord	56, 73, 96
24	bulb sockets, night-light, with power cord	71, 78, 82
36	bulb sockets, small	87, 88, 90, 92
12	bunsen burners	49, 53, 57, 59, 73
12 sets	calcium salts, assorted	73
6 boxes	candles	76
12	cans, soup, empty	1, 5, 54
12	cans, soup, full	5, 54
2	cans, 1-gallon	1
4	capacitors	88
12	cars, toy, with track	7, 20, 21, 25
2 boxes	carbon paper	7, 36
6 pkg	cards (3" × 5")	82
12 sheets	cardboard	78, 79, 83
16 ft	chain, strong (tow variety)	10, 22
2 boxes	chalk	42, 65, 68
1 jar	chalk dust	42
12	champagne bottles	8
12	clamps, C (6-inch)	19
12	clamps, test-tube	21, 23, 25, 26, 36, 38, 57, 60, 64, 65, 71
200	coins (pennies)	98
36	compasses, magnetic, small	88, 93, 94
1–12	computers	5, 6, 7, 12, 13, 4, 16, 20, 24, 28, 38, 40, 44, 51, 54, 57, 58, 60, 64, 65, 66, 71, 94

Quantity	Item	Lab Number
12	cookie sheets (or trays with sides)	41, 85
12 sets	copper salts	73
1–12	data plotting software	40, 65, 71, 78, 82, 85, 98
12	darts, bouncing	19, 20
6	dolls (small)	11
12 boxes	dominoes	3, 99
1	Doppler ball (piezoelectric speaker with 9-volt battery mounted inside 4"-diameter foam ball)	66
12	dowels	36, 67
12	dynamics carts	11, 14, 16, 18
12	electroscopes	86, 96
6	electrostatic kits	96
12	electric dip heaters	51
12	embroidery hoops, 10" or 12" wooden	8
6	extension cords, 50-ft	95
12	eyedroppers	42, 67
24	eye hooks	36
6	fans	15
12	files, large triangular	31
24	film canisters (35 mm)	45
12	flashlights	38, 64, 65, 70, 72
12	flasks, Florence	52, 63
1 set	food coloring	52, 56
12	friction carts	21
12	friction boards	29
24	fulcrums, knife-edge	34
1	funnel	1
12	galvanometers	95, 96
8	handheld hand-cranked generators	89, 91
12	glass plates (8" × 10")	76
12	glass rods	63
1 box	glass tubing (6 mm)	1, 52, 57, 59
12 sheets	glassine paper	79
1 pint	glycerin	55
1–12	*Good Stuff* or equivalent	66, 80
12	graduated cylinders, 10-mL	42
12	graduated cylinders, 100-mL	41
12	graduated cylinders, 250-mL	58
12	graduated cylinders, 500-mL	45, 49, 66
2 boxes	graph paper	12, 13, 26, 27, 28, 34, 43, 48, 49, 96, 97
6	hook masses, 500-g	17, 18, 43, 46
2	hose clamps	60
12	hot plates	44, 50, 52, 60
4	induction coils	73
100 ft	insulated wire	94
1–12	interface boxes (optional)	7, 24, 38, 57, 58, 64
1–12	*LabTools* or equivalent	4, 7, 12, 13, 24, 27, 28, 37, 44, 51, 54, 57, 58, 60, 64, 65, 71, 93
1–12	*Laboratory Simulations* or equivalent	6, 7, 14, 20, 26, 55, 41, 42, 97
5 lb	iron filings	93
10 lb	iron scraps	45
12	jars, glass	55, 93, 97
6	kaleidoscopes	77
6	lamps with 200-watt bulbs	96
6	lasers, helium-neon	83
16	lenses, convex	79, 82, 85

Quantity	Item	Lab Number
12	lenses, diverging	83
12	lever clamps, knife-edge	33
12	light probes	24, 38
12 sets	lithium salts	73
1 roll	magnesium ribbon	94
24	magnets, bar	93, 94
12	magnets, horseshoe	93, 94
100	marbles	4, 27, 95, 97
12 doz	marking pens	5, 17, 63
12 rolls	masking tape	4, 5, 33, 39, 43, 64, 68, 77, 94
24	mass hangers	12, 17, 32, 33
12 sets	masses, slotted	9, 26, 30, 32, 33, 34
1 box	matches	76
4	mercury vapor lamps	96
24	metersticks	2, 3, 4, 7, 12, 13, 14, 19, 20, 21, 22, 24, 25, 26, 27, 28, 29, 31, 32, 33, 34, 54, 55, 56, 63, 64, 65, 66, 69, 70, 78, 81, 82, 83, 95, 97
24	meterstick clamps	32
6	micrometers	41
1 box	microscope slides	84, 85
24	mirrors (4" × 5")	72, 77
12	mirrors, concave, spherical	78
2 lb	modeling clay	46, 77, 78, 82
3 lb	paraffin wax	59, 60
5 lb	nails	36, 49, 50
20 ft	nichrome wire	73
4	nichrome wire apparatus	89
500	nuts, 1/4-inch	8
1 gal	oil, clear vegetable	93
1 liter	oleic acid (linseed oil)	42
6	oscilloscopes, preferably dual-trace	94
12	outlet extenders (3-socket)	71
2 cans	paint, black spray	56
2 boxes	paper clips, small & large	16, 17,18, 37, 43
6 rolls	paper towels	50, 52, 57, 58, 63
6	parallel bulb apparatus	91
12	paraffin blocks	67
2 boxes	pencils	75
2 boxes	pencils, colored	6
12 dozen	pens, felt-tipped, for transparencies	93, 98
6 sets	pendulum apparatus	28
12	pendulum clamps	5, 19, 23, 24, 64, 65
1	phonograph turntable	36
6	photogate timers	24, 38
12 sheets	plastic, clear	92, 93
1 box	plastic wrap	56
1 roll	polarized plastic	72
12 sets	potassium salts sets	73
6	power supplies, variable ac/dc	56, 71, 73, 78, 96
2 dozen	protractors	4, 15, 29, 77
12	pulleys, adjustable bench clamp	17
24	pulleys, free standing	9, 12, 13
24	pulleys, low-friction	18
12	quart containers	49
4	reaction boards, 8 ft × 2 in. × 12 in.	36
2 dozen	resistors, assorted	88, 89, 90
12	resonance tubes (or golf-club tubes)	69

Quantity	Item	Lab Number
6	ripple tanks with wave generators	67
12	rings	63, 57
24	ring stands (sturdy)	5, 17, 19, 21, 23, 25, 35, 36, 38, 57, 60, 63, 64
5 lb	rock salt	44, 61
12	rods, 18-inch	23
6 boxes	rubber bands	9, 17, 18, 37, 43
6 pieces	rubber hose	67
24	rubber stoppers, cut	67
12	rubber stoppers, one-hole (assorted sizes)	54, 55, 63, 75
12	rubber stoppers, two-hole	1, 52
100 ft	rubber tubing	1, 53, 59
12	sailboats, toy	15
24	shoe boxes with lids	41, 98
12	Slinky™ spring toys, long form	66
1 gal	soap solution (dishwashing detergent)	85
12 sets	sodium salts	73
1–12	sonic ranging system	5
12	spectroscopes	73
3 dozen	springs, assorted	43
6	steam generators	53, 57, 59
12	steam traps	59
12 sets	steel balls (or marbles), of different mass	4, 5, 7, 23, 27, 28, 35, 36
12	steel bobs, with center hole	24, 64, 65
300	steel bolts	45
5000	steel pellets (BB's)	41
1 pkg	steel wool	96
1	stereo, portable, with removable speakers	66
6	stopwatches	2, 3, 4, 11, 12, 13, 14, 7, 20, 22, 36, 64, 99
2 boxes	straight pins	81
1 box	straws	16
1 ball	string, lightweight	9, 12, 13, 14, 16, 17, 18, 15, 19, 23, 24, 36, 30, 34, 39, 45, 46, 47, 64, 65
12 sets	strontium salts	73
6	Styrofoam® (plastic foam) balls (1", 2", 4" 8")	44
3 dozen	Styrofoam cups	37, 49, 50, 56, 57, 58, 59, 61
24	switches, knife, double-throw	92
12	temperature probes (optional)	57, 58
12	tennis balls	7, 14, 27
12	test tubes, assorted sizes	55, 59, 60
12	thermometers, range –20°C to 110°C	44, 49, 50, 51, 52, 55, 56, 58, 59, 60, 61, 63
1 box	thumbtacks	39
12	tire pressure gauges	48
12	toy drinking birds	62
4	transistor radios	74
12 sets	tuning forks	66, 69
4	umbrellas	74
12	voltmeters	91
500	washers, 1-inch	37
1 roll	wire, connecting	87, 88, 89, 90, 91
12 pieces	wire, guide or fishing line	16
4 dozen	wire leads, with alligator clips	94
6	wood balls	27
12	wood blocks	4, 9, 29, 46, 55, 94
12	zinc plates (2" × 4")	94

Teacher Notes for Individual Labs

1 Making Hypotheses

1. Since atoms cannot be observed directly, physicists base their model of atomic structure on data collected through experimentation. Inferences are made that best describe the atom in terms of available data. The model of the atom has been changed and refined over the years as more information is collected. This lab encourages students to make inferences that allow them to construct a model of a system that they cannot view internally.
2. Fill the elevated can nearly to the top with water. Can B should have about 2 cm of water in it. Pour water from the beaker into the funnel (or thistle tube). The water level in Can B will rise, creating unequal air pressure in the two cans. Air flowing from Can B to Can A in the rubber tubing increases the air pressure on the surface of the water in Can A. Water is forced through the tubing and fountain, creating a siphon effect.
3. Have the apparatus set up before students come to class. When class starts, pour the water into the funnel to start the siphoning action.
4. Encourage students to ask questions about the setup, such as "Is the liquid water?" or "Is there an electric pump?" but don't disclose the operation of the system. After a few minutes, encourage the students to sketch what they think the insides of the cans look like. Encourage creativity in the explanations they offer. When the students finally agree on an explanation that is feasible, put the apparatus away. Don't let them look at it. Explain that this is how physicists infer models that explain things they can't actually investigate further.

2 The Physics 500

1. This lab will prove an effective introduction to average speed if students understand that the measured speeds are not uniform.
2. Encourage students to show the units used in measured and computed values in the top of each column along with the name of the quantity. Because this laboratory is exploratory, no units were included in Data Table A, and students should insert their own.
3. Students need to be encouraged to find unique races for the timed trials. These computed speeds are average speeds, although the longer the distance traveled, the closer the average speed will be to the speed at the finish line if the speed is constant after the start.
4. *Sample Computation*
   A student runs 35.0 meters in 4.67 seconds.

$$\text{average speed} = \frac{35.0\text{ m}}{4.67\text{ s}} = 7.49\text{ m/s}$$

The equivalent in miles per hour (Question 3) is:

$$(7.49\text{ m/s}) \times \frac{2.24\text{ mi/h}}{1.00\text{ m/s}} = 16.8\text{ mi/h}$$

3 The Domino Effect

1. The amount of friction provided by the table surface can affect the toppling speed of the dominoes.
2. You must exclude the case where all the dominoes are packed one against the next (spacing distance = 0).
3. Students will quickly learn that good science requires a great deal of patience. Encourage students to come up with refinements of this experiment that would yield more accurate results.
4. Some dominoes give a nice maximum speed around a spacing of 0.6 domino lengths, but other dominoes give a minimum at a spacing of 0.5 to 0.6 domino lengths. Using heavy plastic dominoes gives the maximum and light wooden ones give the minimum (results are therefore inconclusive). It is possible that the friction between the domino and the table surface also plays a significant role. Future experimentation required!

4 Merrily We Roll Along!

1. If you have a class set of air tracks or the PASCO dynamics cart and tracks, you might wish to use them instead of the ball on a ramp.
2. Set up the coordinate axes of the distance vs. time graphs of Step 4 for your students ahead of time so that you can compare each group's graphs by overlapping them on the overhead projector.

5 Conceptual Graphing

1. Sonic rangers are available from several sources. See T13 for a list of vendors.
2. You may wish to use string tagged with masking tape every so often if taping the floor is impractical.

3. Encourage students to grasp the quantitative meaning of the graphs displayed on the computer monitor; the actual magnitudes of the velocities and distances are not important here. The motion detector measures the distance between itself and the moving object. When the object is moving *away* from the detector, the distance is increasing, and, thus, the distance vs. time graph slopes upward. When the object is moving *toward* the detector, the distance is decreasing, and, thus, the distance vs. time graph slopes downward. Thus, the slope of the graph conveys information about whether the object is moving away from or toward the detector.

5. Don't *tell* students about taking the height of the can into account; let that be their discovery.

6 Race Track

1. You might want to set up a round-robin elimination tournament.
2. If students overdesign their courses with too many curves, the effect on the maximum speed will be quickly realized.
3. Students enjoy playing the "Race Track" game on the computer, using the *Laboratory Simulations-I* software available from Laserpoint. This game forces them to think—and they enjoy it!

7 Bull's Eye

1. Depending on the level of your class and course design, you may wish to postpone this lab until after your students have learned Newton's laws. This lab makes a great culminating experience for mastery learning.
2. This lab provides a real challenge for your students. Have each group explain to you exactly how they can predict the landing point of the marble. To prevent catastrophic disappointment, don't let them try until you are convinced their method is correct. Ask questions such as "How are you measuring speed?" "How are you measuring time?" and "How are you measuring distance?"
3. A major point of confusion is *time*. There is the time for the ball to roll from one photogate (or light probe) to the next—which is not to be confused with the time it takes the ball to fall to the floor from the ramp. The photogate time is horizontal motion ($v = d/t$), which is *constant velocity,* whereas the projectile motion has a vertical component ($d = \frac{1}{2}at^2$), which is *accelerated* motion.
4. Be sure to caution the students against allowing the marbles to strike the floor when they are measuring their horizontal speeds. No cheating!

8 Going Nuts

1. Practice whisking the hoop from under the nuts ahead of time, so that you can demonstrate to the students what the result is. To keep the nuts from spewing all over the room, you must grab the hoop on the *opposite* side from where your hand is coming so that the hoop is elongated horizontally, allowing the top of the hoop to drop below the nuts before they fall straight down. If you grab the hoop on the near side, it compresses into a vertical oval and ejects the nuts all over the place!
2. Don't spoil the fun by showing the "correct" way to grab the hoop. Let them discover that for themselves!
3. Champagne bottles work really well for this activity.

9 Buckle Up!

1. Have the dynamics carts crash into wood blocks or bricks.
2. Discuss Newton's first law and how it relates to auto seat belts. Talk to the students about air bags and shoulder straps and child restraints.
3. This would be an excellent time to use seat belt safety materials available from the U.S. Department of Transportation.

10 24-Hour Towing Service

1. Some teachers prefer to use this exploratory activity as a demonstration. As an alternative to taking the class outside, heavy chairs or desks can be moved inside if they are attached to a suitable anchor point.
2. The purpose of this lab is to have the students suggest and test various ways for a single person to move a parked car noticeably in the horizontal direction. Because of their small mass, small tires, and tight suspensions, smaller cars work quite well. Park the car on a level surface and have the students try various techniques.
3. When, and if, the students have become exhausted or exasperated without success, you can connect the chain to the strong tree. Make sure that the chain is quite taut. (To tighten the chain, back the car until the chain tightens and then lock the brake. If all efforts fail, release the brake and place the transmission in neutral. A *very small* horizontal force on the chain will then move the car.) This may be a sufficient

clue to evoke new hypotheses. It is most dramatic to choose a smaller person to pull at right angles to the chain at the midpoint between the car and the tree. Ropes that stretch and tires or suspension that "give" with a sideways force will not provide a good demonstration.

4. This same procedure could be used to fell trees, pull stumps, pull dents out of car fenders, and pull loose teeth!
5. Student observations and hypotheses in this exploratory lab will vary depending on their background. Encourage students to form groups and present hypotheses of their observations. After the students have shared their ideas and understandings, you may wish to move them into the following analysis, which is one way to present vector addition.

When a force F_3 is applied by the hand pulling at right angles to the chain's direction, the tension in the chain can be shown as vectors F_1 and F_2.

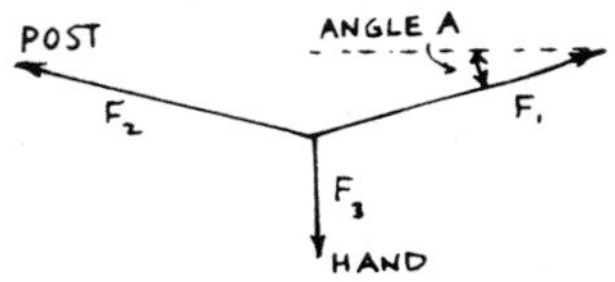

The force F_2 is exerted by the post or tree, and the force F_1 is exerted by the car. (The force F_1 is equal in magnitude but opposite in direction to the force exerted by the chain on the car.) If the point where the force F_3 is applied is not under acceleration, then the sum of the three forces on the point is zero.

$$F_1 + F_2 + F_3 = 0$$

Notice that a relatively small force F_3 is balanced by large forces F_1 and F_2. As angle A approaches zero, then F_1 and F_2 will approach infinity for any given applied force F_3. Some students may incorrectly see this force multiplication as a means of multiplying energy, but most will correctly observe that the small force applied by the hand acts through a large distance and produces a large force that acts through a small distance.

11 Getting Pushy

1. This lab will help students understand that a constant net force does *not* necessarily produce constant speed.
2. Ask students whether a rocket capable of producing 3 pounds (15 N) of horizontal force fastened to a skater could significantly change the skater's motion. (Answer: It would! Any force greater than that of the friction to be overcome will produce acceleration.)
3. The actual distances or forces may have to be adjusted to individual teaching situations.

12 Constant Force and Changing Mass

1. If you have a class set of air tracks, you may wish to use them instead of dynamics carts. However, the dynamics carts and track system from PASCO give excellent results.

13 Constant Mass and Changing Force

1. If you have a class set of air tracks, you may wish to use them instead of dynamics carts. However, the dynamics carts and track system from PASCO give excellent results.

14 Impact Speed

1. The central idea of this experiment is to attach significance to the area under a graph. This lab is very advanced conceptually, but is quite simple experimentally. Don't be discouraged if students do not grasp the full meaning right away.
2. This makes an excellent home experiment.
3. If your students don't like graphing, the program "Impact Speed" on the *Laboratory Simulations-I* software from Laserpoint does it for them. For best effect, have students do at least one graph by hand, and then allow them to marvel while the computer does it! (The heart of this program took first place in the Physics Division of the 1986 California State Science Fair and was recognized in the Honor Group of the 1988 Westinghouse Search for Talent Contest. Written by Jay Obernolte (who now programs for Sega-Gensis), this simulation is based on an idea from Verne Rockcastle of Cornell University (Ithaca, NY). The program generates the two plots by counting the pixels under the first graph ("theoretical" speed line) and then uses the same number of pixels to plot the second graph ("actual" speed line). Clever!
4. It may be very instructive to the students to do the following problem before attempting "Impact Speed."

> A motorcyclist at a stop sign accelerates from rest to 12 m/s in 3 seconds, then maintains constant speed for

10 seconds. Then the cyclist accelerates to 20 m/s in 4 seconds and maintains constant speed for 20 seconds. The cyclist sights another stop sign and comes to rest in 7 seconds. What is the distance between the two stop signs?

Trying to memorize a complicated formula is hopeless! A better approach is to compute the area under the speed vs. time graph. It is an easy task (but tedious!) to compute the area of the triangles, rectangles, and trapezoid under the graph.

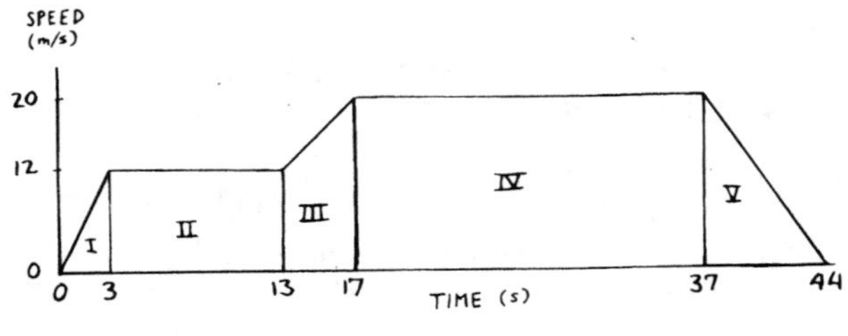

Section	Shape	Area
I	triangle	$\frac{1}{2}$ (height) × (base) $= \frac{1}{2}$ (12 m/s)(3 s) = 18 m
II	rectangle	(height) × (base) = (12 m/s)(10 s) = 120 m
III	trapezoid	$\frac{1}{2}$ (left side + right side) × (width) $= \frac{1}{2}$ (12 m/s + 20 m/s) × (4 s) = 64 m
IV	rectangle	(height) × (base) = (20 m/s)(20 s) = 400 m
V	triangle	$\frac{1}{2}$ (height) × (base) $= \frac{1}{2}$ (20 m/s)(7 s) = 70

The total distance traveled is the total area under the speed vs. time graph:

$$\text{total area} = (18\text{ m}) + (120\text{ m}) + (64\text{ m}) + (400\text{ m}) + (70\text{ m}) = 672\text{ m}$$

5. If your students do this laboratory, "Wrap Your Energy in a Bow" (Experiment 26) makes an ideal follow-up.

15 Riding with the Wind

1. This experiment provides an excellent opportunity for students to learn how to resolve a vector into parallel and perpendicular components.
2. Although the PASCO sailboat gives excellent results, a sailboat can be fabricated by making 4 saw-cuts into a 6-inch-long two-by-four. Two saw-cuts should be perpendicular to each other, and the third and fourth should be at a 45° angle to the first two. These saw-cuts make a nice groove to hold the sail. The two-by-four can be temporarily fastened to a dynamics cart with duct tape.
3. Caution students about the hazards of handling electric fans.

16 Balloon Rockets

1. Allow the students to be creative in their rocket designs. Monofilament fishing line works well as a guide line strung across the room or between any two stable objects.
2. The most efficient designs usually attach the balloon to a straw and slip the guide wire through the straw. (Oblong balloons seem to work best.) Students might design a two-stage rocket to make the return trip.

17 Tension

1. Spring balances *cannot* be substituted for the rubber bands used in this experiment (to indicate equal tension throughout the string) because the weight of a balance will affect the tension. The weight of a rubber band is negligible in comparison.

18 Tug-of-War

1. Students who are the first to say that the content of this lab is obvious are usually the first to be confused about why the tension in the string in Figure F is *not* twice each hanging weight!
2. Ask your students to compare Figure F with a tug-of-war.

19 Go Cart

1. Previous editions of this lab manual included a similar activity titled "Bouncing Dart" in which the dart either stuck or bounced off a block of wood. The activity was revised so that it now *unambiguously* demonstrates that the dart delivers a greater impulse on the cart when it bounces than when it doesn't—evidenced by the fact that the cart rolls a greater distance when struck by the bouncing dart.
2. This seemingly simple activity is a good example of how even the simplest phenomena can actually be quite complicated. The following analysis of "Go Cart" was provided by Ken Ford.

It is interesting to analyze three special cases mathematically. In all three cases, momentum is conserved in the collision. Let the dart have mass m, speed v just before the collision, and speed v_d just after the collision. Let the cart have mass M, zero speed just before the collision, and speed v_c just after the collision. Momentum conservation means that

$$mv = mv_d + Mv_c.$$

Case 1. Perfectly elastic collision (kinetic energy is conserved): The condition of energy conservation can be written

$$\frac{1}{2} mv^2 = \frac{1}{2} mv_d^2 + \frac{1}{2} Mv_c^2.$$

The momentum and energy equations provide two equations for two unknowns, v_d and v_c. Before solving them, it is convenient to divide both equations by m, which changes them to equations involving only the mass ratio M/m, not m or M separately. You can then, for instance, solve the momentum equation for v_d and substitute into the energy equation, solving it for v_c. The results for the two unknowns are

$$v_c = \frac{2}{1 + r} v$$

$$v_c = \frac{1 - r}{1 + r} v,$$

where r is the mass ratio: $r = M/m$.

For the cart much more massive than the dart ($r \gg 1$), the dart rebounds with nearly its initial speed ($v_d \approx - v$) and the cart moves forward with a small speed [$v_c \approx (2/r)v$]. The case of equal masses ($r = 1$) is the interesting special case in which the dart stops and the cart moves forward with the dart's initial speed v (this is equivalent to the straight-on collision of one billiard ball with another and is the condition for maximum energy transfer from the dart to the cart).

Case 2. Perfectly inelastic collision (dart sticks to cart and moves with it): Kinetic energy is not conserved. Since cart and dart move together after the collision, $v_c = v_d$, and the momentum equation can be solved by itself to give

$$v_c = v_d \frac{1}{1 + r} v.$$

The speed of the cart in this case is exactly half what it is for the perfectly elastic collision, and its kinetic energy is one-fourth that of the elastic collision. One would therefore expect to see the cart roll about four times as far after a perfectly elastic collision as after a perfectly inelastic one in which the dart sticks to the cart.

Case 3. The dart stops (no bounce, no stick): Again kinetic energy is not conserved. The final speed of the dart is $v_d = 0$, so the momentum equation reduces to

$$mv = Mv_c,$$

giving

$$v_c = \frac{1}{r} v.$$

(This case is physically possible only for $r \geq 1$. If the dart were more massive than the cart ($r < 1$), the dart could not stop. It would keep moving after the collision.) This case is intermediate between the perfectly elastic and perfectly inelastic cases.

Cases 1 and 3 approximate what actually happens in the activity. It is therefore of interest to find the ratio of kinetic energies given to the cart for these two cases. It is the same as the ratio of the squares of the cart's speed in the two cases and is given by

$$\frac{\text{KE (elastic)}}{\text{KE (dart stops)}} = \frac{4r^2}{(1 + r)^2}$$

This energy ratio should be approximately the same as the ratio of distances moved by the cart in the elastic and dart-stopping cases. Theoretically, it is 1 when the cart and dart have equal mass and 4 when the cart is much more massive that the dart. You can calculate this ratio for the actual masses used in your experiment and compare with the relative rolling distances.

20 Tailgated by a Dart

1. Students should know about momentum conservation before this laboratory.
2. Resist telling your students exactly how to go about finding the speed of the dart. Let them try to figure it out.
3. The total momentum in the system is the same before and after impact. The car's average speed after the impact, the mass of the dart, and the mass of the car must be measured first—before the initial speed of the dart can be found.
4. *Sample Data and Computations*
   Distance the car moved = 1.48 m
   Coasting time = 4.77 s

   Car's average speed = (1.48 m)/(4.77 s)
   = 0.310 m/s

   Since average speed equals (initial speed + final speed)/2, and the final speed is zero, then:

   Speed of car upon impact = 2(0.310 m/s)
   = 0.620 m/s

   Mass of car = 58.37 g

Mass of dart = 1.43 g

Momentum before impact = Momentum after impact

(1.43 g)(initial speed of dart) = (58.37 g + 1.43 g)(0.620 m/s)
Initial speed of dart = 25.9 m/s

5. Students may want to use a light probe and computer to check the car's speed upon impact on a short length of track.

21 Making the Grade

1. The idea here is to let the students explore the fact that work is equal to the force times the distance through which the force acts. It is hoped that they will see that the amount of work done is the same regardless of the path they take. Don't tell them this, but try to guide them so that they will eventually arrive at this conclusion. For all paths, the product of force and distance should be the same.

22 Muscle Up!

1. Encourage creativity as students select activities. Throwing a medicine ball a measured distance straight up, climbing a rope, a leg press, and a pull-up could all be good activities.
2. Sometimes, the output distance of a load is greater than the input distance of the effort. Nautilus equipment commonly found in weight-lifting rooms will be difficult to use because cams and levers produce variable resistances throughout the motion of the exercise.
3. On a bench press, do not measure the distance the weights are lifted; measure the distance the handgrips move.

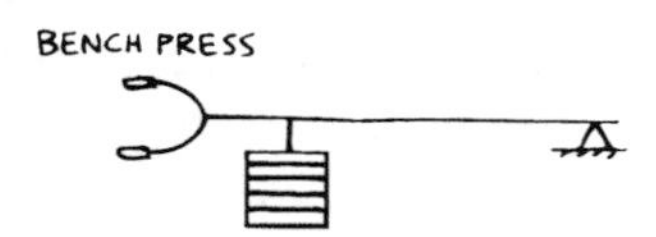

4. *Sample Data and Computations*
   Activity: leg press
   Force = 260 pounds
   Distance = 60 cm
   Time = 1.3 s

   When calculating work in joules, remind students that the force must be in newtons and the distance in meters. To convert a force in pounds to newtons, use the conversion factor of 4.45 newtons per pound.

Force = (260 lb)(4.45 N/lb) = 1160 N

Work = (force) × (distance)
= (1160 N)(0.60 m) = 700 joules

$$\text{Power} = \frac{\text{work}}{\text{time}} = \frac{700 \text{ J}}{1.3 \text{ s}} = 540 \text{ watts}$$

23 Cut Short

1. This exploratory activity is vital and should be done *before* you mention energy conservation.

24 Conserving Your Energy

1. This experiment should be done after "Cut Short" (Activity 23). With care and the use of light probes, students are gratified by the close agreement of PE and KE.

25 How Hot Are Your Hot Wheels®?

1. Both ends of the track should be elevated to a height of about one meter above the floor or lab table.
2. It is important to secure the track to the floor or lab table so that some of the car's energy is not transformed into motion of the track. Any toy car will do, but Hot Wheels toy cars work exceptionally well. Some students may be willing to bring cars from home.

26 Wrap Your Energy in a Bow

1. This experiment is designed to look at the more general case of energy transferred when the force on a system is not constant. The most general way of looking at work done or energy transferred is to compute the area under the force vs. distance graph.
2. This lab makes an excellent follow-up to "Impact Speed" (Experiment 14), which also draws information from the area under a graph.
3. You will need heavy-duty spring scales. The ones calibrated in newtons and available from scientific supply houses are quite expensive. The kind sold in hardware stores, with capacities of 25 lb or 50 lb, are cheaper, and you can have the students convert pounds to newtons. (There are 4.45 newtons in one pound.)
4. An archer's bow often can be borrowed from the physical education department.
5. Make sure that the bow is firmly supported to avoid injury or damage to the bow.

6. The force vs. stretch of a compound bow is very interesting since there is a point where the force decreases with the stretch of the string in the bow.
7. Observe students to determine whether they are increasing the stretch distance in small enough increments so that they catch any variances in the force profile. Too small increments will result in a tedious collection of data that does not add sufficiently to the force profile. Six to ten data points should not be uncommon for this exercise. After the graph has been plotted, there may be interest in a few more readings at critical points if the force pattern is changing.
8. *Sample Data*

Stretch Distance (cm)	Force (N)
0.0	0.0
1.0	11.1
2.0	20.0
3.0	24.5
4.0	29.0
5.0	35.6
6.0	44.5
7.0	46.8
8.0	57.9
9.0	66.8
10.0	75.7
11.0	82.4
12.0	91.3
13.0	95.8
14.0	104.7
15.0	109.1
16.0	115.8
17.0	120.3
18.0	124.7
19.0	127.0

When these data are plotted on a graph of force (in newtons) vs. distance (in meters), the points will not lie right on a straight line. A *smooth* curve should be drawn that has about equal numbers of points off to either side of the curve.

9. The total energy stored by the bow can be approximated by breaking up the area under the curve into sections and counting the number of small squares in each section. Multiply the width of one small square (in meters) by the height of the square (in newtons) to obtain the area of the square in joules. Then, multiply this energy value by the total number of squares under the curve to get the total energy stored by the bow.
10. To find the height that an arrow could be shot, apply the idea that the energy stored by the bow at a certain stretch will be transferred to the arrow, which will gain gravitational potential energy in its ascent.

$$\text{area under graph} = PE_{bow} = mgh$$

where m = mass of arrow
g = acceleration due to gravity
h = height attained by arrow

Say the area under the curve is 53 J and the arrow's mass is 50 grams, or 0.050 kg. Then,

$$h = \frac{PE}{mg} = \frac{53\ \text{J}}{(0.050\ \text{kg})\ (9.8\ \text{m/s}^2)} = 110\ \text{m}$$

When the mass of the arrow is 75 grams, or 0.075 kg, the height is

$$h = \frac{53\ \text{J}}{(0.075\ \text{kg})\ (9.8\ \text{m/s}^2)} = 72\ \text{m}$$

If you have students check the height of the arrow experimentally, be sure that the trajectory of the arrow is completely clear of persons and property subject to damage by a flying arrow. Remember, that to get hit by a flying arrow is similar to being hit by an arrow shot directly from the bow!

11. The speed is found by equating the KE with the PE stored by the bow.

$$KE = \frac{1}{2}mv^2 = PE$$

$$v = \sqrt{\frac{2PE}{m}}$$

$$= \frac{(2)(53\ \text{J})}{(0.050\ \text{kg})} = 46\ \text{m/s}$$

12. The KE of the arrow is not equal to the work done on the bow. It turns out that the bow is quite inefficient. David Wiley, physics professor at the University of Pennsylvania, Johnstown, has pointed out that only about 60–75% the PE of a drawn bow goes into the KE of an arrow, depending on the type of bow—the rest heats the bow.

27 On a Roll

1. This makes an excellent take-home lab.
2. If the classrooms are not carpeted, the library may have a carpeted area where students can perform this lab.
3. The PE of the ball at the release point is all converted to KE of the ball at the bottom of the ramp. The ball comes to a stop on the carpet because of a friction force that we can assume is constant. The ball does work against the friction force until it has no more energy left. The work done equals the friction force F times the

stopping distance *d*. The work done equals the KE at the bottom of the ramp.

$$\text{work} = F \times d = \text{KE}$$

$$d = \frac{\text{KE}}{F}$$

28 Releasing Your Potential

1. The launch speed $v = \sqrt{2gh}$ is *not* valid for the launcher because the launcher is a physical pendulum, more complex than a simple pendulum. It has less rotational inertia than a simple pendulum of the same mass and length. So when the launcher hits the crossbar, the ball at the bottom end of it is moving faster than it would be if it were swinging at the end of a simple pendulum. Therefore, we simply say that the speed calculated for a simple pendulum, $v = \sqrt{2gh}$, is a lower limit: the *minimum* launch speed. The error so introduced is only 5–25%, which justifies not belaboring the distinction between simple and physical pendula at this point of the course.

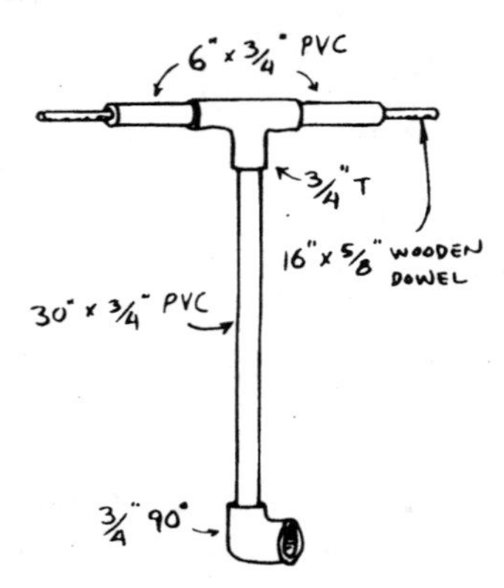

2. A 5/8-inch by 16-inch wood dowel can be used as a bearing. Clamp the dowel to two tall ring stands or lab posts.
3. Attach a crossbar just above where the ball "launcher" will strike it.
4. Make all connections as firm as possible so that when the pendulum bar strikes the crossbar, the ball is jolted or "launched" as cleanly as possible.
5. Adjust the position of the crossbar so that when the pendulum is in the rest position, it is *vertical* to the table. If the ball rolls in or out when carefully placed in the launcher, it probably is *not* level. Readjust the crossbar until the ball will remain in the launcher.
6. It is a good idea to place a strip of particle board along the projected path of the launched ball to prevent marring of the table surface. If you place the entire apparatus on the edge of the lab table, place the particle board on the floor.
7. Place carbon ditto face-down on top of a strip of paper towel where students expect the launched ball to land.
8. Students will need to practice releasing the pendulum bar with the ball in the launcher so that the ball does not fall out *before* the pendulum bar strikes the crossbar. The trick to this is for students to pull away their hands from the ball and the launcher *simultaneously.*
9. Be sure to plan several days to allow your students the time to absorb the key concepts covered in this experiment. After students have had the opportunity to acquire and organize their data, reinforce their efforts with a class discussion. Before they plot their data, give them transparencies with the same scales on the axes so that you can compare their graphs by placing them on an overhead projector.
10. Since energy is conserved, the kinetic energy gained is equal to the potential energy lost.

$$\text{KE}_{\text{gained}} = \frac{1}{2}mv^2 = \text{PE}_{\text{lost}} = mgh$$

$$v^2 \sim h$$

$$v \sim \sqrt{h}$$

11. In Step 6, the mass of the ball has little or no effect on how far it travels downrange.
12. The speed at the bottom is proportional to the square root of the height; the graph of launch speed vs. height (Step 7) will not be a straight line. However, a graph of the *square* of the speed vs. height will be a straight line.
13. The range of the ball equals the horizontal launch speed v times the falling time t.

$$\text{range} = v \times t$$

The falling time depends only on how high the launch position, h, is above the floor. From the equation $h + \frac{1}{2}at^2$, $t = \sqrt{2h/g}$. The range is, thus, proportional to the launch speed v and $\sqrt{h}$, the square root of the height above the floor. A graph of range vs. height (Step 7) is, therefore, not a straight line. A graph of the *square* of the range vs. height *would* be a straight line. This analysis makes an excellent follow-up to "Trial and Error."

$$\text{range} \sim v \sim \sqrt{h}$$

$$\text{range} \sim \sqrt{h}$$

29 Slip-Stick

1. Emphasize the qualitative nature of this experiment.
2. Remind students that when they are measuring friction forces, they should keep the spring balances parallel to the surface.
3. For the "Going Further" section, if your students are mathematically inclined and have had trigonometry, you may want to show the relationship between the coefficient of sliding friction and tan θ in Figure C.

$$\mu_{\text{sliding}} = \frac{F_f}{N} = \tan\theta$$

where F_f is the drag force and N is the normal force.

30 Going in Circles

1. If you do not have a class set of accelerometers or turntables, but can make one setup, this lab makes an excellent class demonstration. A 12-inch accelerometer, attached to the turntable with masking tape, shows the effect of increased acceleration when set at the various speeds.
2. Some students have great difficulty conceptualizing centripetal acceleration and centripetal force. They tend to think it is a separate fundamental force, such as gravity. This lab helps defeat the misconception that water in a bucket swirled overhead is held in by "centrifugal" force. Comparing the appearance of the accelerometer when it is accelerated in a straight line with when it is rotating helps reinforce the concept that the direction of the acceleration (and, hence, the *net* force) is *radially inward.* It is the bucket's running into the water that causes the water to remain in the bucket.

31 Where's Your CG?

1. The use of triangular supports is important.
2. Most bathroom scales have a convex lens cover over the scale. Students can stack boots on the foot platform of the scale so that the reaction board does not contact the convex lens cover.
3. Students should be careful to make sure that the overhangs on each triangular support are equal.
4. You might want to ask your students why the board is called a "reaction" board. The reason is that when they lie on the board, their weight can be considered the action force. The board exerts an equal and opposite reaction force on them.

32 Torque Feeler

1. This is a great exploratory activity to get students to distinguish between *torque* and *force.*
2. This activity should be done *before* you begin discussing or defining torque.

33 Weighing an Elephant

1. Be sure that your students include the mass of the mass hangers. (The mass of the knife-edge fulcrum does not enter the computation because its weight acts above the pivot and contributes nothing to torque.)
2. Have your students use at least 100-gram masses on each side to minimize the effects due to friction in the fulcrum and to make balancing easier.
3. The product of the smaller mass times its distance from the fulcrum should equal the product of the larger mass times its distance to the fulcrum within a few percent.

34 Keeping in Balance

1. Many a bright student who thinks that center of gravity is a simple concept thinks that in Step 6 only 85% of the mass of the meterstick balances the mass hung on the short end! It is important for your students to understand that the *entire* mass of the meterstick acts as if it is concentrated at the center of gravity, not just the 85% of the mass that is on the same side of the fulcrum as the CG. This is a sophisticated concept.

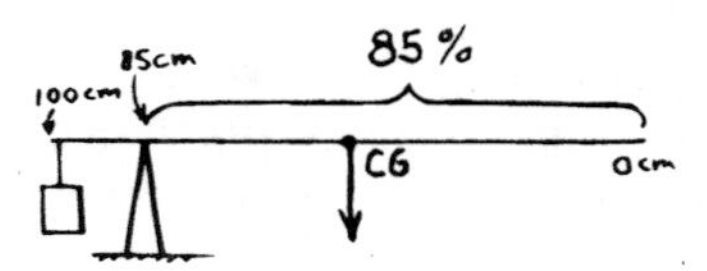

2. Don't let students make the mistake of thinking, in Step 10, that the mass of the rock they tape to their meterstick becomes an *unknown* as in the "Weighing an Elephant" lab (Experiment 33). The idea is rather to make the taped mass part of the meterstick so that the meterstick becomes an asymmetrical object. The students are to find the center of gravity of this asymmetrical meterstick, as they would a baseball bat or any other asymmetrical object.

35 Rotational Derby

1. Guard against using objects that are too small to minimize the effects due to friction. In Steps 4 through 6, different-sized undented food cans work well, provided the contents do not slosh around.

2. There is an important analogy between freely falling and rolling objects you should remember before attempting this lab: All objects fall at the same acceleration g, because they have the *same ratio of force to mass.* An increase in mass is exactly compensated by a proportional increase in gravitational force. Similarly, the accelerations of objects rolling down inclined planes depend on the ratio of torque to rotational inertia. Be careful not to allow your students to conclude that all objects of the same *geometrical shape* (such as all balls, all cylinders, all rings, etc.) have the same *rotational inertia.* They do NOT, for they may have different radii and different masses. What objects of similar shape *do* have is the same *ratio of torque to rotational inertia.*

36 Acceleration of Free Fall

1. You might wish to have students do Part A of "Computerized Gravity" (Experiment 37) at the same time they do this experiment, especially if only one computer is available.
2. If a ring stand is used to suspend the apparatus, be sure to clamp the base of the ring stand firmly to the laboratory table to prevent wobbling.
3. Do *not* attempt to calculate the period of the swinging meterstick using the simple pendulum equation.

$$T = 2\pi\sqrt{L/g}$$

The mass of a simple pendulum is concentrated in the bob, whereas the mass of the meterstick is spread over its entire length.
4. Depending on their apparatus, student results for the acceleration of gravity should be within 5% of 9.80 m/s^2.

37 Computerized Gravity

1. Part A of this experiment can be done at the same time as "Acceleration of Free Fall" (Experiment 36), especially if only one computer is available. Several groups of students can cycle through it in one lab period.
2. The letter "g" for Part A can be cut from 1/8-inch thick cardboard or from a sheet of aluminum (obtainable from Arbor Scientific, or perhaps a shop teacher). Using evenly spaced black electrical tape is not advised.
3. Part B is a great review opportunity to clear up confusion on accelerated motion from Chapter 2. It also reinforces the "Going Further" section of "Merrily We Roll Along!" (Experiment 4) but is *far* easier for students to accomplish. Dropping the picket fence through the light probes is embarrassingly simple compared with rolling a ball down an incline. You might point out how technological advances do benefit humankind—especially physics students!
4. The distance *d* the picket fence drops is proportional to the square of the time *t.*

$$d = \frac{1}{2}gt^2$$

$$d \sim t^2$$

Plotting distance vs. time (Step 7) gives a parabola. A graph of distance vs. the square of the time would give a straight line.
5. The instantaneous velocity *v* of the picket fence is proportional to the time.

$$v = gt$$

$$v \sim t$$

The graph of velocity vs. time (Step 8) will be a straight line. The slope of the graph is the acceleration of gravity, *g.* To measure the slope, students should divide the vertical change between two points on the graph line (in speed units) by the horizontal change between those same two points (in time units). If students come to you with a nonsensical value for *g,* they may have divided the horizontal change by the vertical change.

38 Apparent Weightlessness

1. Students need to select coins of a weight that causes sufficient tension in the rubber band so that when the cup is released, the band pulls them into the cup. If the coins are too heavy, the cup will turn upside down! A practical solution is to use empty soup cans (which have more mass) instead of Styrofoam (plastic foam) or paper cups. The holes for Step 4 can be made in cans ahead of time and taped over with masking tape until they are needed.
2. The water-filled cups can be dropped into a trash can that has a plastic liner. (If you are using soup cans, it is better to drop them into such a trash can than into a sink.)

39 Getting Eccentric

1. This activity is a nice way to give students a hands-on ellipse experience, where they can learn what affects the shape of ellipses.
2. Since Kepler's laws are not mentioned in the text, you might wish to discuss Kepler's laws as they relate to the orbital motion of bodies such as planets and comets.

3. If you plan for your students to do "Trial and Error" (Experiment 40), do NOT mention that the square of the period is proportional to the cube of the average orbit radius (Kepler's third law).

40 Trial and Error

1. The purpose of this activity is for students to:
   (a) Learn the ease and power with which the computer can plot graphs.
   (b) Learn how easily the computer can manipulate data.
   (c) Learn that when their graph is a straight line, then the quantity on the vertical axis is proportional to the quantity of the horizontal axis.
   (d) Learn how to find the functional relationship between variables.
   (e) Have fun doing on the computer what would probably never get done by hand.
2. This activity can be done any time you choose to introduce computers into your course. It is recommended that students make graphs by hand on transparencies through Chapter 2. Once your students discover the power of the computer to plot graphs and manipulate data, it will be difficult to get them to do it any other way!
3. Do NOT tell students that $T^2 \sim R^3$. Allow them to have the satisfaction of discovering it.

41 Flat as a Pancake

1. You might want to do a quick prelab demonstration showing that the diameter of a marble is the same as the thickness of a "pancake" of 20 to 35 marbles—the pancake's volume divided by its top or bottom surface area.
2. Note that the volume of the BB pancake is a bit greater than the volume of the same BB's in the graduated cylinder. This causes the estimated value to be about 5 to 10% higher than the actual diameter. A typical BB measures about 4.5 mm using a micrometer.
3. It takes a volume of about 47 cm^3 of BB's to spread out in a layer with an area of about 100 cm^2. Rimmed cookie sheets work well because students can easily make a square or rectangle "monolayer" and keep the spillage of BB's to a minimum.
4. A nice follow-up activity is to have your students find the density of a rectangular block of aluminum by measuring its mass and volume. Then hand them a piece of aluminum foil and ask them to find its thickness.

   Since they have just computed the density of aluminum, they can measure the mass of the foil and then compute its volume. Since its volume is simply its top or bottom surface area multiplied by its thickness, once they measure its surface area they can compute its thickness by dividing its volume by its surface.

$$\text{density} = \frac{\text{mass}}{\text{volume}}$$

$$\text{volume} = \frac{\text{mass}}{\text{density}}$$

$$\text{thickness} = \frac{\text{volume}}{\text{area}}$$

$$= \frac{(\text{mass})/(\text{density})}{\text{area}}$$

42 Extra Small

1. This investigation provides the student with a very simple method of making a direct measurement of the diameter of a molecule. Although the computations are a bit complicated, the results are well worth the effort.
2. Rimmed cookie sheets or cafeteria trays make excellent trays for this and are inexpensive. Ripple tanks also work well.
3. Preparation of the 0.5% solution of oleic acid is critical. Exacting care must be used or student results will be drastically affected:
   (a) By volume: Measure 5 mL of oleic acid and 95 mL of alcohol (methyl or isopropyl) into a graduated cylinder. Place the solution in a clean bottle and shake well. Then, measure off 5 mL of this solution and mix it with 45 mL of alcohol. The resulting solution is 0.5% oleic acid by volume.
   (b) By mass: Dissolve 450 milligrams oleic acid in 100 mL of alcohol.
4. Inform students that the oleic acid solution they are using is mostly alcohol with a small amount of oleic acid in it. When they add a drop of solution in a tray of water, the alcohol dissolves in the water, but the oleic acid floats on top, just as a drop of oil floats on water.
5. Caution students not to put too much powder on the water surface. If they do, the powder will push back on the acid film and reduce its diameter. That is, the powder will hem in the film.
6. In Step 3, a typical value for the average diameter is 30.0 cm. The radius r is then

$$r = (30.0\ \text{cm})/2 = 15.0\ \text{cm}$$

The area A of the circle is

$$A = \pi r^2$$
$$= (3.14)(15.0\ \text{cm})^2$$
$$= 706\ \text{cm}^2$$
$$= 7.06 \times 10^2\ \text{cm}^2$$

7. In Step 3, the number of drops in 3 cm^3 of 0.5% oleic acid solution is about 114. The volume of one drop is

$$(3\ \text{cm}^3)/(114) = 0.026\ \text{cm}^3$$
$$= 2.6 \times 10^{-2}\ \text{cm}^3$$

8. In Step 4, the volume of acid in a single drop equals 0.005 multiplied by the volume of one drop:

$$(0.005)(2.6 \times 10^{-2}\ \text{cm}^3) = 1.3 \times 10^{-4}\ \text{cm}^3$$

9. In Step 5, the length of an oleic acid molecule equals the volume of oleic acid in one drop divided by the area of the circle:

$$\text{length} = (\text{volume})/(\text{area})$$
$$= (1.3 \times 10^{-4}\ \text{cm}^3)/(7.06 \times 10^2\ \text{cm}^2)$$
$$= 0.18 \times 10^{-6}\ \text{cm}$$
$$= 1.8 \times 10^{-7}\ \text{cm}$$

Good measurements yield values between 1.0×10^{-7} cm and 2.0×10^{-7} cm. Usually, there is not enough time to repeat the procedure. However, if there is time, be sure that students clean the trays thoroughly before making a second measurement.

43 Stretch

1. If you are using as standard Hooke's law apparatus, explain the use of the mirrored surface on which the scale is superimposed. (When the observer has the pointer lined up with its image in such a way that the image is blocked out, parallax is eliminated.)
2. The mass-weight distinction presents a problem here. The springs you will likely use are very fine, and the masses used are 10 g to 150 g. The weights in newtons of these masses are fairly small, since the weight of one *kilogram* is 9.8 N:

$$\text{weight of 10 g} = \text{weight of 0.010 kg}$$
$$= (0.010\ \text{kg})(9.8\ \text{N/kg})$$
$$= 0.098\ \text{N}$$

To circumvent this, you may want to work with "weights" in grams, or "gram weights," and plot "gram weights" vs. stretch in centimeters. Then, the spring constant k will be in gram weights per centimeter. This can then be converted as a last step to newtons per meter.

$$k \text{ in N/m} = (k \text{ in g/cm})(9.8\ \text{N/kg})(0.001\ \text{kg/g})(100\ \text{cm/m})$$
$$= (k \text{ in g/cm})(0.98\ \text{N·cm/g·m})$$

3. The spring constant for two springs connected in series is not the same as the spring constant for each individual spring. For two springs otherwise the same, a long spring has a different spring constant from that of a short spring. The amount of stretch for a given applied force depends on the length of the spring!

44 Geometric Physics

1. Scaling is not covered in most physics textbooks. Yet, it governs physical phenomena as diverse as the relative strengths of ants and other animals and the terminal speed of raindrops.
2. Another activity you might want your students to try is building arches out of small bars of modeling clay (1 cm × 1 cm × 2 cm). They will quickly discover that weight increases faster than strength.

45 Eureka!

1. In this exploratory activity, your students will learn to differentiate among mass, weight, volume, and density. Only density is a characteristic property of a substance. Do *not* spoil the exploratory by telling this ahead of time, however.

46 Sink or Swim

1. Many students believe that the buoyant force acting on a floating object is greater than the buoyant force acting on a submerged object. This is not necessarily true. The buoyant force acting on a 10-kilogram rock is far greater than the buoyant force acting on a floating beach ball.
2. Your students will need to devise a way to measure the volume of water displaced precisely. Overflow cans or graduated cylinders are useful. Our method is to have your students put a piece of masking tape on the side of a beaker and carefully mark the different water levels on the tape. Then, they can go back and measure the difference using a graduated cylinder.

47 Weighty Stuff

1. Tire pressure gauges are available at auto supply stores. Gauges with 1-pound increments are preferable.
2. In Step 4, the mass of the air pumped into the basketball is the mass of the inflated basketball and paper (Step 3) minus the mass of the flattened basketball and paper (Step 1). Typically, this value is 4 or 5 grams.
3. A Pyrex® brand glass casserole dish makes a good "canal" for Step 8.

48 Inflation

1. This activity helps students distinguish between force and pressure.
2. This is a fun lab that does have its limitations. Students may find that the weight of the car nearly equals the pressure times the contact area of the tires (which, for tires of good tread, is typically about 80% the outlined area). However, these coincidental results can be misleading—the role of the sidewalls is significant.

49 Heat Mixes: Part I

1. In this lab, your students will discover that the final temperature of hot water mixed together with cold water depends on *how much* hot water is mixed with *how much* cold water.
2. Your students need to exercise care when marking the water levels of their cups. The large 14-ounce plastic foam cups work better (larger amounts of water) than the small 6-ounce variety.
3. You may note that the constant cooling of water due to evaporation contributes to error.

50 Heat Mixes: Part II

1. In this lab, your students will discover that the final temperature of a mixture of nails and an equal mass of water depends on something besides the amount (or mass): the substance's ability to absorb heat energy, that is, its *specific heat.*
2. Use stubby nails that are completely covered by an amount of water of the same mass. Long skinny nails may protrude above the water.
3. The specific heat of iron is 0.11 cal/g·°C, or about 1/9 that of water. The nails, therefore, will change temperature 9 times as much per calorie of heat transferred.

51 Antifreeze in the Summer?

1. This lab provides you and your students an excellent opportunity to analyze a real-world problem. It's easy for people to think about the effects of one variable, such as specific heat, but often there are many variables that affect the results. Antifreeze is used in radiators for several reasons:
   (a) to act as an antifreeze
   (b) to act as an antiboil
   (c) to maintain the efficiency of the system with anticorrosives.
2. Instead of having everybody find the specific heat of a 50-50 mixture, you might have one group do the experiment with 100% pure antifreeze, and others with 75%, 50%, 25% solutions, and pure water (0%). Pool student results at the end of the lab.
3. To prepare 400 g of a 50% mixture by mass of antifreeze and water, add 200 g water to 200 g pure antifreeze. This is accomplished most easily by adding 200 mL water to 180 mL antifreeze. You can compute the volume of pure antifreeze required to make any percent solution from the density of pure antifreeze, which is 1.11 g/mL.
   Example:
   To make 400 g of a 25% mixture, mass of antifreeze needed is

$$\text{mass} = 25\% \times (400\text{ g}) = 100\text{ g}$$

The volume of this mass of antifreeze is

$$\text{volume} = \frac{\text{mass}}{\text{density}}$$

$$= (100\text{ g})/(1.11\text{ g/mL})$$

$$= 90\text{ mL}$$

Concentration of Antifreeze (%)	T_i (°C)	T_f (°C)	ΔT (°C)	Amount of Mixture (g)	Heating Time (s)	Specific Heat $\left(\frac{\text{cal}}{\text{g·°C}}\right)$	Boiling Temp (°C)
0	20.0	55.8	35.8	400	200	1.0	98
25	20.0	59.4	39.4	400	200	.91	101
50	20.0	63.7	43.7	400	200	.82	105
75	20.0	70.5	50.5	400	200	.71	113
100	20.0	77.8	57.8	400	200	.62	135

4. Immersion heaters are inexpensive and work well *so long as students do not operate them when they are not submerged in liquid.* If you don't have those, there are several other methods by which you can determine the specific heat of a liquid such as antifreeze.

   One is the method of mixtures, where you start with known amounts of antifreeze and water at dissimilar temperatures, and pour them together in an insulated container and note the final temperature, as in the previous lab.

   A second method involves the transfer of heat from a piece of metal, such as an aluminum cylinder, to the antifreeze. The specific heat of the metal would need to be known either by looking up the value or by an experiment. Place the known mass of metal in boiling water for five minutes and then into a known amount of antifreeze at a known temperature in an insulated container. From the initial temperatures of the metal and the antifreeze and the final temperature of the two materials, the specific heat of the antifreeze can be computed.
5. Sample data are given on the previous page.
6. *Sample Computations*

   The quantity of heat transferred to the water (Step 2) is

$$Q = mc\Delta T$$

$$= (400\text{ g})(1.00\text{ cal/g}\cdot{}^\circ\text{C})(35.8^\circ\text{C})$$

$$= 14\,300\text{ cal}$$

   The specific heat of a 50% mixture of antifreeze and water (Step 4) is

$$c = Q/m\Delta T$$

$$= (14\,300\text{ cal})/(400\text{ g})(43.7^\circ\text{C})$$

$$= 0.82\text{ cal/g}\cdot\text{C}$$

52 Gulf Stream in a Flask

1. If time and/or equipment are not available for this lab to be done as a student laboratory, it can be done as a class demonstration.
2. Have the glass tubing installed in the stoppers before the students come to lab. Note that the tube that sticks out of the stopper needs to extend only about 3 inches above the stopper. The other tube should go to the bottom of the flask.
3. Make sure not to tell the students the results at the outset. Let them discover what happens.
4. Extreme care should be exercised when handling the flask with the hot water, and when inserting the stopper assembly. Some hot water may "leak" out when inserting the stopper assembly. For safety reasons, be sure to watch students very closely when they handle the flask of hot water.

53 The Bridge Connection

1. This is a sophisticated experiment that requires a high level of mathematical computation. With reasonable care, the results are fairly reliable and clearly demonstrate an engineering application of thermal expansion. Be certain of the operation of the apparatus before students attempt using it. If your students have particularly low math skills or your equipment does not operate well, you are advised to skip it.
2. There are at least three different varieties of thermal expansion apparatus: the lever form, the micrometer form, and the roller form. This manual describes use of the roller form because it may be adapted from existing IPS equipment and because use of the axle-pointer requires proportional reasoning.
3. You can reduce time by having the water in the steam generators heated before your students arrive.
4. Caution your students about handling hot expansion rods; they should allow them to cool first.
5. You may want to go over with your students how to use a micrometer or vernier caliper. If you do not have a class set, you might train an able group of students to tutor other students as you help the rest of the class. Or, you can simply measure the diameter of the axle of the pointer yourself and put it on the chalkboard.
6. Watch for the following:
   (a) The groove in the expansion rod should be firmly placed on the fulcrum.
   (b) The axle should be free to rotate, and the pointer should not rub against the dial.
   (c) The double-wheel rollers should be properly adjusted with an Allen wrench so they roll freely with a minimum of friction.
7. Some coefficients of linear expansion for common materials are given in the table.

Material	**Coefficient of Linear Expansion (°C^{-1})**
Lead	29×10^{-6}
Aluminum	26×10^{-6}
Brass	19×10^{-6}
Concrete	12×10^{-6}
Iron or steel	11×10^{-6}
Platinum	9.0×10^{-6}
Ordinary glass	8.5×10^{-6}
Pyrex glass	3.3×10^{-6}

54 Cooling Off

1. The temperature of a cooling object decreases exponentially with time, as shown in the graph. (The equation for this graph is given in the answer to Question 9.)

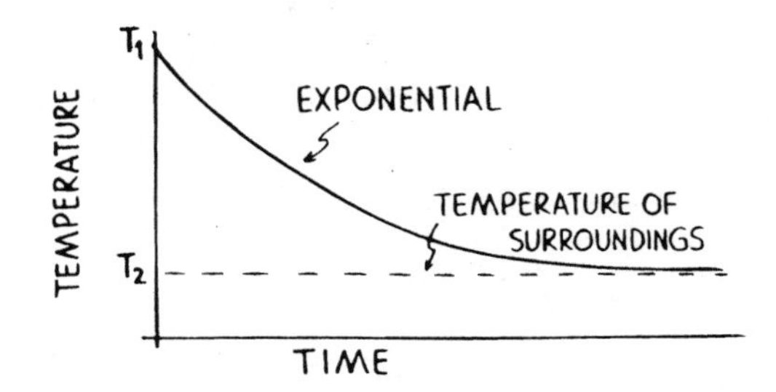

If your students are mathematically inclined, you might want to try to explain the meaning of exponential increase or decrease. Refer your students to Appendix E, "Exponential Growth and Doubling Time," of the text.

2. If you have data plotting software or Vernier Software's "Graphical Analysis" program, your students may want to do Step 7, in which they can analyze their data using logarithms. Also, if you have it, see page 30 of Vernier's "Temperature Plotter" documentation.

55 Solar Equality

1. Use jars with tight-fitting lids, such as peanut butter or mustard jars. Obtain rubber stoppers, and drill a hole in each jar lid large enough to hold a stopper.
2. Spray paint the metal strips beforehand to save time.
3. The metal strips and glass jars may be saved from year to year.
4. There are two environmental conditions that are important in obtaining accurate results: the day should be clear, with minimal cloud cover; and the indoor and outdoor temperatures should be as close as possible.
5. Introduce this laboratory by asking students to guess the power output of the sun. Review the idea that a watt is a common unit of power, and define it in terms of common wattages, such as those of lightbulbs or appliances.
6. To avoid broken thermometers, have the students come to you for guidance in putting the thermometers through the stopper. Use the glycerin as a lubricant.
7. *Sample Data and Computations*
   Maximum temperature recorded in sunlight: 60°C
   Distance from lightbulb to foil strip when temperature reading is same: 9.5 cm, or 0.095 m

$$\text{sun's wattage} = \frac{(\text{bulb's wattage}) \times (\text{sun's distance})^2}{(\text{bulb's distance})^2}$$

$$= \frac{(100\ \text{W})(1.5 \times 10^{11}\ \text{m})^2}{(0.095\ \text{m})^2}$$

$$= 2.5 \times 10^{26}\ \text{W}$$

The number of 100-watt lightbulbs needed to equal the sun's power is

$$\frac{(\text{sun's wattage})}{100\ \text{W}} = \frac{2.5 \times 10^{26}\ \text{W}}{100\ \text{W}}$$

$$= 2.5 \times 10^{24}$$

It would be impossible for the earth to produce enough electricity to perform this feat. The table comparing power values will help students see the reason.

Wattage Value (watts)	Item
10^{26}	Power output of sun
10^{17}	Solar power incident on the earth
10^{15}	Solar power striking the United States
10^{13}	Total world power consumption
10^{12}	Total U.S. power consumption
10^{9}	Output of one large generating plant
10^{4}	Solar power striking a house roof
10^{3}	Output of electric stove or clothes dryer
10^{2}	Output of one lightbulb
10^{0}	Output of electric clock or flashlight
10^{0}	Output of a sedentary human

56 Solar Energy

1. A combination of blue and green food coloring will make a very dark solution. However, there are other combinations you might try.
2. This activity works best on a very sunny day.
3. The activity is short, so you might want students to run several trials and average their data.
4. *Sample Data and Computations*
   Volume of water: 140 mL
   Mass of water: 140 g
   Initial water temperature: 23°C
   Final water temperature: 26°C
   Temperature difference: 3°C
   Typical top diameter of Styrofoam (plastic foam) cup: 6.9 cm

The surface area of the top of a typical Styrofoam cup is

$$\text{area} = \pi(\text{diameter}/2)^2$$

$$= 3.14(3.5 \text{ cm})^2 = 38 \text{ cm}^2$$

The energy collected in the cup was

$$\text{energy} = mc\Delta T$$

$$= (140 \text{ g})(1.0 \text{ cal/g·°C})(3\text{°C})$$

$$= 400 \text{ cal}$$

The solar energy flux was

$$\text{solar energy flux} = \frac{\text{energy}}{(\text{area}) \times (\text{time})}$$

$$= \frac{400 \text{ cal}}{(38 \text{ cm}^2)(10 \text{ min})}$$

$$= 1 \text{ cal/cm}^2\text{·min}$$

57 Boiling Is a Cooling Process

1. If available, Franklin's flasks are convenient because crushed ice will stay in the crater of the bottom when they are turned upside down. Even with Florence flasks, ice can be placed on the flat part of the Florence flask.
2. Have the students go through the dry run outlined in Steps 2 and 3. Caution them about safety in handling the flask of hot water in Step 5.
3. With most test tube clamps, your students can simply rotate the flask upside down in position, minimizing the need to handle it.
4. The graph of temperature vs. time for the water in the flask (Step 7) will have the form of the one shown.

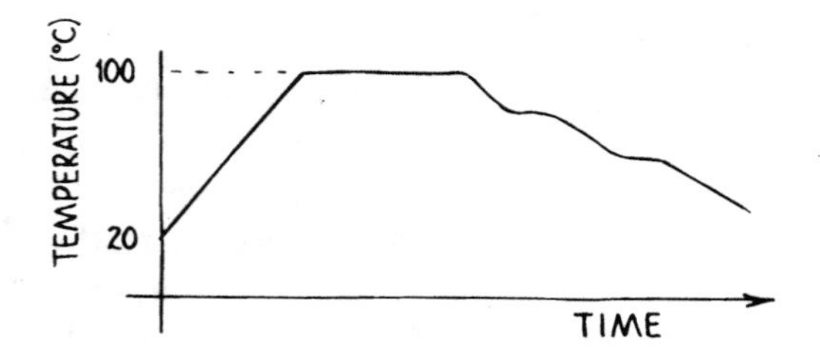

58 Melting Away

1. Place a bowl of ice cubes out before the lab begins to allow them to come up to a slippery wet 0°C. Cubes right out of the freezer are probably below 0°C and dry.
2. Remember that once an ice cube melts, the 0°C ice water that the ice turns into is then warmed by the surrounding warmer water to the final temperature of the mixture. Be sure to have your students start off with enough water so that the water doesn't drop to 0°C before the ice melts.
3. If subfreezing ice cubes are used, the energy used to warm the ice cube from its initial subfreezing temperature up to 0°C will make a contribution to the measured "heat of fusion." This contribution will be equal to the specific heat of ice, about 0.5 cal/g·°C, multiplied by that temperature change.

59 Getting Steamed Up

1. Have your students observe the appearance of the steam as it leaves the tubing. If you have adjusted things correctly, you should not see anything next to the tubing, yet a cloud develops nearby, looking like our standard version of "steam." The cloud is tiny droplets of liquid water condensing as it hits the "cool" room air, but then evaporating again. Steam itself is invisible.
2. The part of the steam that does manage to bubble up through the water does not condense in the cup nor does it heat the water. Only the steam that condenses in the cup heats the water.
3. Steam will condense in the tubing that connects the steam generator to the cup. This water will drain into the cup and increase the mass of the water in the cup significantly. Using a steam trap eliminates this problem. If you don't have commercial steam traps, you can make one from a 250-mL flask and a two-hole stopper, as shown in Figure A on page xxx. Keep the tubing from the steam trap to the cup as short as possible.
4. Don't forget that the 100°C water that condenses from steam cools down to the final temperature of the water in the cup.

60 Changing Phase

1. By comparing the cooling curves of water and of naphthalene in the test tubes, your students should be able to associate the flattened portion of the temperature graph with the energy released by the paraffin in changing state.
2. Make sure that the students understand that they are to heat the paraffin in the water bath until the paraffin has completely liquefied.
3. This is a good lab to use the thermometer probe(s) to show the advantage in using a computer to do much of the detailed monotonous tasks of collecting data and displaying it graphically. Use two temperature probes, and take data every 30 seconds.

4. Previous editions of this lab manual used naphthalene (moth flakes) instead of paraffin wax. Napthalene has the advantage of having a relatively convenient melting/freezing point of 79°C. However, it has the disadvantage of not being easily removed from the thermometers or temperature probes. If not thoroughly cleaned, the naphthalene may penetrate the tip of the probe, causing it to fail.

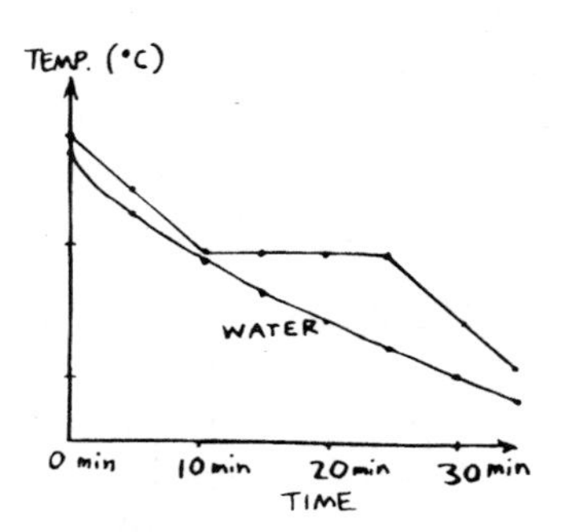

61 Work for Your Ice Cream

1. The instructional goal of this activity can easily be lost. It is up to the individual teacher to determine how much scientific rigor to maintain. There appears to be an adverse relationship between class excitement and the ability to analyze data *numerically*.
2. Several energy transfers are involved. Watch for them. In this lab, careful thought needs to be given to what information is required to make the calculations. Consider assigning the determination of certain energy transfers to groups of students. It is recommended that you introduce this lab to your class a day before they do it because students will need time to formulate methods to evaluate many variables.
3. More on the answer to Question 3. (1) Salt can dissolve not only in liquid water, but also in *solid* water (ice)! Once salt is dissolved in ice, the mixture is above its melting point (which can be as low as –20°C). It "wants" to be liquid. It extracts energy from its environment to provide the energy needed to melt the ice. Then the slushy salt water is actually colder than the ice was. (2) The melting point of ice is lowered by pressure. (That's what makes ice slippery and enables ice skaters to skate.) When a tire rolls over ice, it melts a thin layer. This thin water layer can then dissolve salt. In an ice cream freezer, the salt-ice slush is formed mostly by the first process and becomes cold enough to make ice cream.

Sample Data and Computations

4. The following data were collected with a particular ice cream mix and a dedicated group of physics teachers.

Force on Crank (N)	Number of Turns
8.8	0
8.8	100
8.8	200
10.8	300
11.8	400
13.7	500
17.6	600
23.5	700
33.3	800
41.1	900

The distance the force moves equals the number of turns times the circumference of the circle, which in this case was 1.13 m. A force vs. distance graph was plotted. The energy transferred was estimated from the area under the curve, which was approximately 17 400 J, or 4160 cal.

5. The specific heat of the ice cream mix, by using the method of mixtures of water and the liquid mix, was found to equal 0.80 cal/g·°C.
6. The heat lost by the mix as it cooled from the original temperature of 14°C to –10°C for the 3.05 kg of mix was 58 600 cal.
7. The heat of fusion was found to be 45 cal/g by measuring the decrease in temperature of water as some ice cream melted in the water, thus causing the 3.05 kg of ice cream mix to lose 137 000 cal.
8. The total energy expended and lost by the ice cream mix was approximately 200 kcal.
9. The 2.4 kg of ice melted and absorbed 192 cal.

62 The Drinking Bird

1. Probably the ultimate dissertation ever written on the drinking bird is "The Hydro-Thermal-Dynamical Duck" by Henry A. Bent and Harold J. Teague, *Journal of College Science Teaching*, Vol. VIII, Number 1, September 1978, p. 18.
2. Push the neck of the bird up and down until the fulcrum is positioned so that the bird can tip over and return.

63 The Uncommon Cold

1. In this experiment, the pressure on some trapped air is assumed to remain constant, and the volume is assumed to be proportional to the temperature. The volume and temperature are measured at the temperatures of boiling water and of melting ice. The temperature for zero volume of air is extrapolated from a straight-line graph of volume vs. temperature drawn through the two data points. This is an accurate procedure because the direct proportionality of volume to temperature is a valid law in the temperature range in which you will be working.

2. Caution students to use a paper towel or pot holder to hold the hot clamp and flask after the air in the flask has been heated.
3. Be sure to use a one-hole stopper. A two-hole stopper will give very poor results.
4. Great care should be taken in not allowing air to enter the flask as the flask is placed in the ice bath and the glass rod is removed from the stopper.
5. If the flask and stopper are not totally dry inside, evaporated water in the flask will condense and draw in a much larger volume of water from the ice bath.
6. Do not allow students to assume that a 250-mL flask has a capacity of 250 mL. Have them use the graduated cylinder to measure how much water it can hold.
7. *Sample Data and Computations*
   Temperature of boiling water: 98.5°C
   Volume of trapped air in boiling water bath: 335.0 mL
   Temperature of ice bath: 2°C
   Volume of water in the flask from ice bath: 87.5 mL
   Volume of trapped air in the ice bath: 247.5 mL
   A graph of volume of trapped air vs. temperature can then be made by drawing a straight line through the two data points. Where the graph line intersects the x-axis, the volume would be zero and the temperature would be at absolute zero.

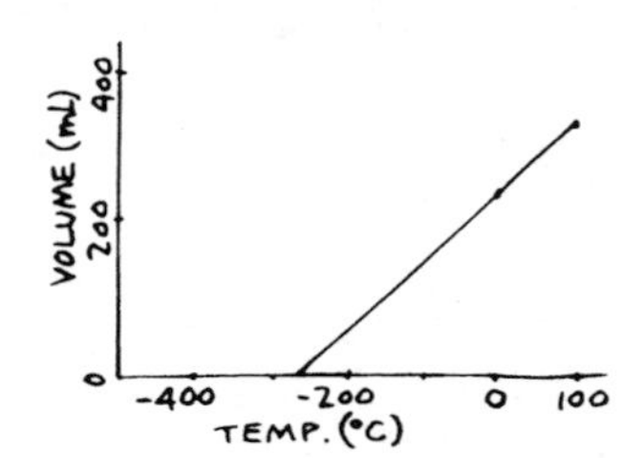

8. The accuracy with which the points are plotted is very critical. It should be emphasized that extrapolating data is a high-risk procedure. Not only is the accuracy of predicted values in doubt, but we can't be sure that nature behaves the same outside of collected data points as between collected data points. Student values for absolute zero may easily vary up to 50°C from the accepted value of –273°C.

64 Tick-Tock

1. Provide your students with a variety of pendulum bobs and string, and allow them to discover that the period depends solely on its length.
2. Consider having a student contest to see what group is the first to find the correct length, which is about 25 cm.

65 Grandfather's Clock

1. This lab is an extension of "Tick-Tock" (Activity 64). Now that the students know that the period of a pendulum depends on its length, they will investigate the mathematical relationship in more detail.
2. This laborious and tedious endeavor is transformed into a delightful and pleasurable one for students who use data plotting software.
3. It is far more important for your students to learn that a plot whose graph is a straight line shows direct proportion than to find a particular equation for a pendulum.

66 Catch a Wave

1. This exploratory lab introduces the notions of constructive and destructive interference, beats, traveling and standing waves, the Doppler effect, and sonic booms, and distinguishes transverse and longitudinal waves.
2. If only one computer is available, you might allow students to do this lab while others work on another assignment. Alternatively, this lab is easily adapted as a teacher demonstration.

67 Ripple While You Work

1. This lab is easily adapted to a teacher demonstration.
2. If students will be doing the lab, demonstrate the operation of a ripple tank before the students begin the lab.

68 Chalk Talk

1. The purpose of this activity is to demonstrate that vibrational motion is a source of sound energy that can be transmitted through a medium, and that there are a number of variables that affect the frequency.
2. The most pure musical sounds will probably be produced as the "instrument" is pulled toward the "musician" with the chalk on the leading edge. Allow students only a brief time to experiment, and stop the activity while interest is still high.

69 Mach One

1. Golf tubes can be purchased at most local sporting goods shops. You might shorten them a bit so that they are a bit less unwieldy.

2. Do not dwell on the reason for adding 0.4 times the diameter of the tube to the measured length of the air column.
3. There are other common examples of resonance: pushing (or pumping) a swing at the proper frequency, a car that rattles at just one speed, the throbbing from two engines that are run at slightly different RPM's, the TV commercial in which a glass is shattered, and so on.
4. Strike the tuning fork with a rubber heel, soft wood, or a rubber hammer. Tuning forks between 256 Hz and 500 Hz are ideal. Don't let the students touch the graduated cylinder with the tuning fork. It may damage the fork or the cylinder.
5. At a room temperature of 20°C, the speed of sound is about 344 m/s. If you use a tuning fork with a frequency of 440 Hz, then the wavelength is

$$\text{wavelength} = \frac{\text{speed}}{\text{frequency}}$$

$$= \frac{344\ \text{m/s}}{440\ \text{Hz}} = 0.78\ \text{m} = 78\ \text{cm}$$

This means you can expect the first resonance to occur at 1/4 of 78 cm, or 19.5 cm. The second resonance will occur at 3/4 wavelength, the third at 5/4 wavelength, and so forth.
6. In Step 7, students can compute the speed of sound in air by multiplying the wavelength obtained in Step 6 by the frequency of the tuning fork recorded in Step 2.
7. Don't scrimp on antacid tablets; for best results use two tablets per tube.

70 Shady Business

1. Shadows are such a common phenomena that your students probably do not fully understand them.
2. After the lab, you might wish to explain the formation of eclipses, using the terminology *umbra* and *penumbra*. Don't tip off your students by using these words before the lab, however. (Eclipses are described in the textbook.)
3. The formation of the penumbra is primarily the result of straight light rays emanating from a *non*-point source—not the bending (diffraction) of light around the opaque edges of the object.

71 Absolutely Relative

1. It is necessary to darken the room as much as possible to perform this lab; the light probes are *extremely* sensitive!
2. Black construction paper or black cloth on the table helps eliminate almost all reflection so that the lab can be performed without ring stands.
3. Calibrating the light probes eliminates the need for discussing the unit of luminous intensity, the candela. All measurements are expressed as a percentage of the intensity at the calibrated distance. This parallels methods used in astronomy.
4. Although the inverse-square relationship can be obtained by manual graphing techniques, it is far more efficient to use the computer.
5. As the graphs on this page indicate, on a graph of intensity vs. distance, the intensity decreases rapidly as the distance increases, but never quite reaches zero. The graph of intensity vs. $1/(\text{distance})^2$ is a straight line.

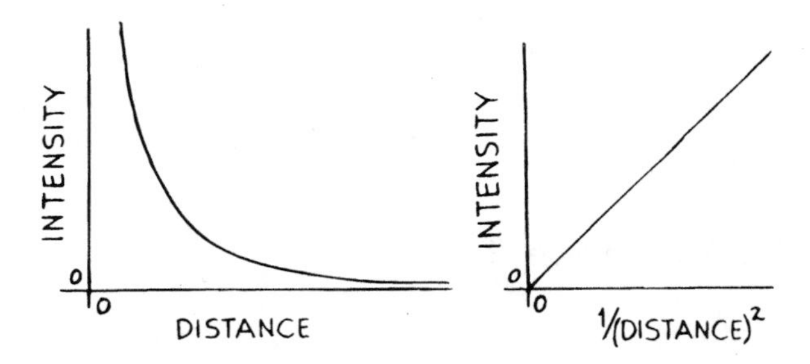

72 Shades

1. Inexpensive polarizing sheets can be obtained from Edmund Scientific Co.
2. Although the explanation for the "Going Further" makes a great review of vectors, it is quite sophisticated and should not be required of all students. Appendix D of the text introduces the vector nature of light and how light is polarized. Polarized light that is passed through the filter closest to the light source has a component parallel to the axis of the second filter (which is at 45° to the axis of the first filter). This component gets through. *It* has a component parallel to the axis of the third filter (which is also at 45° to the axis of the second filter), so that component gets through.
3. Demonstrating polarizing filters on an overhead projector makes a great follow-up class demonstration. You might cap your demonstration by rotating a single polarizing filter above a clear transparency containing many overlapping strips of clear, shiny tape (not matte-finish, permanent tape).

73 Flaming Out

1. You can assemble your own materials for this laboratory, but much time and effort can be saved by purchasing commercial spectra-of-the-elements kits.

2. Emphasize the importance of *not* mixing the test wires.
3. If you assemble your own spectroscopes, you will need to purchase plastic replica grating material. This can be inexpensively purchased from Edmund Scientific Co.
4. Share this quote with your students. Even great minds can be wrong!

   "More than a century ago, Auguste Comte also decried as absolute nonsense the idea of attributing a chemical composition to a distant star. It may have made sense to speak of the chemical composition of the Sun, he was willing to admit; but certainly not the composition of a star to which there is not the slightest possibility of anyone ever traveling. Of course, in the meantime, half a dozen ways have been found to get at the composition of a star, and many a satisfactory check has been made of one method against another." (A. Comte, "Cours de philosophie positive," Paris, 1835.)

74 Satellite TV

1. Allow the students to experiment with several weak stations.
2. You may even have students use the foil-lined umbrella to try to block out radio waves by placing the radio outside the umbrella, at the tip.

75 Images

1. Modeling clay or rubber bands and small pieces of wood can be used to support the mirrors.
2. Do remember that this is an exploratory activity designed to arouse student curiosity. Don't be concerned about "right" answers at this time. Let the students explore image formation.
3. Encourage the students to make diagrams depicting how they think light travels from the object pencil to their eyes in order to form an image. The second pencil should be placed behind the mirror, but allow students to grapple with this placement problem *before* giving them hints.

76 Pepper's Ghost

1. You can read about John Henry Pepper in Edwin A. Bawes, *Great Illusionists* (New York: Shotwell Books, 1979).
2. To support the thick (approximately 3/8-inch) plate of glass, use several 3/4-inch thick wooden supports with a shape similar to that shown below. (Two thin sheets of glass will work as a substitute for a thick sheet.)

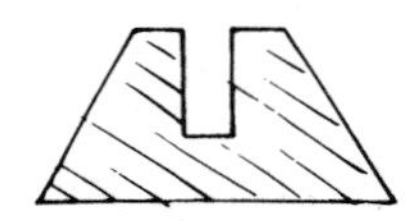

3. Encourage the students to use diagrams to help them think of possible explanations about the images they see.

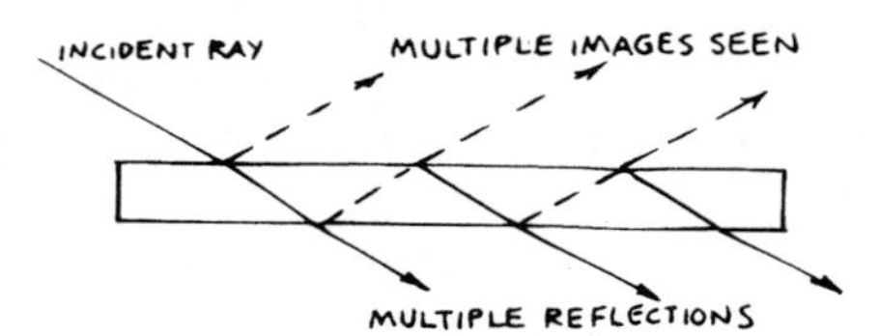

77 The Kaleidoscope

1. There are interesting uses of multiple reflections. They are often seen at a carnival fun house. Another use is the kaleidoscope, which is discussed here.
2. As the mirrors become closer to 0°, the number of images increases enormously. Two mirrors placed face to face could have an unlimited number of images as the reflection is bounced back and forth. A simple example is looking through a small mirror at a large mirror with your back to the large mirror.
3. An interesting additional activity is to obtain two cheap kaleidoscopes: one for students to tear apart and one for viewing the total effect.

78 Funland

1. The biggest stumbling block students will encounter with this lab is that for this analysis d_i and d_o is the distance from the object or image to the focal point. The chief advantage of this technique is that the data graphs nicely as a hyperbola, indicating an inverse relationship between d_i and d_o. Students have the natural tendency to measure d_i and d_o from mirror to the object or image.
2. This lab is a mirror image conceptually of Experiment 82, "Bifocals." If time is short, you might have your students do either one or the other, or have half the class do one lab while the other half does the other.
3. Mirrors of diameter 38 mm and focal length 25 cm work well.
4. Glass mirrors can be protected and numbered by mounting them on a square piece of cardboard. This also makes them easier to manipulate.

5. The filament of a clear 7-watt night-light bulb makes an excellent object. With careful observation, students can ascertain whether the image is inverted or erect.
6. If the sun is out, students can measure the focal lengths of their mirrors outdoors.
7. Be sure that students understand that in Data Table B they are to record the distances from the focal point, not from the mirror surface.
8. If students multiply image distance d_i by object distance d_o, for different locations of an object, the product is nearly constant, typically within 20%. This is because $d_o\, d_i = f^2$.
9. At the end of the lab, you may wish to derive the equation

$$d_o\, d_i = f^2.$$

In the ray diagram for a concave mirror, the light gray right triangle near the mirror has a base very close to f and a height equal to the object height, h_o. The darker gray right triangle near the mirror has a base very close to f and a height equal to the object height, h_o. The darker gray right triangle near the mirror also has a base very close to f and a height equal to the object height, h_o. The darker gray right triangle near the mirror also has a base very close to f and a height equal to the image height, h_i.

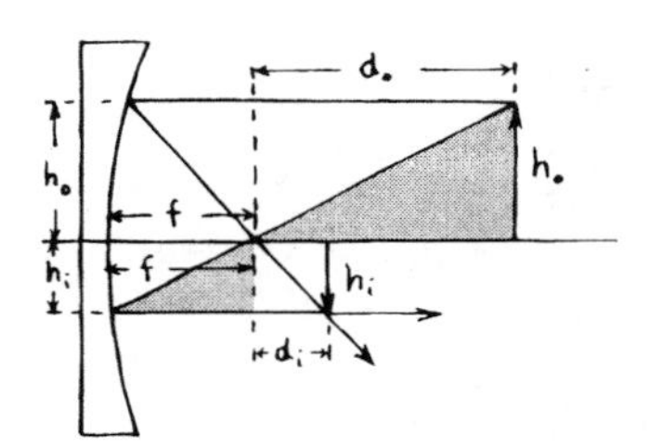

Since the two darker gray triangles are similar, the ratio of their heights equals the ratio of their bases:

$$\frac{h_o}{h_i} = \frac{d_o}{f}$$

Also, since the two light gray triangles are similar:

$$\frac{h_o}{h_i} = \frac{f}{d_i}$$

Thus,

$$\frac{h_o}{h_i} = \frac{d_o}{f} = \frac{f}{d_i}$$

$$d_o d_i = f^2$$

79 Camera Obscura

1. This activity is an extension of Activity 1 on page 478 of the textbook.
2. The camera shown in Figure A can be made from a covered shoe box or any other box that has a long cover. Cut one end off the box. Use a convex lens of diameter 25 mm to 38 mm and a focal length less than the length of the box. Cut an opening for the lens in the other end of the box. Tape the edges of the lens to the box. Make a pinhole next to the lens. Make a screen by taping glassine paper to the inside of the box at one focal length from the lens. Cover the lens with a flap of aluminum foil.
3. It may be of interest to your students that the word *camera* is derived from the Greek word *kamara,* meaning "vaulted room." Royalty in the sixteenth century were entertained by the camera obscura.
4. A lens bends light for the same reason a prism does: the faces are not parallel. Only at the center of a lens are both faces parallel to each other. As a result, light that passes through the center of a lens undergoes the least bending.

   A pinhole acts like the center of a lens. A pinhole image can be formed at *any* distance from the pinhole, so that on a fixed screen, objects at varying distances from the pinhole are all in focus.
5. For further information, contact:
   Eric Renner's Pinhole Resource, Star Route 15, Box 1355, San Lorenzo, NM 88041. Phone: (505) 536-9942.

80 Thin Lens

1. It is best that students acquire a qualitative introduction to the formation of images with convex and concave lenses *before* proceeding to the quantitative experiment, "Bifocals."

81 Lensless Lens

1. This is a simple activity to do and reinforces the point made in Activity 79, "Camera Obscura," that a pinhole, unlike a lens, focuses equally well on objects at all distances.
2. Since the light not passing through the center cannot be focused on the retina within a certain fixed distance, the image appears blurry. The reason a pinhole acts as a magnifying lens is that it obscures all the light except that which passes through the center of the lens of the eye. This allows you bring the object (the printed page) closer in focus albeit dimmer.

82 Bifocals

1. This lab is a mirror image conceptually of Experiment 78, "Funland." If time is short, you might have your students do either one or the other, or have half the class do one lab while the other half does the other. As is the case with "Funland," the biggest stumbling block students will encounter with this lab is that for this analysis d_i and d_o is the distance from the object or image to the focal point. The chief advantage of this technique is that the data graphs nicely as a hyperbola, indicating an inverse relationship between d_i and d_o. Students have the natural tendency to measure d_i and d_o from mirror to the object or image.
2. The analysis of a convex lens is very similar to that of a concave mirror. If time is in short supply, you might choose to do one in lab and lecture about the other.
3. Lenses of diameter 38 mm with focal lengths between 25 and 40 cm work well. One inexpensive way to keep your lenses organized is to keep them in an egg carton.
4. If the sun is out, students can measure the focal lengths of their lenses outdoors.
5. The filament of a clear 7-watt night-light bulb makes an excellent object. With careful observation, students can ascertain whether the image is inverted or erect.
6. If students multiply d_o and d_i for different locations of an object, the product is nearly constant, typically within 20%. This is because $d_o\, d_i = f^2$.
7. It is better for your students to discover the inverse relationship between image and object distances than for them to memorize or verify formulas they are likely to forget.
8. In Step 10, the students are being asked to form a telescope with two converging lenses. In a telescope, the objective lens (the lens closer to the object) forms a real image at the focal point of the eyepiece (the lens closer to the eye). This image becomes the *object* of the eyepiece, which forms a magnified, virtual image. The magnification equals the focal length of the objective divided by the focal length of the eyepiece, so the lens with the longer focal length should be the objective.
9. The telescope was invented in Holland, but some controversy exists over the actual inventor—according to *Encarta Encyclopedia.* The invention is usually ascribed to Hans Lippershey (1587–1619), a Dutch spectacles maker, about 1608. Galileo exhibited, in 1609, the first telescope on record. The German astronomer Johannes Kepler discovered the principle of the astronomical telescope with two convex lenses. (Lippershey and Galileo used one converging and one diverging lens.) This idea was actually employed in a telescope constructed by the German Jesuit astronomer Christoph Scheiner (1579?–1650) about 1630. Because of the difficulties caused by spherical aberration, astronomical telescopes had to be of considerable focal length, some of them up to 200 ft (60 m).
10. The relationship $d_o d_i = f^2$ can be derived geometrically for a lens as well as for a concave mirror. (See teacher notes for Experiment 78, "Funland," and use similar reasoning with the ray diagram of the convex lens shown here.)

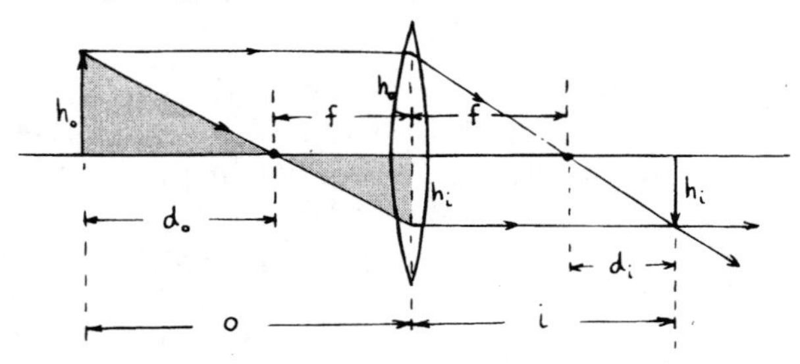

The relationship can also be derived algebraically from the thin-lens equation

$$\frac{1}{o} + \frac{1}{i} = \frac{1}{f}$$

where o = distance from lens to object
i = distance from lens to image

Since d_o and d_i are distances from the *focal point* to the object and image, then for objects and images beyond the focal point

$$o = f + d_o$$
$$i = f + d_i$$

Thus, the thin-lens equation becomes

$$\frac{1}{f + d_o} + \frac{1}{f + d_i} = \frac{1}{f}$$

$$f(f + d_i) + f(f + d_o) = (f + d_o)(f + d_i)$$

$$f^2 + fd_i + f^2 + fd_o = f^2 + fd_i + fd_o + d_o + d_i$$

$$f^2 = d_o d_i$$

11. Here are some interesting tidbits about the history of lenses from *Encarta Encyclopedia.* In 1268 the English philosopher Roger Bacon recorded the earliest statement about the optical use of lenses. Possibly as early as the tenth century, however, the Chinese had made use of magnifying glasses placed in frames.

Eyeglasses were first used in Europe in Italy, and some portraits dating from the Middle Ages depict persons wearing eyeglasses. With the invention of the printing press in the fifteenth century, the demand for eyeglasses increased and by 1629 was large enough for a charter to be granted to a guild of spectacle makers in England. The first bifocal glasses were made for Benjamin Franklin at his suggestion about 1760. In early times the only eyeglasses having spherical lenses were manufactured to correct nearsightedness and farsightedness. Not until the end of the nineteenth century did the cylindrical lens for the correction of astigmatism come into common use.

83 Where's the Point?

1. Students will have some difficulty in measuring the diameter of the laser beam because the beam decreases in intensity rapidly near the edges, making the edge fuzzy. For this reason, when one wishes to strive for better precision, the beam intensity is measured with a photometer at the beam center. Then, the half-power points—where the beam is half as bright—are located at several places around the circumference. The beam diameter is then defined as the diameter between half-power points.
2. If you don't have a photometer, the results will be more of an estimate. Shorter focal lengths are the most difficult to measure.

84 Air Lens

1. Students need to know that light travels at different speeds through different materials, and that the relative difference in speeds determines the direction of refraction at the boundary between two materials.
2. The air lens is an interesting exception to common lenses, which are material lenses in an air environment. It illustrates that when air and glass are interchanged, a convex (air) lens diverges light, and a concave (air) lens converges light.

85 Rainbows Without Rain

1. A wonderful soap-bubble solution can be made from 2/3 cup Dawn® brand dishwashing detergent and 3 tablespoons glycerin for every gallon of water. It is best to let this solution stand overnight.
2. If students hold the soap-film frame vertical long enough, they should notice that the top of the frame appears dark. The light wave that reflects from the first surface, the air-to-water boundary, reflects with a half-wavelength (180°) phase shift—just like the Slinky® spring toy reflection from a rigid support or the reflection of a light spring from the junction with a heavy spring. The light wave reflecting from the back side of the film, or water-to-air boundary, reflects with no phase shift—just like the reflection on a heavy spring from a junction with a light spring. These two reflected light waves are then out of phase and produce destructive interference.

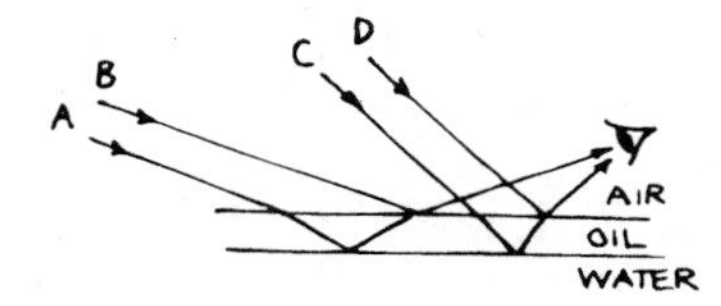

If the wedge of soap film gets to be 1/4 wavelength thick, then the path-length difference of the two reflected waves is 1/2 wavelength. With the 1/2 wave shift, there is constructive interference for that particular wavelength. The longer the wavelength, the thicker the wedge will need to be before the paths will interfere constructively. As the wave path lengths increase, there will be alternate destructive and constructive interference for all wavelengths. So the first color band to appear is blue.

3. Since this is an exploratory activity, you should not expect your students to produce fully developed explanations of their observations. However, it is hoped that the students will be able to relate some of their understandings of properties of waves to these self-demonstrations and will be able to give some qualitative explanations of the phenomena.
4. The observations of oil films floating on water also show the effect of interference of light waves. In the diagram of the oil film on water, Waves C and D will interfere constructively for a shorter wavelength than Waves A and B if the path lengths vary in each pair of waves by a given number of wavelengths.

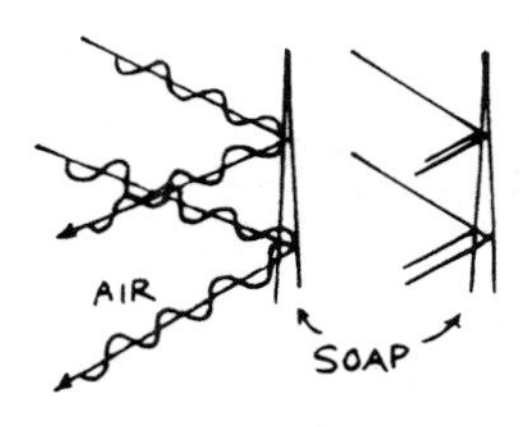

You may find many bands of a given color as the thickness of the oil film varies by multiples of 1/2 the wavelength. In this case, there will be a 180° shift at *both* reflecting surfaces since, at both surfaces, the interface is from a less optically dense material to a more optically dense material.

86 Static Cling

1. Many different types of materials found in physics storerooms will produce electrostatic charges. The most common ones are fur and a hard rubber rod (when rubbed together, they produce a negative charge on the rubber rod) and silk and glass (when rubbed together, they produce a positive charge on the glass rod).
2. The plastic golf club tubes become highly charged even under poor atmospheric conditions. It is difficult to get the tubes to transfer a charge directly to the electroscope. However, the electroscope will charge extremely well by induction.
3. In Step 6, the plastic foam peanuts seem to jump around in a very delightful and haphazard fashion. Some of the peanuts will acquire a charge from the tube and some of them will be repelled, but most remain attracted to the tube. The students might wish to try other things such as balloons, paper, and other pieces of plastic.

87 Sparky, the Electrician

1. In Step 1, encourage the students to try a number of arrangements. They will learn through experimentation, or by imitating others, that both ends of the battery, the threaded portion of the bulb, and the base of the bulb must all be involved in successful attempts to light the bulb.
2. Step 3 presents a unique challenge for the students to be creative and innovative. Remember to encourage the students to explore. Some students may actually hit upon parallel arrangements for the bulbs and battery and may get more than five bulbs operating at once.
3. Steps 4 through 7 introduce the student to the concepts of series and parallel circuits. Through this activity, the students should be able to come up with their own descriptions or definitions of series and parallel circuits.

88 Brown Out

1. This lab was adapted by Sheila Cronin from the *Electricity Visualized* CASTLE program as developed by Mel Steinberg of Smith College. It uses capacitors, batteries, and bulbs for an intuitive approach to electric circuits and the operation of components. The goal of the lab is to develop an understanding of how a resistor (and capacitor, too) function in a circuit. Qualitative exploration, rather than computation, is emphasized. The computational approach to resistance is developed in the following lab, "Ohm Sweet Ohm." Encourage students to visualize current flow in the circuit.
2. This lab also introduces students to the operation and use of capacitors.
3. It is important to use a voltage source between 3.6 and 4.5 volts.

89 Ohm Sweet Ohm

1. The Genecon® (a handheld, hand-cranked generator) and the nichrome wire apparatus provide a way to *see* how current varies with resistance via the brightness of a bulb. It's very hands-on, very conceptual. However, there are two drawbacks to keep in mind: (1) the resistance of the bulb changes with temperature, and (2) using the bulb lead as a current indicator may introduce an unwanted branch to your otherwise simple circuit. It is, therefore, very important that students follow the instructions provided in the lab manual carefully or they may unknowingly complicate their circuits.
2. The two major concepts that you want your students to learn by doing this lab are: (1) *resistance* varies with the length and cross-sectional area of wire, and (2) *current* varies directly with voltage and inversely with resistance. The second statement is *Ohm's Law.*
3. Electricity and magnetism are abstract to many people because they cannot "see" or "feel" electrons "flow" in a wire. Encourage your students to use the Genecon to get an intuitive feel for these two concepts as much as possible before attempting "Going Further."

90 Getting Wired

1. This lab was adapted by Sheila Cronin from the *Electricity Visualized* CASTLE program as developed by Mel Steinberg of Smith College. This program uses capacitors, batteries, and bulbs to build an intuitive approach to electric circuits. Encourage students to try to visualize *how* and *why* the charge is moving in the circuits.
2. Encourage students to distinguish the *amount* of compass needle deflection from the *direction* of the compass needle deflection.
3. The deflection of a compass by an electric current was discovered by accident by Ostered during a demonstration.

91 Cranking Up

1. This is a nice activity that really allows students to *feel* the work involved in keeping the circuits energized. The torque required to crank the

Genecon increases with the addition of each bulb in the circuit. More power is required to energize more bulbs.

92 3-Way Switch

1. Many an electrician has miswired a 3-way switch. Although one of the most common of household circuits, this circuit will provide a real challenge for your students.
2. After the students have constructed their circuits, compare the two settings on a single-pole double-throw switch to the "up" and "down" settings on a household 3-way switch. The "up" setting closes the circuit between the switch and one travel wire (wire between the two switches). The "down" setting closes the circuit between the switch and the other travel wire.

93 3-D Magnetic Field

1. Prepare a jar of clear vegetable oil and iron filings as follows. A one-pint jar with flat sides works well. Instant coffee often comes in such containers. Fill the jar to the neck with clear vegetable oil and add one teaspoon of iron filings. Place the lid tightly on the jar.
2. A strong horseshoe magnet works best with the jar of filings in oil. Using the jar allows students to develop the idea that a field is three-dimensional. When the mixture in the jar is shaken and put close to the magnet, expect to see some interesting results. The iron filings will align along the magnetic field lines like lots of fingers reaching outward.
3. The small, inexpensive compasses work well for Steps 3 and 4.
4. For Steps 5 and 6, caution your students about keeping the iron filings off the magnet.

94 You're Repulsive

1. The primary purpose of this activity is to allow students to explore the effect of a magnetic field on a moving charge. Initially, the expectations should be for the students to discover that there is a deflection of the electron beam in the presence of a magnetic field and that the deflection is at right angles to the magnetic field.
2. Don't be determined to teach the students the left-hand rule at this time. However, you could ask them whether they notice any pattern to the deflections and whether they would be able to predict the direction of the deflection given the orientation of a magnet and the direction of the electron beam.
3. As is indicated in the lab, if you don't have an oscilloscope, you may use an Apple II Series computer and monitor with the given program. You should find reasonably strong magnets to get nicely detectable deflections of the electron beam.
4. You should test in advance whether you can get deflections on your galvanometer or ammeter by moving a magnet quickly through a three-loop coil (Step 5). If not, you might try to get a deflection by either adding loops to the coil (make it a ten-loop coil, for example) or finding stronger magnets.
5. Allow the students some time to discover the motion that is needed to induce a current. Ask what variables they think influence the amount of current induced.

95 Jump Rope Generator

1. The rotational speed is more important than the direction. The measured output is usually almost the same in a north-south alignment as in an east-west alignment because magnetic field lines are still being cut by a component of the wire's velocity.

Sample Data

2. Output from a 20-foot twirling cord is about 5×10^{-6} A.
3. The voltage induced is about 2 mV, so the total power output is about

$$\begin{aligned} \text{power} &= \text{voltage} \times \text{current} \\ &= (2 \times 10^{-3}\ \text{V})(5 \times 10^{-6}\ \text{A}) \\ &= 1 \times 10^{-8}\ \text{W} \end{aligned}$$

The energy input to produce this power is *most* probably too small to sense. (The back EMF does produce a small force that would resist the motion of the wire, but it is very small.)

96 Particular Waves

1. Electrostatics kits have been distributed by Sargent-Welch for years. Other suppliers may carry them, also.
2. Electrostatic results are optimum when the air is cool and dry.
3. If the probes of your electroscopes are round, bend the zinc plates or magnesium ribbon so that they will remain on the probes.
4. Have students buff the zinc or magnesium vigorously with steel wool until it shines brightly just prior to performing the activity. The layer of zinc oxide that rapidly forms prevents the emission of photoelectrons.
5. The chief results of this lab are:

(a) Weak ultraviolet light discharges a negatively charged electroscope while bright infrared (or white) light does not. This is because excess electrons on a zinc surface in contact with the electroscope are easily dislodged by the higher-energy photons of ultraviolet light, and are mutually repelled into the surrounding air. The lower-energy photons of infrared or white light are unable to dislodge electrons.

(b) A positively charged electroscope cannot be discharged by ultraviolet light because dislodged electrons are attracted back to the positively charged zinc surface and, thus, are not removed from the electroscope.

6. Some ultraviolet lamps designed to induce the phosphorescence of minerals have two illumination settings—long and short wavelengths. The long wavelength setting will not discharge the electroscope, but the short wavelength setting will—nice!

97 Nuclear Marbles

1. This activity is an analogy of the Rutherford scattering experiment in which Rutherford determined the order of magnitude of the nucleus (Figure 17.8 of the text).
2. This lab could be initiated by telling the students to imagine that, in the front of the classroom, there is a very dense opaque cloud (not surrounding the teacher, of course!) with a massive ball suspended in this cloud. They can't get inside the cloud to measure the size of the ball directly, but are to suggest as many ways as they can think of to get some idea of the size of the ball. Help the students make the transition to the problem of determining the size of nuclei in a very thin foil.
3. Be sure the students roll the marbles randomly. To do this, have them close their eyes or look elsewhere as they roll the marble toward the target.
4. *Sample Data and Computations*
   Number of hits $H = 48$
   Width of target area $L = 60$ cm
   Total number of trials $T = 200$
   Number of target marbles $N = 6$
   The marble diameter is

$$\begin{aligned} d &= HL/2TN \\ &= 48(60\text{ cm})/2(200)(6) \\ &= 1.2\text{ cm} \end{aligned}$$

98 Half-Life

1. Depending on your resources, several different items can be used to represent radioactive atoms in this activity. Instead of pennies, the students can use M&M's® candies (about 4 medium-sized bags) or soft-drink bottle caps.
2. A box that is lower and flatter than a shoe box would also be appropriate to use.
3. The students need to keep an accurate count of the pieces removed after each shake.
4. The graph should resemble a radioactive decay curve.
5. A decay curve, like a cooling curve (see Activity 54, "Cooling Off"), decreases exponentially with time. To make the graph in Step 6 come out a straight line, students will need to plot the *logarithm* of the number remaining vs. the time.
6. *Sample Results*
   For pennies, M&M's, or bottle caps, approximately half (50%) will be removed each time. For the brass paper fasteners, approximately 15% will be removed each time, because there is only a 15% chance that a fastener will be lying on its head.

99 Chain Reaction

1. This simulation will provide a visual chain-reaction model that can serve as a springboard for understanding nuclear fission.
2. Playing cards may be substituted for the dominoes. Stand the cards on end in pairs tepee fashion. Line up the tepees in a formation similar to that in Figure A. This method works best on a carpeted surface.
3. If a free neutron interacts with the nucleus of a U-235 atom, the nucleus splits in two, or fissions. The resulting two large pieces are nuclei of new atoms, such as barium and krypton. In addition, several neutrons and a large amount of energy are released. The neutrons can each cause the fission of another U-235 atom.
4. The example of U-235 fission shown in Figure 40.2 of the text should be introduced *after* the student activity.

100 You're on Your Own

Many of the labs in this manual are intended to give you experience in performing certain specific tasks. As such, they often contain very specific directions. Sometimes you may know the "answer" to a lab before you do it. That's okay—because the answer is secondary to the important practice you gain in gathering, organizing, and interpreting data.

The following labs are open-ended, and do not entail specified steps for you to follow. The purpose of these open-ended investigations is to give you experience at attempting to find the solutions to problems for which the answer is *not* known. In these cases you must devise your own method. A list of such labs is shown below.

Choose one of the investigations from the list or come up with one of your own. It is important that you define exactly what your investigation will and will not accomplish. Formulate a hypothesis. As a safety precaution, be sure to have your instructor approve your experimental procedure and apparatus before you begin experimenting. Acquire and organize your data—graphically, if appropriate. Specify as many factors as you can that influenced your experimental outcome. Does your experiment affirm or refute your hypothesis? Remember, some of the most famous experiments in physics had *null* results—for example, the Michelson and Morely experiment—so do not despair if your experimental results refute your hypothesis.

Do and *enjoy!*

- Have your students photograph examples of "real world physics" in their daily lives. Photos may range from sports to model trains. Then have the students analyze and describe the physics illustrated in their photos.
- Find a relationship between the area covered by jigsaw pieces that are uncovered and the same pieces when connected. Predict the area needed to lay out pieces for a puzzle that is known to have a certain area when put together.
- Determine which factors influence the toppling speed of dominoes. What is the maximum speed for a chain of dominoes at least one meter long?
- Devise and build an accelerometer. Calibrate the accelerometer so that it measures acceleration in m/s^2. Explain its operation.
- Devise an experiment that shows how the applied force (or net force) and mass of a system are related to its acceleration.
- Devise a method to measure free fall acceleration.
- Devise a method to measure the speed of an object, such as a Wiffle® ball or tennis ball, as it strikes the ground.
- Devise a simulation using *Interactive Physics*™ (requires software available from Knowledge Revolution).
- Create a *Hypercard* stack that teaches a physics principle (requires a Macintosh computer). Use *Quicktime*® and *QuickTake*® to incorporate both audio and video to your presentation.
- Devise an experiment that demonstrates a) constant speed, b) constant velocity, c) constant acceleration, and d) variable acceleration (jerks). If available, use a sonic ranger.
- Devise an experimental procedure that measures the important variables of a damped harmonic oscillator, such as a loaded spring. If available, use a sonic ranger.
- What is the vertical temperature gradient over a 24-hour period in your physics classroom during a) a school day and b) the weekend? Compare the temperature gradients between days during both warm and cold seasons. How thermally efficient is the classroom? A computer with multiple temperature probes may be very helpful.
- Estimate the average force you exert on a baseball with your hand when pitched. Express your estimate in terms of the distance the ball is thrown and the time of flight—both quantities you can measure.
- Devise a method of measuring the coefficient of rolling friction for a Hot Wheels car.
- Use the CASTLE Kit (available from PASCO) to devise an experiment that investigates the effect different numbers of bulbs have on the charging/discharging times of a capacitor.
- Use the "Millikan Oil Drop" simulation (available from Vernier Software) to investigate the charge on an electron (requires Apple II or IBM compatible computer).
- Devise an experiment in chaos. You may wish to refer to Vernier Software's book *Chaos in the Laboratory and 13 Other Science Projects for the Apple II.*
- Use the "Pulse Mode Timing" of *Super Sonic Plus* to devise an experimental apparatus that will measure the speed of sound in a gas other than air.
- Devise an investigation using a spreadsheet (requires computer and spreadsheet program). You may wish to refer to *Wondering About Physics* by Dewey Dykstra and Robert Fuller (John Wiley and Sons, 1988). This neat little book can be used with a laser videodisc entitled "Physics Vignettes... A Series of 16 Physics What If?...Video Scenarios."
- Calculate the incline of a hill, θ, that you can cycle on at constant speed as a function of the gear radii. Test your predictions by cycling on a nearby incline.
- Segments of a short-span bridge are observed to deflect downward as vehicles cross it. Likewise, rails on a railroad track are deflected as a train rolls by. Devise a technique that will enable you to measure the deflection of the roadbed or the train track.
- Other ideas of your own invention!

Appendix

An excellent software program about significant figures, uncertainty, accuracy, precision, and good laboratory techniques, written for the Apple II Series computer, is a set of programs called *Physics Lab Help-Ware,* available from HelpWare Education Materials.

Teacher Feedback and Evaluation Form

Your feedback on this lab manual will be very helpful in making revisions in future editions. Please tear out this form and use it to let me know how these lab activities and experiments can be improved. Suggestions for new labs are welcome. Your ideas and suggestions will be credited. The author promises a personal reply.

Date: ____________________

Dear Paul,

I want to tell you what I thought of your *Conceptual Physics Laboratory Manual* (© 2002).

I liked certain things about the manual, including:

____

____

____

____

However, I disliked the following:

____

____

____

____

I do feel that the manual could have been improved in the following ways:

____

____

____

____

There are some other things I wish were in the manual, such as:

____

____

____

____

Here is something that happened in my class when I used an idea from your manual:

____

____

____

☐ Enclosed is an alternative learning activity that has worked well with my classes. You may wish to include it in future revisions.

Sincerely,

Name: ____

School: ____

School Address: ____

City, State, Zip: ____

School Phone: (____) ____–______

Please mail this form to:

Paul Robinson, Author
424 Quartz Street
Redwood City, CA 94062

Or, send e-mail to Paul at:

laserpablo@aol.com

CONCEPTUAL PHYSICS

Laboratory Manual

Paul Robinson
San Mateo High School
San Mateo, California
Illustrated by Paul G. Hewitt

Paul G. Hewitt

Needham, Massachusetts
Upper Saddle River, New Jersey
Glenview, Illinois

Contributors

Roy Unruh
University of Northern Iowa
Cedar Falls, Iowa

Tim Cooney
Price Laboratory School
Cedar Falls, Iowa

Clarence Bakken
Gunn High School
Palo Alto, California

Consultants

Kenneth Ford
Germantown Academy
Fort Washington, Pennsylvania

Jay Obernolte
University of California
Los Angeles, California

Cover photograph: Motor Press Agent/Superstock, Inc.

ISBN 0-13-054257-1
Teacher's Edition ISBN 0-13-054258-X
1 2 3 4 5 6 7 8 9 10   04 03 02 01

Acknowledgments

Most of the ideas in this manual come from teachers who share their ideas at American Association of Physics Teachers (AAPT) meetings that I have attended since my first year of teaching. This sharing of ideas and cooperative spirit is a hallmark of our profession.

Many more individuals have contributed their ideas and insights freely and openly than I can mention here. The greatest contributors are Roy Unruh and Tim Cooney, principal authors of the PRISMS (Physics Resources and Instructional Strategies for Motivating Students) Guide. I am especially thankful to them and others on the PRISMS team: Dan McGrail, Ken Shafer, Bob Wilson, Peggy Steffen, and Rollie Freel. For contributions and feedback to the first edition, I am grateful to Brad Huff, Bill von Felten, Manuel Da Costa, and Clarence Bakken, as well as the students of Edison Computech High School who provided valuable feedback. I am especially indebted to my talented former student Jay Obernolte, who developed computer software that originally accompanied this manual.

For helpful lab ideas I thank Evan Jones, Sierra College; Dave Wall, City College of San Francisco; David Ewing, Southwestern Georgia University; and Sheila Cronin, Avon High School, CT, for her adaptations of CASTLE curriculum. Thanks go to Paul Tipler; Frank Crawford, UC Berkeley; Verne Rockcastle, Cornell University; and the late Lester Hirsch for their inspiration. I am especially grateful to Ken Ford, who critiqued this third edition and to my talented and spirited students at San Mateo High School who constantly challenge and inspire me.

For suggestions on integrating the computer in the physics laboratory, I am grateful to my AAPT colleagues Dewey Dykstra, Robert H. Good, Charles Hunt, and Dave and Christine Vernier. Thanks also to Dave Griffith, Kevin Mather, and Paul Stokstad of PASCO Scientific for their professional assistance. I am grateful to my computer consultant and long time friend Skip Wagner for his creative expertise on the computer.

For production assistance I thank Lisa Kappler Robinson and Helen Yan for hand-lettering all the illustrations. Love and thanks to my parents for their encouragement and support, and to my children—David, Kristen, and Brian—and my dear Ellyn—for being so patient and understanding!

Most of all, I would like to express my gratitude to Paul Hewitt for his illustrations and many helpful suggestions.

Paul Robinson

Contents

YOU NEED TO KNOW THE RULES OF A GAME BEFORE YOU CAN FULLY ENJOY IT. LIKEWISE WITH THE PHYSICAL WORLD---TO FULLY APPRECIATE NATURE YOU NEED TO KNOW ITS RULES---WHAT PHYSICS IS ABOUT. READ YOUR TEXTBOOK AND ENJOY!

ALSO LIKE A GAME, TO FULLY UNDERSTAND PHYSICS, YOU NEED TO KNOW HOW TO KEEP SCORE. THIS INVOLVES OBSERVING, MEASURING, AND EXPRESSING YOUR FINDINGS IN NUMBERS. THAT 'S WHAT THIS LAB MANUAL IS ABOUT. <u>DO</u> PHYSICS AND UNDERSTAND!

Goals

The laboratory part of the *Conceptual Physics* program should

1) Provide you with hands-on experience that relates to physics concepts.
2) Provide training in making measurements, recording, organizing, and analyzing data.
3) Provide you with the experiences in elementary problem solving.
4) Show you that all this can be interesting and worthwhile—*doing* physics can be quite enjoyable!

This lab manual contains 61 activities and 38 experiments. Most activities are designed to provide you with hands-on experience that relates to a specific concept. Experiments are usually designed to give you practice using a particular piece of apparatus. The goals for activities and experiments, while similar, are in some ways different.

The chief goal of activities is to acquaint you with a particular physical phenomenon which you may or may not already know something about. The emphasis during an activity is for you to observe relationships, identify variables, and develop tentative explanations of phenomena in a qualitative fashion. In some cases, you will be asked to design experiments or formulate models that lead to a deeper understanding.

Experiments are more quantitative in nature and generally involve acquiring data in a prescribed manner. Here, a greater emphasis is placed on learning how to use a particular piece of equipment, making measurements, identifying and estimating errors, organizing your data, and interpreting your data.

Graphing

When you look at two columns of numbers that are related some way, they probably have little meaning to you. A *graph* is a visual way to see how two quantities are related. You can tell at an instant what the stock market has been doing by glancing at a plot of the Dow Jones Industrial Average plotted as a function of time.

Often you will take data by changing one quantity, called the *independent variable,* in order to see how another quantity, called the *dependent variable,* changes. A graph is made by plotting the values of the independent variable on the *horizontal axis,* or *x-axis* with values of the dependent variable on the *vertical axis,* or *y-axis.* When you make a graph, it is always important to label each of the axes with the quantity and units used to express it. The graph is completed by sketching the best smooth curve or straight line that fits all the points.

To eliminate confusion and increase learning efficiency, your teacher will guide you before each experiment as to how to label the axes and choose a convenient scale. Often you will work in groups and graph your data *as you perform the experiment.* This has the chief advantage of providing immediate feedback if an erroneous data point is made. This gives you time to adjust the apparatus and make the necessary adjustments so that your data are more meaningful. Everyone in the class can be easily compared by simply overlapping the graphs on an overhead projector. This method also has the added benefit of eliminating graphing as a homework assignment!

Use of the Computer

The computer provides a powerful tool in collecting and analyzing your data and displaying it graphically. Data plotting software enables you to input data easily and to plot the corresponding graph in minutes. With the addition of a printer, your graphs can be transferred to paper and included with your lab report.

Because of the computer's ability to calculate rapidly and accurately, it can help you analyze your data quickly and efficiently. The interrelationships between variables become more apparent than if you had to plot the data by hand.

In this course you will be encouraged to use the computer (if it is available in your lab room) as a laboratory instrument that can measure time and temperature and sense light. You can convert the computer into a timer, a light sensor, and a thermometer by connecting to it one or more variable resistance probes. Though not required for any of the labs, the use of probeware greatly facilitates data collection and processing.

Lab Reports

Your teacher may ask that you write up a lab report. Be sure to follow your teacher's specific instructions on how to write a lab report. The general guideline for writing a lab report is: Could another student taking physics at some other school read your report and understand what you did well enough to replicate your work?

Suggested Guidelines for Lab Reports

Lab number and title Write your name, date, and period in the upper right-hand corner of your report. Include the names of your partners underneath.

Purpose Write a brief statement of what you were exploring, verifying, measuring, investigating, etc.

Method Make a rough sketch of the apparatus you used and a brief description of how you planned to accomplish your lab.

Data Show a record of your observations and measurements, including all data tables.

Analysis Show calculations performed, any required graphs, and answers to questions. Summarize what you accomplished in the lab.

Safety in the Physics Laboratory

By following certain common sense in the physics lab, you can make the lab safe not only for yourself but for all those around you.

1. Never work in the lab unless a teacher is present and aware of what you are doing.
2. Prepare for the lab activity or experiment by reading it over first. Ask questions about anything that is unclear to you. Note any cautions that are stated.
3. Dress appropriately for a laboratory. Avoid wearing bulky or loose-fitting clothes or dangling jewelry. Pin or tie back long hair, and roll up loose sleeves.
4. Keep the work area free of any books and materials not needed for what you are working on.
5. Wear safety goggles when working with flames, heated liquids, or glassware.
6. Never throw anything in the laboratory.
7. Use the apparatus only as instructed in the manual or by your teacher. If you wish to try an alternate procedure, obtain your teacher's approval first.
8. If a thermometer breaks, inform your teacher immediately. Do not touch either the mercury or the glass with your bare skin.
9. Do not force glass tubing or thermometer into dry rubber stopper. The hole and the glass should both be lubricated with glycerin (glycerol) or soapy water, and the glass should be gripped through a paper towel to protect the hands.
10. Do not touch anything that may be hot, including burners, hot plates, rings, beakers, electric immersion heaters, and electric bulbs. If you must pick up something that is hot, use a damp paper towel, a pot holder, or some other appropriate holder.
11. When working with electric circuits, be sure that the current is turned off before making adjustments in the circuit.
12. If you are connecting a voltmeter or ammeter to a circuit, have your teacher approve the connections before you turn the current on.
13. Do not connect the terminals of a dry cell or battery to each other with a wire. Such a wire can become dangerously hot.
14. Report any injuries, accidents, or breakages to your teacher immediately. Also report anything that you suspect may be malfunctioning.
15. Work quietly so that you can hear any announcements concerning cautions and safety.
16. Know the locations of fire extinguishers, fire blankets, and the nearest exit.
17. When you have finished your work, check that the water and gas are turned off and that electric circuits are disconnected. Return all materials and apparatus to the places designated by your teacher. Follow your teacher's directions for disposal of any waste materials. Clean the work area.

* *See page xiv for descriptions of the safety symbols you will see throughout this laboratory manual.*

Safety Symbols

These symbols alert you to possible dangers in the laboratory and remind you to work carefully.

Emergency Procedures

Report all injuries and accidents to your teacher immediately. Know the locations of fire blankets, fire extinguishers, the nearest exit, first aid equipment, and the school's office nurse.

Situation	Safe Response
Burns	Flush with cold water until the burning sensation subsides.
Cuts	If bleeding is severe, apply pressure or a compress directly to the cut and get medical attention. If cut is minor, allow to bleed briefly and wash with soap and water.
Electric Shock	Provide fresh air. Adjust the person's position so that the head is lower than the rest of the body. If breathing stops, use artificial resuscitation.
Eye Injury	Flush eye immediately with running water. Remove contact lenses. Do not allow the eye to be rubbed.
Fainting	See Electric Shock.
Fire	Turn off all gas outlets and disconnect all electric circuits. Use a fire blanket or fire extinguisher to smother the fire. Caution: Do not cut off a person's air supply. Never aim fire extinguisher at a person's face.

Name Period Date

1 Making Hypotheses

Purpose

To practice using observations to make hypotheses.

Setup: <1
Lab Time: 1
Learning Cycle: exploratory
Conceptual Level: easy
Mathematical Level: none

Required Equipment/Supplies

2 1-gallon metal cans
2 2-hole #5 stoppers
glass funnel or thistle tube
glass tubing
rubber tubing
500-mL beaker

Adapted from PRISMS.

Activity

Discussion

Science involves asking questions, probing for answers, and inventing simple sets of rules to relate a wide range of observations. Intuition and inspiration come into science, but they are, in the end, part of a systematic process. Science rests on observations. These lead to educated guesses, or *hypotheses.* A hypothesis leads to predictions, which can then be tested. The final step is the formulation of a theory that ties together hypotheses, predictions, and test results. The theory, if it is a good one, will suggest new questions. Then the cycle begins again. Sometimes this process is brief, and a successful theory that explains existing data and makes useful predictions is developed quickly. More often, success takes months or years to achieve. Scientists must be patient people!

Procedure 

Step 1: Observe the operation of the mystery apparatus (shown in Figure A) set up by your teacher.

Observe apparatus.

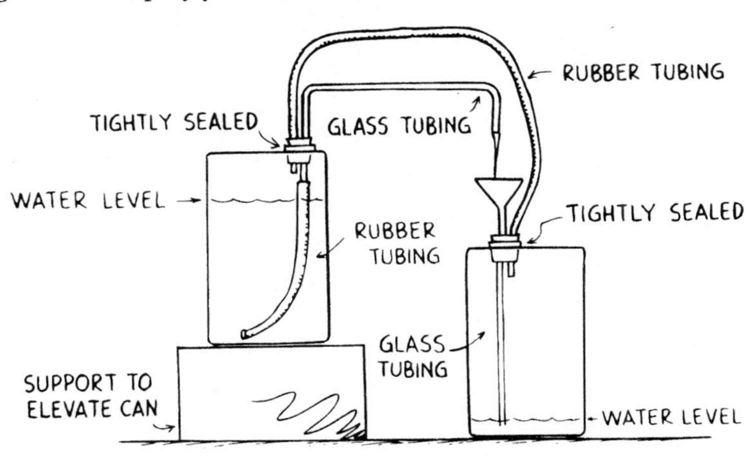

Fig. A

Propose explanations.

SCIENTIFIC PEOPLE ARE CURIOUS...

Step 2: Attempt to explain what is happening in the mystery apparatus and how it works. Write a description of how you think it works.

Students will be expected to offer an explanation as to why the apparatus worked the way it did. Explanations may vary from one to several paragraphs.

Step 3: Report your findings to the rest of the class. The class should reach a consensus about how it works. Record the consensus here.

The water in the funnel increases the air pressure in the can, forcing liquid up the tube.

2 The Physics 500

Purpose

To compute the average speed of at least three different races and to participate in at least one race.

Setup: <1
Lab Time: 1
Learning Cycle: exploratory
Conceptual Level: easy
Mathematical Level: easy

Required Equipment/Supplies

meterstick
stopwatch
string
equipment brought by students for their races

Adapted from PRISMS.

Discussion

In this activity, you will need to think about what measurements are necessary to make in order to compute the average speed of an object. How does the average speed you compute compare with the maximum speed? How could you find the maximum speed of a runner or a car between stoplights?

Activity

Procedure 

Measure average speed.

Step 1: Work in groups of about three students. Select instruments to measure distance and time. Develop a plan that will enable you to determine speed. Two students race each other in races such as hopping on one foot, rolling on the lawn, or walking backward. The third student collects and organizes data to determine the average speed of each racer. Repeat this process until each member of your group has a chance to be the timer. For the race in which you are the timer, record your plan and the type of race.

____

____

____

When measurements are to be made in an experiment, a good experimenter organizes a table showing *all* data, not just the data that "seem to be right." Record your data in Data Table A. Show the units you used as well as the quantities. For each measurement, record as many digits as you can read directly from the measuring instrument, plus one estimated digit. Then calculate the average speed for each student.

ACTIVITY	DISTANCE	TIME	SPEED

Data Table A

Compute unknown distance.

Step 2: Upon completing Step 1, report to your teacher. Your teacher will then ask you to perform one of your events over an unknown distance. Then, compute the distance covered (such as the distance across the amphitheater or the length of a corridor), using the average speed from Step 1.

event: ______________________________

average speed = ______________________

distance = ___________________________

Analysis

1. How does average speed relate to the distance covered and the time taken for travel?

   The average speed is the total distance divided by the total time.

2. Should the recorded average speed represent the maximum speed for each event? Explain.

   No. Average means the average of the high, the middle, and the low values.

3. Which event had the greatest average speed in the class in miles per hour (1.00 m/s = 2.24 mi/h)?

   Answers will vary. The maximum should be about 20 miles per hour.

4. Does your measurement technique for speed enable you to measure the fastest speed attained during an event?

   No, this technique measures the average speed for the event, not the fastest speed attained during it.

Name Period Date

3 The Domino Effect

Purpose

To investigate the ways in which distance, time, and average speed are interrelated by maximizing the speed of falling dominoes. To become familiar with elementary graphing techniques.

Setup: <1

Lab Time: 1

Learning Cycle: concept development

Conceptual Level: easy

Mathematical Level: easy

Required Equipment/Supplies

approximately 50 dominoes
stopwatch
meterstick

Thanks to Clarence Bakken for his ideas and suggestions for this lab.

Activity

Discussion

A central property of motion is *speed*—the rate at which distance is covered. By rate, we mean how much or how many of something per unit of time: how many kilometers traveled in an hour, how many feet moved in a second, how many raindrops hitting a roof in a minute, how much interest earned on a bank account in a year. When we measure the speed of an automobile, we measure the rate at which this easily seen physical thing moves over the ground—for instance, how many kilometers per hour. But when we measure the speed of sound or the speed of light, we measure the rate at which energy moves. We cannot see this energy. We can, however, see and measure the speed of the energy pulse that makes a row of dominoes fall.

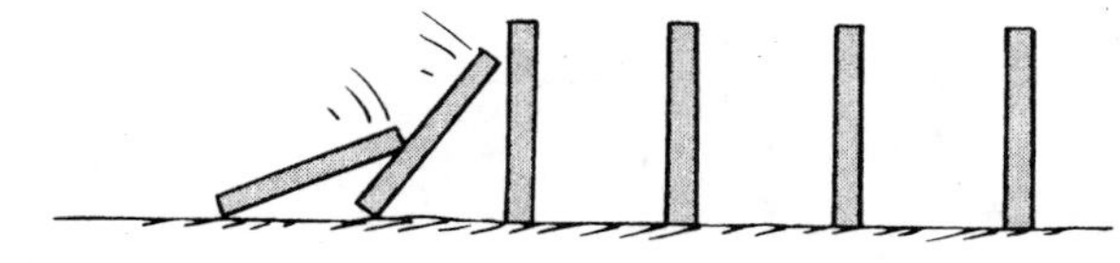

Procedure

Step 1: Set up 50 dominoes in a straight row, with equal spacing between them. The dominoes must be spaced at *least* the thickness of one domino apart. Your goal is to maximize the speed at which a row of dominoes falls down. Set the dominoes in a way you think will give the greatest speed.

Set up string of dominoes.

Step 2: Measure the total length of your row of dominoes.

length = ______________________

(1) Student results may be inconclusive—stress importance of method rather than getting the "answer."

(2) If students are using the computer photogate system, caution them to position the light probes several dominoes from either end.

Step 3: Compute the average spacing distance between dominoes by measuring the length from the middle of the first domino to the middle of the last one, and divide this by the number of domino spacings.

average distance between dominoes = ______________

Step 4: Measure the length of a domino.

length of domino = ______________

spacing distance = ______________ domino lengths

Step 5: Measure the time it takes for your row of dominoes to fall down.

time = ______________

Compute average toppling speed.

Step 6: Compute the average toppling speed for your row of dominoes.

average speed = ______________

Repeat for different spacings.

Step 7: Repeat Steps 5 and 6 for at least three more spacings. Include a spacing that is about as small as you can make it and still produce toppling and a spacing that is about as large as you can make it and still produce toppling. Record your data (including data for the first trial) in Data Table A.

Graph data.

Step 8: Using a separate piece of graph paper, make a graph of your data by sketching a smooth curve through your data points. Identify the point on the curve where the speed is maximum or minimum (this need not be exactly at one of your measured points).

TRIAL	LENGTH	AVERAGE SPACING	TIME	SPEED

Data Table A

Name Period Date

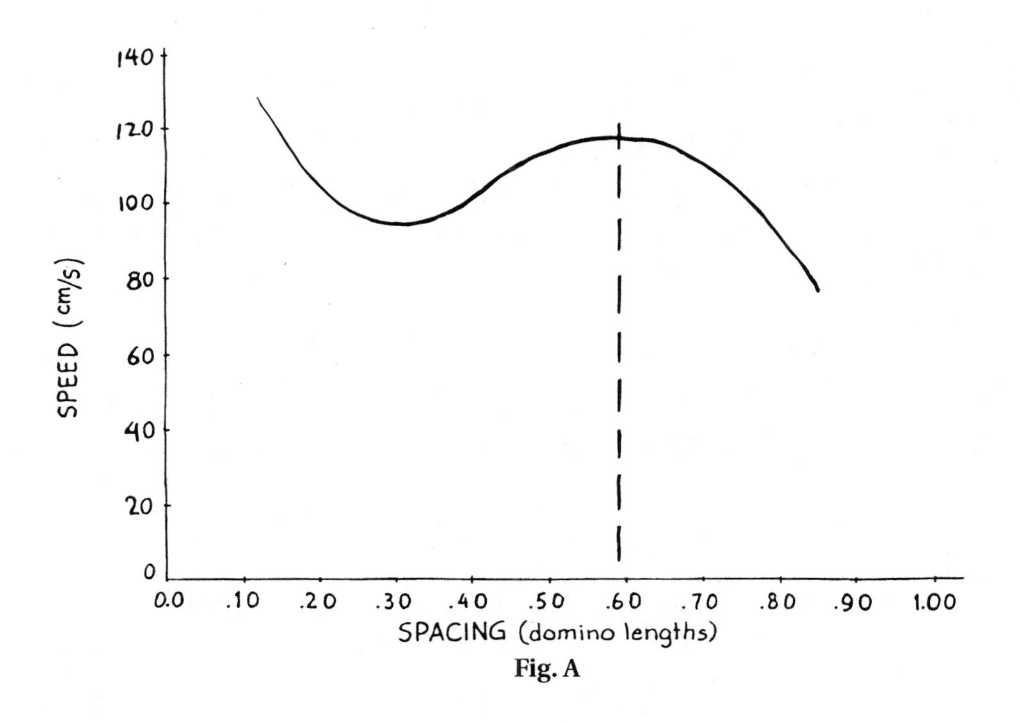

Fig. A

Analysis

1. What is a definition of average speed?

Average speed is the total distance divided by the total time.

2. What are the factors that affect the speed of falling dominoes?

The factors include the spacing; table surface; and height, uniformity, and composition of the dominoes.

3. Why do we use *average speed* for the pulse running down the dominoes rather than *instantaneous speed?*

(a) Dominoes can't be set up with exactly equal spacing. Speed therefore varies somewhat as the pulse travels along. (b) The smallest interval for which a speed can be defined is one domino spacing. Speed cannot be defined at a smaller "instant" than the time for one domino to topple.

Some dominoes give a nice maximum speed around a spacing of 0.6 domino lengths, but other dominoes give a minimum at a spacing of 0.5 to 0.6 domino lengths. This is a real effect of using heavy plastic dominoes that gave a maximum and light wooden ones that gave a minimum; results are therefore inconclusive. It is possible that the friction between the domino and the table surface also plays a significant role. Future experimentation required!

4. From your graph, what is the maximum or minimum toppling speed?

Heavy plastic dominoes show maximum speeds around 100–110 cm/s.

Light wooden dominoes show minimum speeds around 80 cm/s.

5. What spacing between dominoes do you predict would give the maximum or minimum speed? What is the ratio of this spacing to the length of a domino?

For heavier dominoes giving a maximum speed, the maximum occurs typically at a spacing of around 0.55–0.60 domino lengths. For lighter dominoes giving a minimum speed, the minimum occurs typically at a spacing of 0.3 to 0.5 domino lengths.

6. At the maximum or minimum toppling speed of the row of dominoes, how long a row of dominoes would be required to make a string that takes one minute to fall?

Substituting the speed v and the time $t = 60$ s into the equation $d = vt$ gives the distance, which will likely be in the range 50 to 65 m.

4 Merrily We Roll Along!

Purpose

To investigate the relationship between distance and time for a ball rolling down an incline.

Setup: 1

Lab Time: >1

Learning Cycle: concept development

Conceptual Level: difficult

Mathematical Level: moderate

Required Equipment/Supplies

2-meter ramp
steel ball bearing or marble
wood block
stopwatch
tape
meterstick
protractor
overhead transparencies

Optional Equipment/Supplies

computer
3 light probes with interface
3 flashlights or other light sources
6 ring stands with clamps

Discussion

Measurement of the motion of a freely falling object is difficult because the speed increases rapidly. In fact, it increases by nearly 10 m/s every second. The distance that the object falls becomes very large, very quickly. Galileo slowed down the motion by using inclined planes. The component of gravity acting along the direction of the inclined plane is less than the full force of gravity that acts straight down—so the change of speed happens more slowly and is easier to measure. The less steep the incline, the smaller the acceleration. The physics of free fall can be understood by first considering the motion of a ball on an inclined plane.

This experiment will require you to make many timing measurements using a stopwatch or, if available, a computer. If you use a stopwatch, develop good timing techniques so as to minimize errors due to reaction time. The computer-photogate system is a powerful stopwatch because it not only eliminates reaction time, but can be used in a variety of timing modes. Since the diameter of the ball can be measured directly, its *speed* can be found by dividing the diameter by the amount of time it

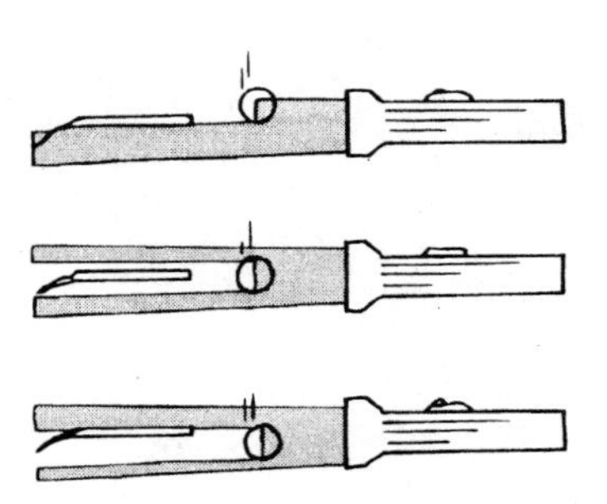

takes to pass through the beam striking the light probe. Strictly speaking, since the ball's speed is increasing as it rolls through the light beam, it exits the beam *slightly* faster than it enters the beam. The width of the ball divided by the time the ball eclipses the light beam gives the *average* speed of the ball through the light beam, not the *instantaneous* speed. But once the ball has picked up speed partway down the incline, its percentage change in speed is small during the short time that it eclipses the light. Then its measured average speed is practically the same as its instantaneous speed.

Procedure

Set up ramp.

Step 1: Set up a ramp with the angle of the incline at about 10° to the horizontal, as shown in Figure A.

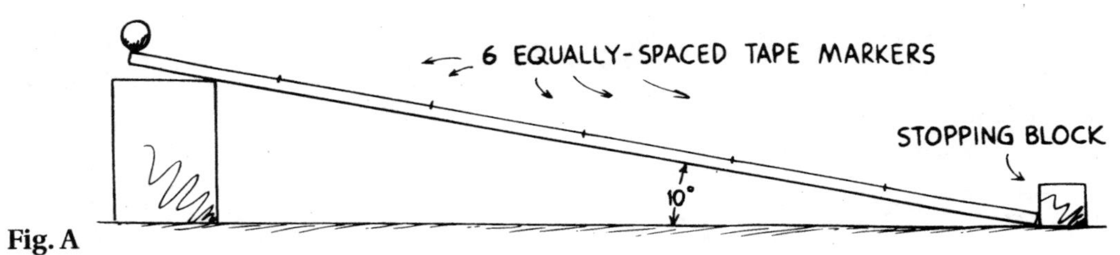

Fig. A

Divide ramp into equal parts.

Step 2: Divide the ramp's length into six equal parts and mark the six positions on the board with pieces of tape. These positions will be your release points. Suppose your ramp is 200 cm long. Divide 200 cm by 6 to get 33.33 cm per section. Mark your release points every 33.33 cm from the bottom. Place a stopping block at the bottom of the ramp to allow you to hear when the ball reaches the bottom.

Data Table A

DISTANCE (cm)	TIME (s)			
	TRIAL 1	TRIAL 2	TRIAL 3	AVERAGE

Time the ball down the incline.

Step 3: Use either a stopwatch or a computer to measure the time it takes the ball to roll down the ramp from each of the six points. (If you use the computer, position one light probe at the release point and the other at the bottom of the ramp.) Use a ruler or a pencil to hold the ball at its starting position, then pull it away quickly parallel to incline to release the ball uniformly. Do several practice runs with the help of your partners to minimize error. Make at least three timings from each position, and record each time and the average of the three times in Data Table A.

Graph data.

Step 4: Graph your data, plotting distance (vertical axis) vs. average time (horizontal axis) on an overhead transparency. Use the same scales on the coordinate axes as the other groups in your class so that you can compare results.

Step 5: Repeat Steps 2–4 with the incline set at an angle 5° steeper. Record your data in Data Table B. Graph your data as in Step 4.

Change the tilt of the ramp and repeat.

Students should discover that balls rolled from equally spaced distances do not result in multiples of the "natural" time unit but rather the ball covers each additional distance in less additional time.

DISTANCE (cm)	TIME (s)			
	TRIAL 1	TRIAL 2	TRIAL 3	AVERAGE

Data Table B

1. What is acceleration?

Acceleration is the rate at which velocity changes.

2. Does the ball accelerate down the ramp? Cite evidence to defend your answer.

Yes, the ball covers each additional distance in less additional time.

3. What happens to the acceleration if the angle of the ramp is increased?

The acceleration increases with larger angles. The component of gravitational force along the ramp is greater.

Step 6: Remove the tape marks and place them at 10 cm, 40 cm, 90 cm, and 160 cm from the stopping block, as in Figure B. Set the incline of the ramp to be about 10°.

Reposition the tape markers on ramp.

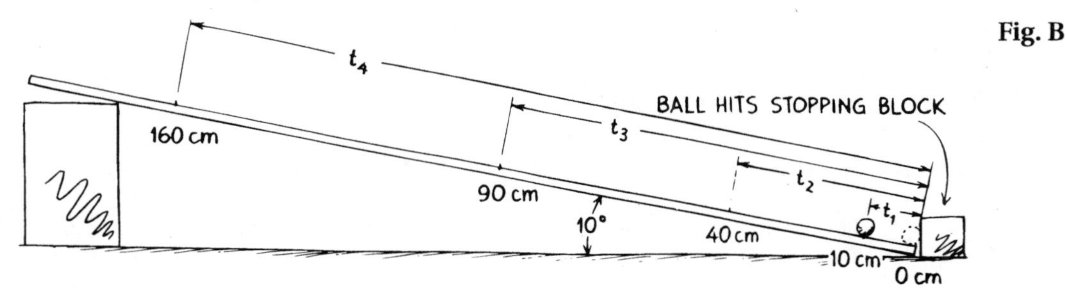

Fig. B

Time the ball down the ramp.

Students should discover that the progressively larger distances result in (more or less) equal intervals of time.

Step 7: Measure the time it takes for the ball to roll down the ramp from each of the four release positions. Make at least three timings from each of the four positions and record each average of the three times in Column 2 of Data Table C.

COLUMN 1	COLUMN 2				COLUMN 3	COLUMN 4
DISTANCE TRAVELED (cm)	ROLLING TIME (s)				TIME DIFFERENCES BETWEEN SUCCESSIVE INTERVALS (s)	TIME IN "NATURAL UNITS"
	TRIAL 1	TRIAL 2	TRIAL 3	AVERAGE		
10				$t_1 =$		1
40				$t_2 =$	$t_2 - t_1 =$	
90				$t_3 =$	$t_3 - t_2 =$	
160				$t_4 =$	$t_4 - t_3 =$	

Data Table C

Graph data.

Step 8: Graph your data, plotting distance (vertical axis) vs. time (horizontal axis) on an overhead transparency. Use the same coordinate axes as the other groups in your class so that you can compare results.

Study your data.

Step 9: Look at the data in Column 2 a little more closely. Notice that the *difference* between t_2 and t_1 is approximately the same as t_1 itself. The difference between t_3 and t_2 is also nearly the same as t_1. What about the difference between t_4 and t_3? Record these three time intervals in Column 3 of Data Table C.

Step 10: If your values in Column 3 are slightly different from one another, find their average by adding the four values and dividing by 4. Do as Galileo did in his famous experiments with inclined planes and call this average time interval one "natural" unit of time. Note that t_1 is already listed as one "natural" unit in Column 4 of Table C. Do you see that t_2 will equal—more or less—two units in Column 4? Record this, and also t_3 and t_4 in "natural" units, rounded off to the nearest integer. Column 4 now contains the rolling times as multiples of the "natural" unit of time.

4. What happens to the *speed* of the ball as it rolls down the ramp? Does it increase, decrease, or remain constant? What evidence can you cite to support your answer?

The speed increases. The ball takes the same time to roll down each (longer) section.

Step 11: Overlay your transparency graph with other groups in your class and compare them.

5. Do balls of different mass have different accelerations?

No. This is clearly evident when student graphs are compared on the overhead projector.

Step 12: Investigate more carefully the distances traveled by the rolling ball in Table D. Fill in the blanks of Columns 2 and 3 to see the pattern.

COLUMN 1	COLUMN 2	COLUMN 3
DISTANCE TRAVELED (cm)	FIRST FOUR INTEGERS	SQUARES OF FIRST FOUR INTEGERS
10	1	1
40	2	4
90	3	9
160		

Table D

6. What is the relation between the distances traveled and the squares of the first four integers?

The distances are proportional to the squares of the first four integers.

Step 13: You are now about to make a very big discovery—so big, in fact, that Galileo is still famous for making it first! Compare the distances with the times in the fourth column of Data Table C. For example, t_2 is two "natural" time units and the distance rolled in time t_2 is 2^2, or 4, times as great as the distance rolled in time t_1.

7. Is the distance the ball rolls proportional to the square of the "natural" unit of time?

Yes, the distance is proportional to the square of the "natural" unit of time.

The sizes of your experimental errors may help you appreciate Galileo's genius as an experimenter. Remember, there were no stopwatches 400 years ago! He used a "water clock," in which the amount of water that drips through a small opening serves to measure the time. Galileo concluded that the distance *d* is proportional to the square of the time *t*.

$$d \sim t^2$$

Step 14: Repeat Steps 6–10 with the incline set at an angle 5° steeper. Record your data in Data Table E.

Increase tilt of ramp.

8. What happens to the acceleration of the ball as the angle of the ramp is increased?

The acceleration increases with increasing angle.

COLUMN 1	COLUMN 2				COLUMN 3	COLUMN 4
DISTANCE TRAVELED (cm)	ROLLING TIME (s)				TIME DIFFERENCES BETWEEN SUCCESSIVE INTERVALS (s)	TIME IN "NATURAL UNITS"
	TRIAL 1	TRIAL 2	TRIAL 3	AVERAGE		
10				$t_1 =$		1
40				$t_2 =$	$t_2 - t_1 =$	
90				$t_3 =$	$t_3 - t_2 =$	
160				$t_4 =$	$t_4 - t_3 =$	

Data Table E

9. Instead of releasing the ball along the ramp, suppose you simply dropped it. It would fall about 5 meters during the first second. How far would it freely fall in 2 seconds? 5 seconds? 10 seconds?

In 2 seconds it would fall 2^2 as far, or 4 x (5 m) = 20 m. In 5 seconds, it would fall 5^2 as far, or 25 x (5 m) = 125 m. In 10 seconds it would fall 10^2 as far, or 100 x (5 m) = 500 m (if air resistance is neglected).

Going Further: Investigating the Speed of the Ball Down the Ramp

Set up computer with light probes.

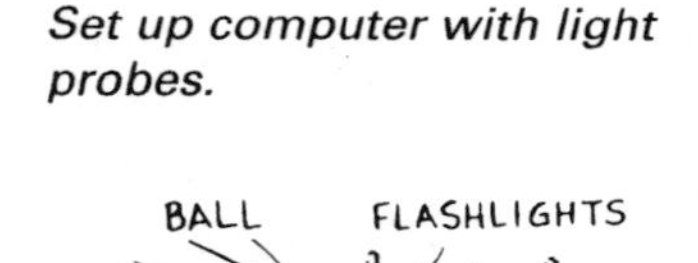

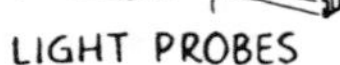

You can use the computer to investigate the speed the ball acquires rolling down the ramp. This can be accomplished several different ways using light probes. One way is to position three light probes 10 cm, 40 cm, and 90 cm from the release point on the ramp. The speed of the ball as it passes a probe can be determined by measuring the time interval between when the front of the ball enters the light beam and the back of the ball leaves it. The distance traveled during this time interval is just the diameter of the ball. Measure the diameter of the ball.

diameter of ball = ________________

With the incline set at an angle of about 10°, measure the respective eclipse times for each probe, and record your timings in Data Table F. To approximate the instantaneous speed of the ball at the three positions, divide the diameter of the ball by the eclipse time. Record the speeds in the table. Also record the average rolling time it took for the ball to travel each distance from the release point, from the information recorded in Data Table C.

Make a plot of instantaneous speed (vertical axis) vs. rolling time (horizontal axis).

Repeat with the ramp at an angle 5° steeper. Use the rolling times recorded in Data Table E.

Students should discover that the larger angle results in shorter times and proportionately greater accelerations.

Name Period Date

	TOTAL DISTANCE TRAVELED (cm)	ECLIPSE TIME (s)	INSTANTANEOUS SPEED (cm/s)	AVERAGE ROLLING TIME (s)
FIRST ANGLE	10			
	40			
	90			
SECOND ANGLE	10			
	40			
	90			

Data Table F

10. How do the slopes of the lines in your graphs of speed vs. time relate to the acceleration of the ball down the ramp?

The slopes of the speed vs. time graphs are greater when the accelerations of the balls are greater.

11. When you determined the speed of the ball with the light probes, were you really determining instantaneous speed? Explain.

No, the light probe is used to measure the time between when the ball enters the first part of the light beam and the time when the ball leaves it. Students are actually determining the average speed over that short time interval.

12. As the angle of the ramp increased, the acceleration of the ball increased. Do you think there is an upper limit to the acceleration of the ball down the ramp? What is it?

Yes, the upper limit is the acceleration of gravity, *g*.

5 Conceptual Graphing

Purpose

To make qualitative interpretations of motions from graphs.

Setup: <1
Lab Time: >1
Learning Cycle: concept development
Conceptual Level: moderate
Mathematical Level: none

Required Equipment/Supplies

sonic ranger
computer
masking tape
marking pen
ring stand
large steel ball
soup can
board
pendulum clamp

Thanks to Bob Tinker and Ron Thornton for their ideas and suggestions for this lab.

Activity

Discussion

Have you ever wondered how bats can fly around in the dark without bumping into things? A bat makes squeaks that reflect off walls and objects, return to the bat's head, and are processed by its brain to give clues as to the location of nearby objects. The automatic focus on some cameras works on very much the same principle. The sonic ranger is a device that measures the time that ultra-high-frequency sound waves take to go to, and return from, a target object. The data are fed to a computer, where they are graphically displayed on the monitor. Data plotting software can display the data in three ways: distance vs. time, velocity vs. time, and acceleration vs. time.

Procedure

Step 1: Your teacher will set up the sonic ranger and the computer for you. Check to see that the sonic ranger appears to be functioning properly.

Check sonic ranger.

Step 2: Place the sonic ranger on a desk or table so that its beam is chest high. A floor stand is very useful for this purpose, if available. Affix 5 meters of masking tape to the floor in a straight line from the sonic ranger, as shown in Figure A.

An alternative method is to use string instead of tape as a rule.

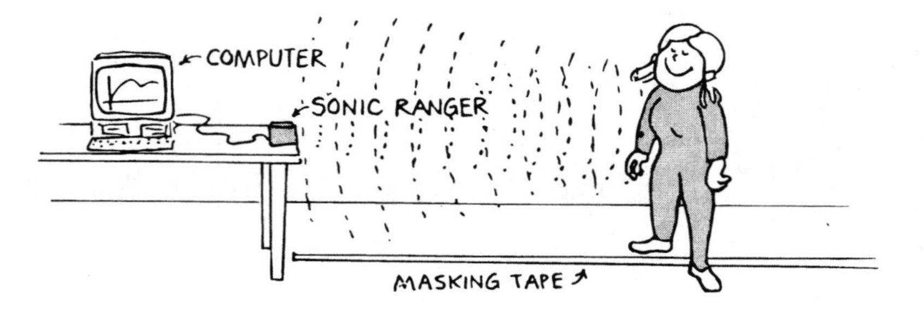

Fig. A

Calibrate the floor.

Step 3: Adjust the sonic ranger software so that it displays a distance vs. time plot. Point the sonic ranger at a student standing at the 5-meter mark of the tape. Calibrate the tape by marking where the computer registers 1 m, 2 m, and so on.

Observe graphs on computer monitor.

Step 4: Stand on the 1-meter mark. Always face the sonic ranger and watch the monitor. Back away from the sonic ranger slowly and observe the graph made. Repeat, backing away from the sonic ranger faster, and observe the graph made.

1. How do the graphs compare?

Both distance vs. time graphs should be straight lines that slope upward to the right. Walking faster away from the sonic ranger gives a graph with a steeper slope.

Step 5: Stand at the far end of the tape. Slowly approach the sonic ranger and observe the graph plotted. Repeat, walking faster, and observe the graph plotted.

2. How do the graphs compare?

Both distance vs. time graphs should be straight lines that slope downward to the right. Walking faster toward the sonic ranger gives a graph with a steeper downward (more negative) slope.

Step 6: Back away from the sonic ranger slowly; stop; then approach quickly.

3. Sketch the shape of the resulting graph.

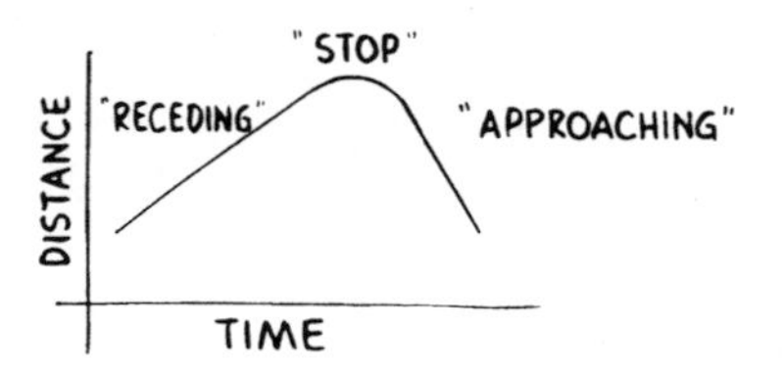

Name Period Date

4. Describe what walking motions result in the graphs shown in Figures B, C, D, and E.

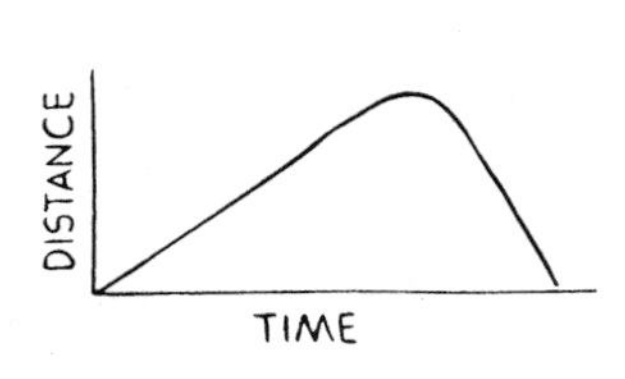

Fig. B

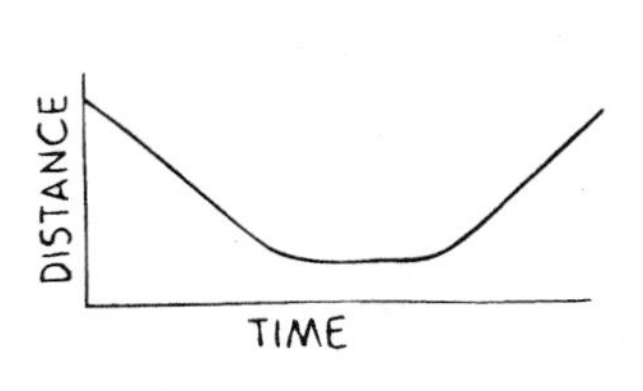

Fig. C

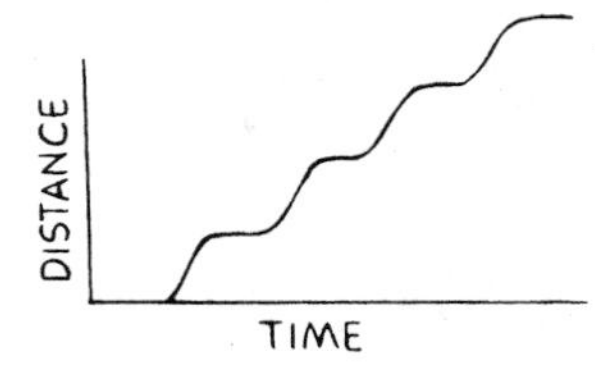

Fig. D

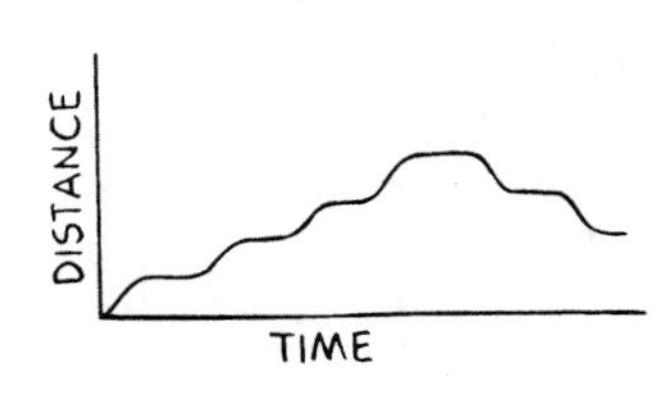

Fig. E

Figure B: Receding from the sonic ranger; then approaching at a greater speed than before.

Figure C: Approaching; then standing still; then receding.

Figure D: Standing still; receding; standing still; receding; etc.

Figure E: Receding; standing still; etc.; approaching; standing still; etc.

Step 7: Repeat Step 4, but use a velocity vs. time plotter.

5. How do the graphs of velocity vs. time compare?

Both velocity vs. time graphs are horizontal straight lines (after an initial start-up period). Walking away faster gives a line farther above the time axis.

Step 8: Repeat Step 5, using the velocity vs. time plotter.

6. How do the two new graphs compare?

Both velocity vs. time graphs are horizontal straight lines (after an initial start-up period) below the time axis. Walking away faster gives a line farther below the time axis.

Step 9: Repeat Step 6, using the velocity vs. time plotter.

7. Sketch the shape of the resulting graph.

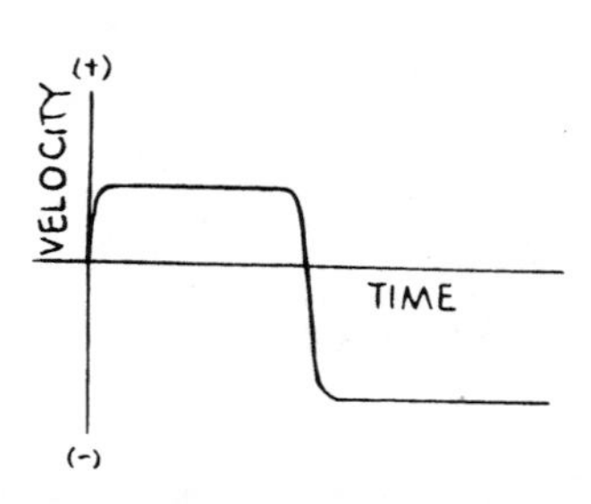

Analyze motion on an incline.

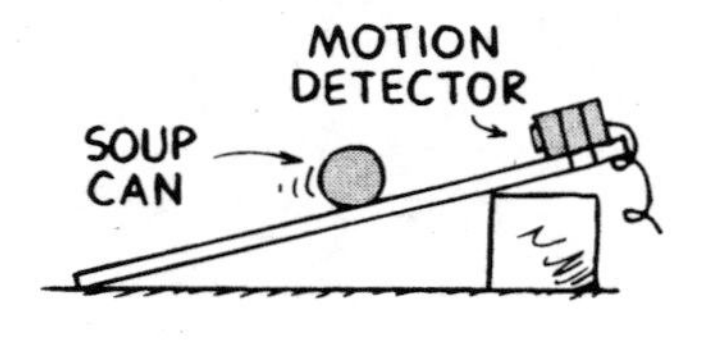

Fig. F

Going Further

Step 10: Set up the sonic ranger as shown in Figure F, to analyze the motion of a can or a large steel ball rolled up an incline. Initially position the can at least 40 cm (the minimum range) from the sonic ranger. Predict what the shapes of the distance vs. time and velocity vs. time graphs will look like for the can or ball rolled up and down the incline.

For best results, be sure to use a smooth laminated board and a new can of Bean with Bacon soup with no dent. It is much easier to roll up the incline in a straight line than a ball.

8. Sketch your predicted shapes in the following space.

Predictions will vary.

Now observe the two graphs, distance vs. time and velocity vs. time.

9. Sketch the shape of the distance vs. time graph.

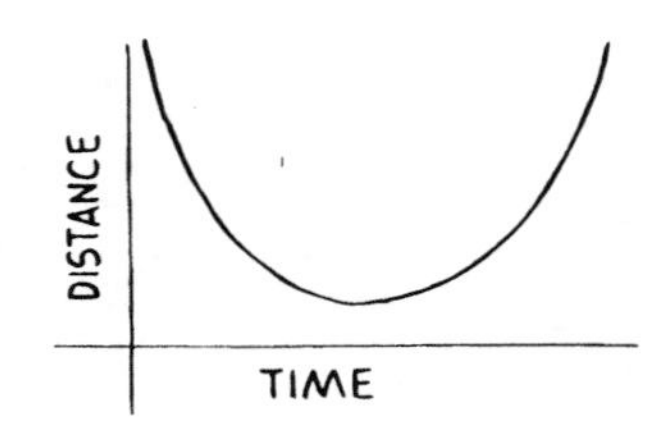

10. Sketch the shape of the velocity vs. time graph.

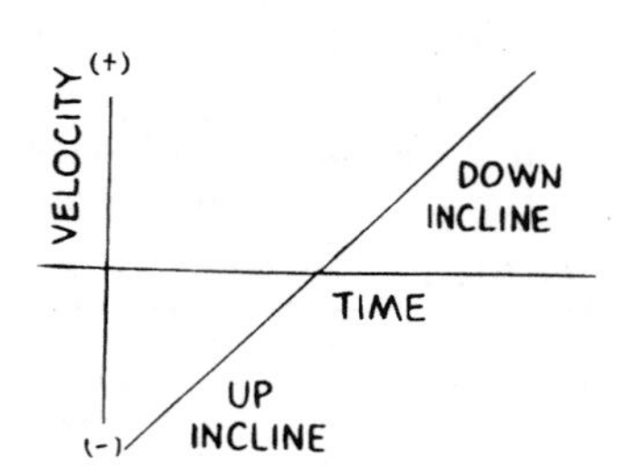

Step 11: Devise an experimental setup that uses the sonic ranger to analyze the motion of a pendulum based on Figure G. Perform your experiment and answer the following questions.

Analyze motion of a pendulum.

Fig. G

11. Sketch the shape of the distance vs. time graph.

The distance vs. time graph will look like a sine or cosine curve (sine if the zero of time is put at zero displacement).

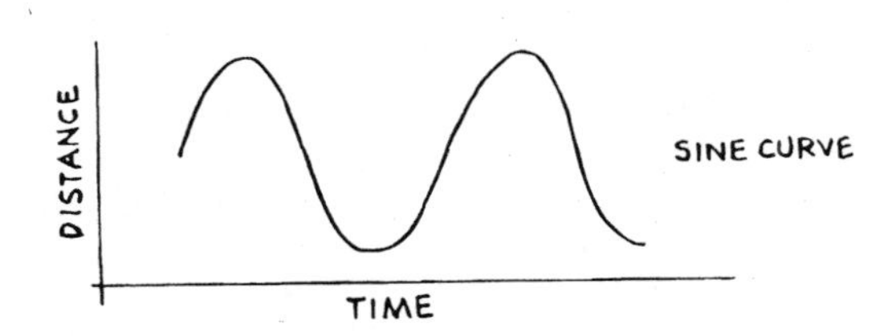

12. Sketch the shape of the velocity vs. time graph.

The velocity vs. time graph will look like a cosine or sine curve (cosine if the zero of time is put at maximum velocity). If these last two graphs are overlapped, students can compare the distance and velocity simultaneously.

An old billiard ball or golf ball works well for a pendulum bob.

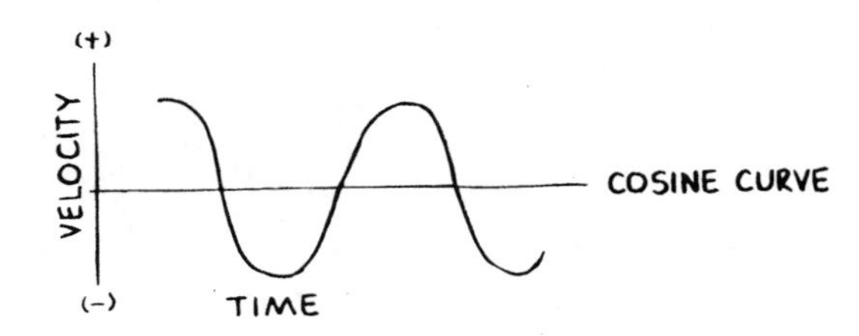

13. Where is the speed greatest for a swinging pendulum bob?

The speed is greatest at the bottom of the swing (for motion in either direction).

14. Where is the speed least for a swinging pendulum bob?

The speed is least at the top of the swing (at either end).

6 Race Track

Purpose

To introduce the concept of constantly changing speed.

Setup: <1

Lab Time: 1

Learning Cycle: concept development

Conceptual Level: easy

Mathematical Level: none

Required Equipment/Supplies

race grid in Figure A
colored pencil or pen
graph paper (optional)

Discussion

A car can accelerate to higher speed at no more than a maximum rate determined by the power of the engine. It can decelerate to lower speed at no more than a maximum rate determined by the brakes, the tires, and the road surface. This simple but fun game will quickly teach you the meaning of fixed acceleration.

Race Track is a truly remarkable simulation of automobile racing. Its inventor is not known. It is described in the column "Mathematical Games" in the journal *Scientific American*, January 1973, p. 108.

Activity

Procedure

Getting started.

Step 1: Each contestant should have a pencil or pen of a different color. Use the race grid in Figure A. Each player draws a tiny box just below a grid point on the starting line. The players then move in order.

Rules of the road.

Step 2: Contestants must obey the following rules.

(1) The first move must be one square forward: horizontally, vertically, or both (diagonally).

(2) On each succeeding move a car can maintain its latest velocity *or* increase or decrease its speed by *one* square per move in the horizontal or vertical direction or *both*. For example, a car going 2 squares per move vertically can change to 2 squares per move vertically *and* 1 square to the left or right; 1 or 3 squares per move vertically; or 1 or 3 squares per move vertically *and* 1 square to the left or right.

(3) The new grid point and the straight-line segment joining it to the preceding grid point must lie entirely within the track.

(4) No two cars may simultaneously occupy the same grid point. That is, no collisions are allowed!

This game is particularly challenging. You may want to have students set up a tournament for extra credit.

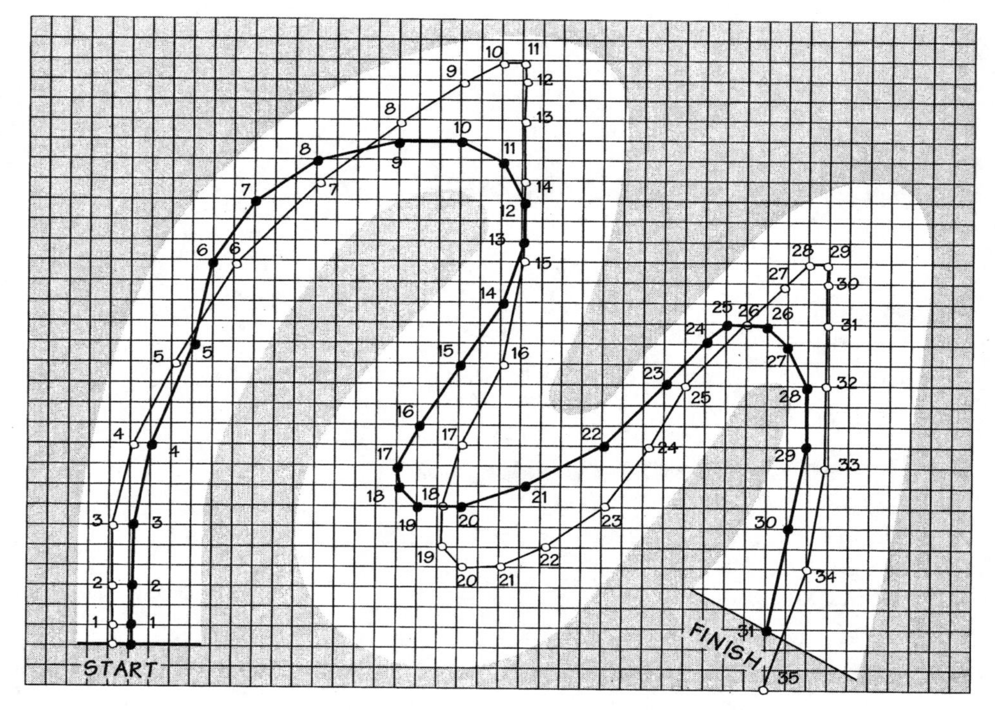

Fig. A

(5) The first player to cross the finish line wins!

To summarize, the speed in either the x or y direction can change by no more than one each turn.

Lay out new course.

Step 3: If you want to play the game again, you can draw a new race grid on a piece of graph paper. The width of the track can vary but should be at least 3 squares wider than the number of cars. To make the game interesting, the track should be strongly curved. Draw a start/finish line at a straight portion of the track.

This challenging game can also be played on a computer using the "Race Track" program on *Laboratory Simulations-I.*

Analysis

1. What was the fastest any player ever went during the course of the race?

   Answers will vary, but it is possible to get up to speeds of 6 or 7 squares per move before crashing.

2. Did that player win the race?

   Probably not. Generally winners are drivers who maximize their average speed—not their instantaneous speed.

3. Did anybody crash? If so, why do you think they did?

   Possibly. Drivers who crash do not take into account that it is as important to slow down as it is to speed up.

Name Period Date

7 Bull's Eye

Experiment

Purpose

To investigate the independence of horizontal and vertical components of motion. To predict the landing point of a projectile.

Setup: 1

Lab Time: >1

Learning Cycle: application

Conceptual Level: difficult

Mathematical Level: difficult

Using a stopwatch instead of photogates or light probes greatly reduces the accuracy students can achieve and likewise diminishes their ability to land the ball in the can.

There are two advantages of having students do the computer simulation: (1) it tutors students needing both conceptual help and help with the math; (2) you can then require students to first acquire their data and get a "Bull's Eye" on the computer *before* attempting to land the ball in the can.

Required Equipment/Supplies

ramp or Hot Wheels® track
1/2-inch (or larger) steel ball
empty soup can
meterstick
plumb line
stopwatch, ticker-tape timer, or
computer
light probes with interface
light sources

Discussion

Imagine a universe without gravity. In this universe, if you tossed a rock where there was no air, it would just keep going—forever. Because the rock would be going at a constant speed, it would cover the same amount of distance in each second (Figure A). The equation for distance traveled when motion is uniform is

$$x = vt$$

The speed is

$$v = \frac{x}{t}$$

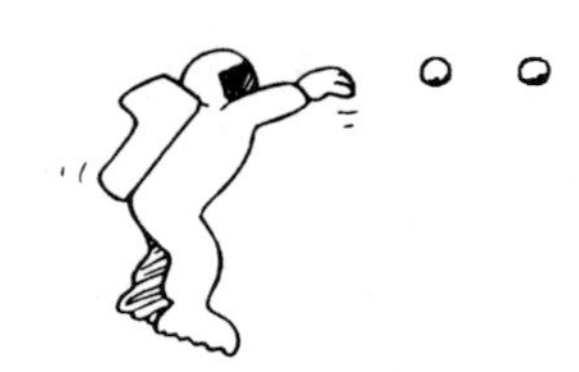

Fig. A

Coming back to earth, what happens when you drop a rock? It falls to the ground and the distance it covers in each second increases (Figure B). Gravity is constantly increasing its speed. The equation of the vertical distance y fallen after any time t is

$$y = \frac{1}{2}gt^2$$

where g is the acceleration of gravity. The falling speed v after time t is

$$v = gt$$

Fig. B

In this experiment, the horizontal projection speed of the ball is estimated by having the ball eclipse either one or two light probes, then dividing the diameter of the ball by the eclipse time. If two probes are used, the total time is $T_2 + T_3$. This is because you are measuring the time from when the first light probe is re-illuminated by the passing ball to when the second probe is re-illuminated. If the ramp is level and the friction is negligible, the ball should roll at a constant speed and T_1 and T_3 should be nearly the same (don't expect them to be identical, however).

Fig. C

What happens when you toss the rock sideways (Figure C)? The curved motion that results can be described as the combination of two straight-line motions: one vertical and the other horizontal. The vertical motion undergoes the acceleration due to gravity, while the horizontal motion does not. The secret to analyzing projectile motion is to keep two separate sets of "books": one that treats the horizontal motion according to

$$x = vt$$

and the other that treats the vertical motion according to

$$y = \frac{1}{2}gt^2$$

It's easy for students to confuse the time it takes the ball to land after leaving the table with the time it takes to roll between the two photogates.

Horizontal motion

- When thinking about how *far*, think about $x = vt$.
- When thinking about how *fast*, think about $v = x/t$.

Vertical motion

- When thinking about how *far*, think about $y = (1/2)\,gt^2$
- When thinking about how *fast*, think about $v = gt$.

Your goal in this experiment is to predict where a steel ball will land when released from a certain height on an incline. The final test of your measurements and computations will be to position an empty soup can so that the ball lands in the can the *first* time!

Procedure

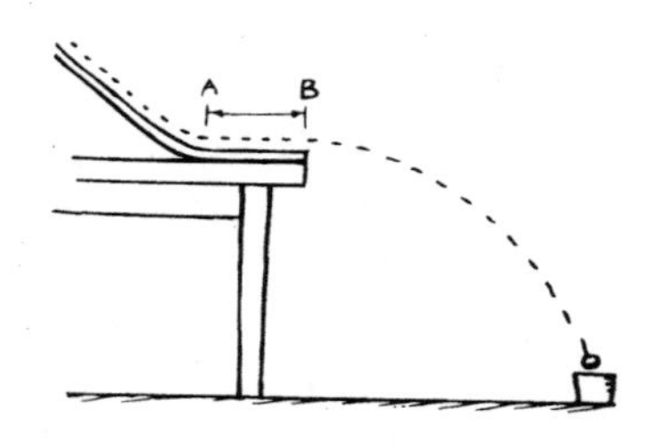

Fig. D

Step 1: Assemble your ramp. Make it as sturdy as possible so the steel balls roll smoothly and reproducibly, as shown in Figure D. The ramp should not sway or bend. The ball must leave the table *horizontally*. Make the horizontal part of the ramp at least 20 cm long. The vertical height of the ramp should be at least 30 cm.

Compute the horizontal speed.

Step 2: Use a stopwatch or light probe to measure the time it takes the ball to travel, from the first moment it reaches the level of the tabletop (point A in Figure D) to the time it leaves the tabletop (point B in Figure D). Divide this time interval into the horizontal distance on the ramp (from point A to point B) to find the horizontal speed. Release the ball from the same point (marked with tape) on the ramp for each of three runs.

Do *not* permit the ball to strike the floor! Record the average horizontal speed of the three runs.

horizontal speed = ______________

Name Period Date

Step 3: Using a plumb line and a string, measure the vertical distance h the ball must drop from the bottom end of the ramp in order to land in an empty soup can on the floor.

Measure the vertical distance.

1. Should the height of the can be taken into account when measuring the vertical distance h? If so, make your measurements accordingly.

The height of the can makes a very *big* difference—particularly if the horizontal speed of the ball is large. The vertical height h should be measured from the track to the *top* of the can or the ball will more than likely hit the side of the can.

$h =$ ______

Step 4: Using the appropriate equation from the discussion, find the time t it takes the ball to fall from the bottom end of the ramp and land in the can. Write the equation that relates h and t.

equation for vertical distance: ______

Show your work in the following space.

$t =$ ______

Step 5: The range is the horizontal distance of travel for a projectile. Predict the range of the ball. Write the equation you used and your predicted range.

Predict the range.

equation for range: ______

predicted range $R =$ ______

Place the can on the floor where you predict it will catch the ball.

Analysis

2. Compare the actual range of the ball with your predicted range. Compute the percentage error. (See Appendix 1 on how to compute percentage error.)

Answers will vary; typically a percentage error of 5% or less is attainable.

3. What may cause the ball to miss the target?

Misses can be caused by air friction, misalignment of the track, and error in measuring the vertical distance.

4. You probably noticed that the range of the ball increased in direct proportion to the speed at which it left the ramp. The speed depends on the release point of the ball on the ramp. What role do you think air resistance had in this experiment?

The ball is more likely to miss the target the farther downrange the target is located, because air friction increases with increasing speed. This is the same reason that line-drive golf and baseballs tail off from their ideal parabolic trajectories.

Going Further

Horizontally launch ball.

Students can either roll a ball from a ramp or use a ballistic pendulum to launch the ball, if available.

Suppose you don't know the firing speed of the steel ball. If you go ahead and fire it, and then measure its range rather than predicting it, you can work backward and calculate the ball's initial speed. This is a good way to calculate speeds in general! Do this for one or two fired balls whose initial speeds you don't know.

8 Going Nuts

Purpose

To explore the concept of inertia.

Setup: <1
Lab Time: 1
Learning Cycle: exploratory
Conceptual Level: easy
Mathematical Level: none

Required Equipment/Supplies

12-inch wooden embroidery hoop
narrow-mouth bottle
12 1/4-inch nuts

Discussion

Have you ever seen magicians on TV whip a tablecloth out from underneath an entire place setting? Can that really be done, or is there some trick to it?

Procedure 

The game is simple. Carefully balance an embroidery hoop vertically on the mouth of a narrow-mouth bottle. Stack the nuts on the top of the hoop. The idea is to get as many nuts as possible into the bottle by touching the hoop with only one hand.

Activity

Analysis

1. Describe the winner's technique.

The best technique is to face the hoop so that you see the full circle. Swing (say) your right arm horizontally across your body, grabbing the *far* side (e.g., left side) of the hoop while your arm is in motion and flicking your wrist to whisk the hoop out from under the nuts.

2. Explain why the winner's technique was successful.

The motion of the hoop is not transferred to the nuts because when the side of the hoop is pulled outward sharply at the start of the swing, the part of the hoop just below the nuts goes *down,* so there is almost no friction between the nuts and the hoop. Inertia keeps the nuts over the mouth of the bottle. With the hoop no longer in place, they drop straight down into the bottle.

9 Buckle Up!

Purpose

To demonstrate how Newton's first law of motion is involved in collisions.

Setup: <1

Lab Time: 1

Learning Cycle: application

Conceptual Level: easy

Mathematical Level: none

Adapted from PRISMS.

Required Equipment/Supplies

4 m of string
2 dynamics carts
2 200-g hook masses
rubber band
2 small dolls
2 pulleys
2 wood blocks

Activity

Discussion

Newton's first law of motion states that an object in motion keeps moving with constant velocity until a force is applied to that object. Seat belts in automobiles and other vehicles are a practical response to Newton's first law of motion. This activity demonstrates in miniature what happens when that important law is ignored.

Procedure

Crash doll on cart with and without "seat belts."

Step 1: Attach 2 m of string to each of two small dynamics carts. Attach a 200-g mass to the other end of each of the strings. Attach the pulleys to the table edge and hang the masses over them with the masses on the floor and the carts on the table. Place a wood block on the table in front of each pulley.

Step 2: Place one doll on each cart. Use a rubber band to serve as a seat belt for one of them.

Step 3: Pull the carts back side by side and release them so they accelerate toward the table's edge.

Analysis

1. What stopped the motion of the doll without a seat belt when the cart crashed to a stop?

   The doll without the seat belt kept moving until it hit the floor.

2. Was there any difference for the doll with a seat belt?

   The doll with the seat belt remained on the cart.

10 24-Hour Towing Service

Purpose

To find a technique to move a car when its wheels are locked.

Setup: <1
Lab Time: 1
Learning Cycle: exploratory
Conceptual Level: difficult
Mathematical Level: moderate

Required Equipment/Supplies

7 m to 10 m of chain or strong rope
tree or other strong vertical support
automobile or other large movable mass

A heavy chain with a large hook (3 inches or so) on each end is recommended.

Discussion

You can exert a force on a parked automobile if you push or pull on it with your bare hands. You can do the same with a rope, but with more possibilities. Even without using pulleys, you can multiply the forces you exert. In this activity, you will try to show that you can exert a far greater force with brains than with brawn.

This is a popular activity with students whether it be done as a student lab or as a class demonstration. Try it and see!

Activity

Procedure

Step 1: Park a car on a level surface with a tree in front of it, the brakes locked, and the gear selector set on "park" or in first gear.

Devise technique to move parked car.

Step 2: Your goal is to move the car closer to the tree. You will do this by exerting force on a rope, chain, or cable tied to the car's front end. How and where the force is exerted is up to your imagination. Your own body is the only energy source you can use. Make a sketch of your method. Show the applied force and the other forces with arrows.

Analysis

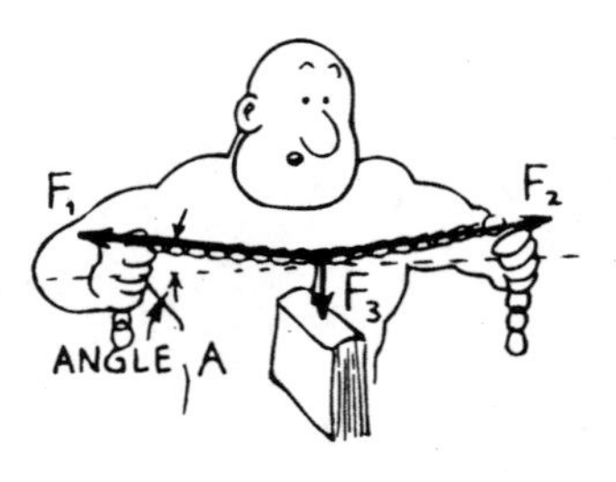

Fig. A

1. Look at Figure A. Suppose a force F_3 is applied to a chain at right angles to the horizontal. Tension in the chain can then be shown as vectors F_1 and F_2. Since the system is not accelerating, all forces must add up to zero. The force F_1 is the tension in the chain and the force on whatever it is attached to, in this case, the right hand of the strongman. The same is true for F_2. The force F_3, in this case the weight of the book, is small, while F_1 and F_2 are large. As the angle A becomes smaller, the forces F_1 and F_2 become larger. This idea is explained further in Chapter 4 of your text. Use a vector diagram to explain how a small sideways force can result in a large pull on the car.

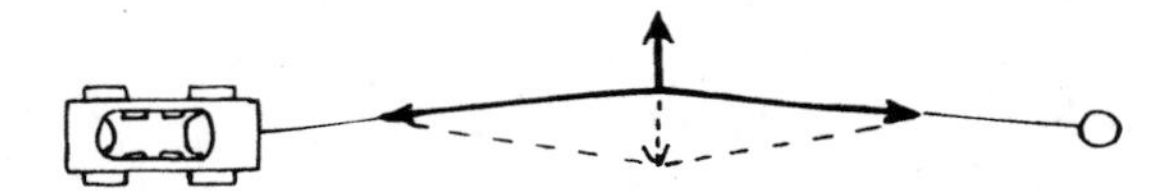

2. List other situations that could use this technique for "force multiplication."

Bow and arrow, string tighteners (for tennis rackets, etc.), dental appliances, beams in construction, etc.

3. This method for making a large force is used to fell trees, pull stumps, straighten dents in car fenders, and pull loose teeth! Explain how this might be possible.

For example, instead of connecting a large chain from a tractor to a tree stump and attempting to pull it out directly, a much greater force can be exerted on the stump by a vertical force on the chain as in Figure A.

Name Period Date

Getting Pushy

Purpose

To investigate the relationship between mass, force, and acceleration.

Setup: <1
Lab Time: >1
Learning Cycle: exploratory
Conceptual Level: moderate
Mathematical Level: easy

Adapted from PRISMS.

Required Equipment/Supplies

roller skates or skateboard
spring balance
stopwatch
meterstick
tape

Discussion

Most of us have felt the acceleration of a car as it leaves a stop sign or the negative acceleration when it comes to a stop. We hear sportscasters describe a running back as accelerating through the defensive line. In this activity, you will investigate some variables that influence acceleration.

Procedure

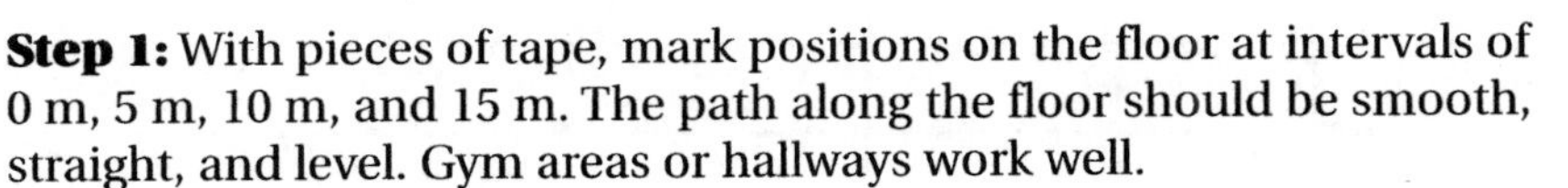

Step 1: With pieces of tape, mark positions on the floor at intervals of 0 m, 5 m, 10 m, and 15 m. The path along the floor should be smooth, straight, and level. Gym areas or hallways work well.

Mark off distances.

Step 2: A student must put on the skates and stand on the 0-m mark. Another student must stand behind the 0-m mark and hold the skater. The skater holds a spring balance by its hook.

Step 3: A third student must grasp the other end of the spring balance and exert a constant pulling force on the skater when the skater is released.

Pull on skater. . . catch skater.

The puller must maintain a constant force throughout the distance the skater is pulled. Do not pull harder to "get going." Time how long it takes to get to the 5-m, 10-m, and 15-m marks, and record this data in Data Table A along with the readings on the spring balance.

TRIAL	DISTANCE (m)	FORCE (N)	TIME (s)
1	5		
	10		
	15		
2	5		
	10		
	15		
3	5		
	10		
	15		

Data Table A

Step 4: Repeat the experiment twice, using different skaters to vary the mass, but keeping the force the same. If the results are inconsistent, the skater may not be holding the skates parallel or may be trying to change directions slightly during the trial.

Step 5: Repeat with the puller maintaining a *different* constant force throughout the distance the skater is pulled, but using the same three skaters as before. Record your results in Data Table B.

TRIAL	DISTANCE (m)	FORCE (N)	TIME (s)
1	5		
	10		
	15		
2	5		
	10		
	15		
3	5		
	10		
	15		

Data Table B

Name Period Date

Analysis

1. Until the time of Galileo, people believed that a constant force is required to produce a constant speed. Do your observations confirm or reject this notion?

   Student observations should cause them to reject this notion. The skater goes faster and faster when pulled.

2. What happens to the speed as you proceed farther and farther along the measured distances?

   The speed increases.

3. What happens to the rate of increase in speed—the acceleration—as you proceed farther and farther along the measured distances?

   The acceleration is probably the same, but this is too subtle to show without measurements.

4. When the force is the same, how does the acceleration depend upon the mass?

   The greater the mass is, the smaller the acceleration is.

5. When the mass of the skater is the same, how does the acceleration depend upon the force?

   The greater the force is, the greater the acceleration is.

6. Suppose a 3-N force is applied to the skater and no movement results. How can this be explained?

   There must be another 3-N force *opposing* the 3-N force applied.

Name Period Date

12 Constant Force and Changing Mass

Experiment

Purpose

To investigate the effect of increases in mass on an accelerating system.

Setup: <1

Lab Time: >1

Learning Cycle: concept development

Conceptual Level: moderate

Mathematical Level: difficult

Required Equipment/Supplies

meterstick
2 Pasco dynamics carts and track
4 500-g masses (2 masses come with Pasco carts)
1 50-g hook mass
pulley with table clamp
triple-beam balance
string
paper clips or small weights
masking tape
graph paper or overhead transparencies
stopwatch, ticker-tape timer, or
 computer
 light probes with interface
 light sources

The Pasco dynamics carts and track system give excellent results. The carts also have the added feature of having about the same mass as the 500-g additional masses that come with the carts. The traditional ball-bearing dynamics carts can easily be substituted for the Pasco carts by using a single cart loaded with 5 identical masses (either 500-g or 1-kg masses work well).

DECREASE MASS BUT KEEP THE SAME APPLIED FORCE

Discussion

Airplanes accelerate from rest on the runway until they reach their take-off speed. Cars accelerate from a stop sign until they reach cruising speed. And when they come to a stop, they decelerate. How does mass affect these accelerations?

In Activity 11, "Getting Pushy," you discovered that less massive people undergo greater acceleration than more massive people when the same force is applied to each. In this experiment you will accelerate a dynamics cart. You will apply the same force to carts of different mass. You will apply the force by suspending a weight over a pulley. The cart and the hanging weight comprise a *system* and accelerate together. A relationship between mass and acceleration should become evident.

Procedure

Set up pulley-and-cart system.

Step 1: Fasten a pulley over the edge of the table. The pulley will change the direction of the force from a downward pull on the mass into a sideways pull on the cart.

Step 2: Mark off a distance on the tabletop slightly shorter than the distance the mass can fall from the table to the floor.

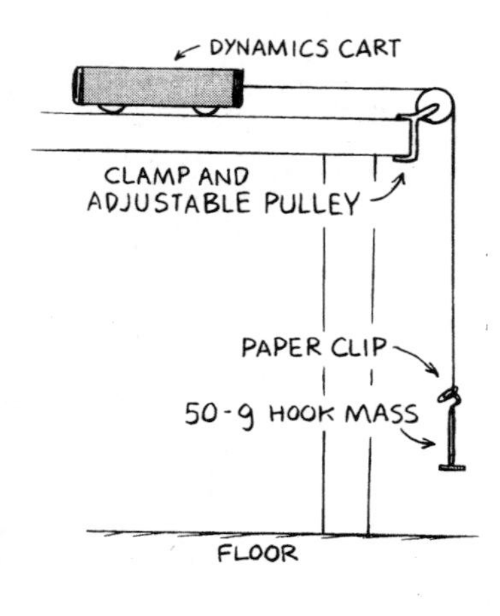

Fig. A

Step 3: Use a triple-beam balance to determine the mass of the carts. Record the total mass of the cart(s) and the four additional masses in Data Table A. Do not include the 50-g hanging mass or the mass of the paper clip counterweight.

Step 4: Nest the two dynamics carts on top of one another, and stack the 500-g masses on the top cart. It may be necessary to secure the masses and carts with masking tape. Tie one end of the string to the cart and thread it over the pulley, as shown in Figure A. To offset frictional effects, add enough paper clips or small weights to the other end of the string so that when the cart is pushed slightly it moves at a *constant speed.* (Give your teacher a break! Stop the cart before it crashes into the pulley during the experiment by adjusting the stop bar on the dynamics track.)

Step 5: Practice accelerating the cart a few times to ensure proper alignment. Add a 50-g hook mass to the paper clip counterweight. Keep this falling weight the same at all times during the experiment.

Set up timing system.

Remind students to adjust the stopping bar on the Pasco track so that the carts don't collide with the pulleys. If not using the Pasco setup, position a piece of foam or even a small wood block to protect the pulley.

Step 6: There are a variety of ways to time the motion of the cart and the falling weight. You could use a stopwatch, a ticker-tape timer (ask your teacher how to operate this device), or a computer with light probes. The distance is the distance between two light probes. Position the cart so that as it is released, it eclipses the first light probe and starts the timer. The timer stops when the second light probe is eclipsed.

Repeat three times, recording the times in Data Table A. Compute and record the average time.

TIME TO COVER SAME DISTANCE				ACCELERATION (m/s^2)
TRIAL 1	TRIAL 2	TRIAL 3	AVG	

Data Table A

Remove masses from the cart and repeat.

Step 7: Repeat Step 5 five times, removing a 500-g mass from the cart each time (the last time removing the top cart). Record the times in Data Table A, along with the average time for each mass.

Compute the acceleration.

Step 8: Use the average times for each mass from Steps 5 and 6 to compute the accelerations of the system. To do this, use the equation for an accelerating system that relates distance d, acceleration a, and time t.

$$d = \frac{1}{2}at^2$$

Rearrange this equation to obtain the acceleration.

$$a = \frac{2d}{t^2}$$

The cart always accelerates through the same distance d. Calculate the acceleration, a. Provided you express your mass units in kilograms and your distance in meters, your units of acceleration will be m/s^2.

Graph your data.

Record your accelerations in Data Table A.

Step 9: Using an overhead transparency or graph paper, make a graph of acceleration (vertical axis) vs. mass (horizontal axis).

Analysis

1. Describe your graph of acceleration vs. mass. Is it a straight-line graph or a curve?

The graph should look like some kind of hyperbola. Any graph that suggests that and shows the acceleration decreasing as the mass increases is acceptable.

2. Share your results with other class members. For a constant applied force, how does increasing the mass of an object affect its acceleration?

For constant applied force, acceleration decreases as mass increases.

Name Period Date

13 Constant Mass and Changing Force

Purpose

To investigate how increasing the applied force affects the acceleration of a system.

Setup: <1

Lab Time: >1

Learning Cycle: concept development

Conceptual Level: difficult

Mathematical Level: difficult

The Pasco dynamics cart and track system give excellent results. However, the traditional ball-bearing dynamics cart can easily substituted for the Pasco cart by using a single cart loaded with 50-g, 100-g, and 200-g masses.

If you don't have 6 hooked masses, any 20-g mass substitutes (such as slotted masses) will do. In that event, have students improvise a method to attach the masses to the string.

Required Equipment/Supplies

Pasco dynamics cart and track
masking tape
pulley with table clamp
6 20-g hook masses
string
paper clips
graph paper or overhead transparency
stopwatch, ticker-tape timer, or
    computer
    light probes with interface
    light sources

Discussion

KEEP THE MASS THE SAME WHILE INCREASING THE APPLIED FORCE

Have you ever noticed when an elevator cage at a construction site goes *up* that a large counterweight (usually made of concrete) comes *down*? The elevator and the counterweight are connected by a strong cable. Thus, the elevator doesn't move without the counterweight moving the same amount. Since the electric motor can't move one without moving the other, the two so connected together form a *system*.

In Experiment 12, "Constant Force and Changing Mass," you learned how the acceleration of a dynamics cart was affected by increasing the mass of the cart. But the cart is really part of a *system* consisting of the *cart and the falling weight*—just as the elevator and the counterweight together form a system. The cart doesn't move without the falling weight moving exactly the same amount. By adding mass to the cart, you were actually adding mass to the *cart-and-falling-weight system.*

Adding mass to the cart rather than to the falling weight was *not* accidental, however. In doing so, you changed only one variable—mass—to see how it affects the acceleration of the entire system.

In this experiment, you will investigate how increasing the applied force on a cart-and-falling-weight system affects its acceleration while keeping the mass of the system constant. The applied force is increased without changing the mass *by removing mass from the cart and placing it on the hanging weight.*

ONLY THE WEIGHT OF THE FALLING MASS ACCELERATES THE SYSTEM . . .
BUT THE TOTAL MASS OF BOTH THE FALLING WEIGHT AND THE LOADED CART RESIST ACCELERATION

The same general procedure used in Experiment 12 will be used here. In Experiment 12, you measured the total time the cart accelerated to compute the acceleration using the formula:

$$a = \frac{2d}{t^2}$$

Procedure

Set up cart-and-hanging-weight system.

Step 1: Set up your apparatus much the same as you did in Experiment 12, except that you should load the dynamics cart with six 20-g masses. Masking tape may be required to hold the masses in position.

Step 2: Attach one end of the string to the cart, pass the other over the pulley, and tie a large paper clip to the end of the string. To offset frictional effects, place just enough paper clips or other small weights on the end of the string so that when the cart is moved by a small tap, it rolls on the track or table with constant speed. Do not remove this counterweight at any time during the experiment.

Set up timing system.

Students should release the cart so that it starts the time just as it is released.

Step 3: You can time the system in a variety of ways. You could use a stopwatch, a ticker-tape timer (your teacher will show you how this device works), or a computer with light probes. The distance is the distance between two light probes. Position the cart so that as it is released, it eclipses the first light probe and starts the timer. The timer stops when the second light probe is eclipsed.

Step 4: For the first trial, remove one of the hooked 20-g masses from the cart and hang it on the end of the string.

Measure the acceleration time.

If using a ball-bearing cart, students must remove the 50-g mass from the string and place it back on the cart and take a 100-g mass from the cart and place it on the end of the string, so as to increase the mass of the hanging weight by 50 grams each trial. Repeat Step 6 for 150-g, 200-g, and 250-g masses on the string. Some trials will require two masses on the end of the string. Be *sure* students return each mass back on the cart after using it in order to keep the mass constant.

Step 5: With your timing system all set, release the cart and measure the time it takes to accelerate your cart toward the pulley. Catch the cart *before* it crashes into the pulley and spews your masses all over the floor and damages the pulley! Repeat each trial at least twice. Record your data and compute the average of your three trials in Data Table A.

Data Table A

MASS OF FALLING WEIGHT	TIME TO COVER SAME DISTANCE				ACCELERATION (m/s^2)
	TRIAL 1	TRIAL 2	TRIAL 3	AVG	
20 g					
40 g					
60 g					
80 g					
100 g					

Increase the applied force.

Step 6: Remove another 20-g mass from the cart and place it on the end of the string with the other 20-g mass. Make several runs and record your data in Data Table A.

Name Period Date

Step 7: Repeat Step 5 five times, increasing the mass of the falling weight by 20 grams each time. Make several runs and record your data in Data Table A.

Step 8: With the help of your teacher, calculate the accelerations and use an overhead transparency to make a graph of acceleration (vertical axis) vs. force (horizontal axis). Since the cart always accelerates through the same distance *d*, the acceleration equals $2d/t^2$. Express your distance in meters and your acceleration will have units of m/s^2. Call the force caused by the single 20-g mass "F," the force caused by two 20-g masses "2F," and so on.

Graph your data.

Analysis

1. Describe your graph of acceleration vs. force. Do your data points produce a straight-line graph or a curved graph?

   The graph is a straight-line graph.

2. Does the acceleration of the cart increase or decrease as the force increases?

   The acceleration increases with increasing force.

3. Was the mass of the accelerating system (cart + falling weight) the same in each case?

   Yes; at least it should be!

4. In Experiment 12, you learned (or should have!) that when a constant force is applied, the acceleration of the system *decreases* as its mass increases. The acceleration is *inversely proportional* to the mass of the system. Here in Experiment 13, the acceleration of the system increases as the force applied to it (the weights of the falling masses) increases. That is, the acceleration is *directly proportional* to the applied force. Combine the results of Experiments 12 and 13 to come up with a general relationship between force, mass, and acceleration.

   The experimental findings should show that the acceleration is directly proportional to the force and inversely proportional to the mass ($a \sim \frac{F}{m}$).

Name Period Date

14 Impact Speed

Experiment

Purpose

To estimate the speed of a falling object as it strikes the ground.

Setup: <1
Lab Time: >1
Learning Cycle: application
Conceptual Level: difficult
Mathematical Level: difficult

Required Equipment/Supplies

stopwatch
string with a rock tied to one end
object with a large drag coefficient, such as a leaf or feather
Styrofoam® plastic foam ball or Ping-Pong® table-tennis ball

Thanks to Verne Rockcastle for this lab.

Discussion

The area of a rectangle is its height multiplied by its base. The area of a triangle is its height multiplied by half its base. If the height and base are measured in meters, the area is measured in square meters. Consider the area under a graph of speed vs. time. The height represents the speed measured in meters per second, m/s, and the base represents time measured in seconds, s. The area of this patch is the speed times the time, expressed in units of m/s times s, which equals m. The speed times the time is the distance traveled. *The area under a graph of speed vs. time represents the distance traveled.* This very powerful idea underlies the advanced mathematics called integral calculus. You will investigate the idea that the area under a graph of speed vs. time can be used to predict the behavior of objects falling in the presence of air resistance.

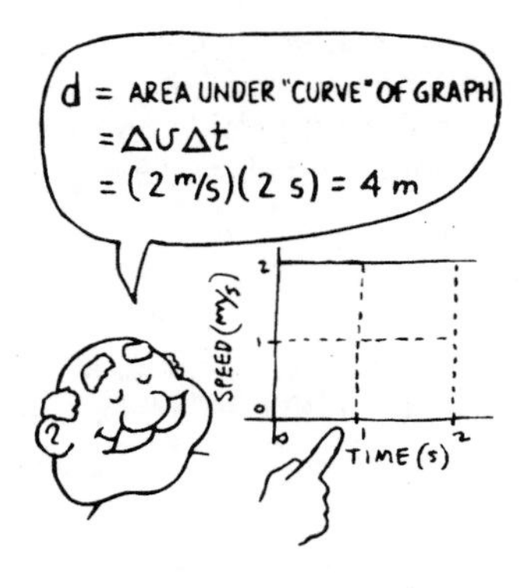

This lab is very sophisticated conceptually but very elegant and relatively simple to perform. Challenge your best students to try it!

If there were no air friction, a falling tennis ball or Styrofoam ball would fall at constant acceleration g so its change of speed is

$$v_f - v_i = gt$$

where v_f = final speed

v_i = initial speed

g = acceleration of gravity

t = time of fall

A graph of speed vs. time of fall is shown in Figure A, where $v_i = 0$. The y-axis represents the speed v_f of the freely falling object at the end of any time t. The area under the graph line is a triangle of base t and height v_f, so the area equals $\frac{1}{2} v_f t$.

To check that this does equal the distance traveled, note the following. The *average* speed $\bar{v}$ is half of v_f. The distance d traveled by a con-

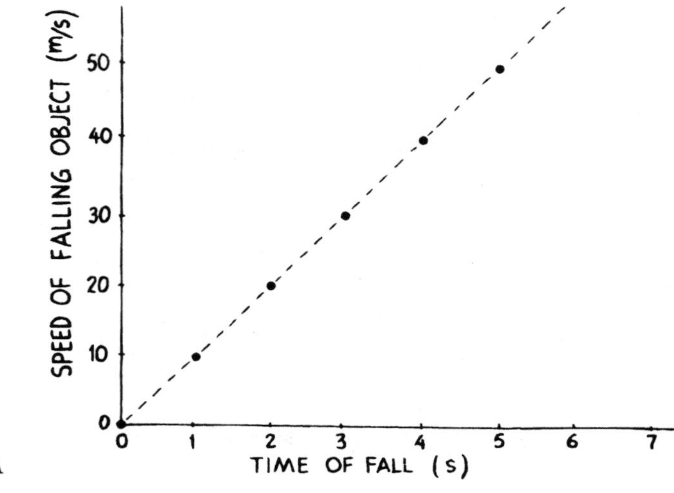

Fig. A

stantly accelerating body is its average speed $\bar{v}$ multiplied by the duration t of travel.

$$d = \bar{v}t \quad \text{or} \quad d = \frac{1}{2}v_f t$$

This lab is greatly enhanced by the computer simulation by the same title programmed by Jay Obernolte. The heart of this program won first prize in the Physics Division of the California State Science Fair in 1986, and Jay was recognized by the Westinghouse Search for Talent Contest in 1988.

If you time a tennis ball falling from rest a distance of 43.0 m in air (say from the twelfth floor of a building), the fall takes 3.5 s, *longer* than the theoretical time of 2.96 s. Air friction is *not negligible* for most objects, including tennis balls. A graph of the actual speed vs. the time of fall looks like the curve in Figure B.

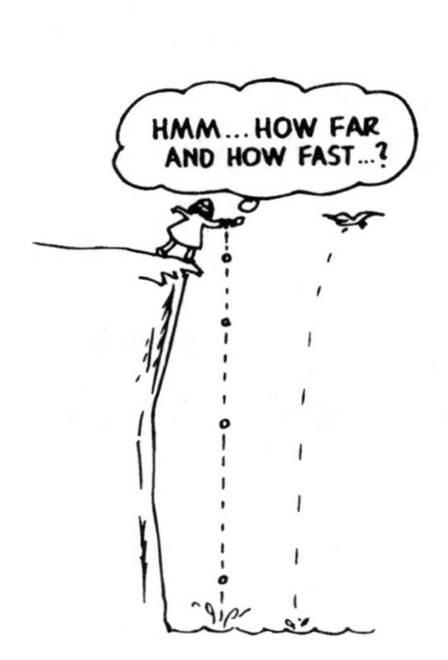

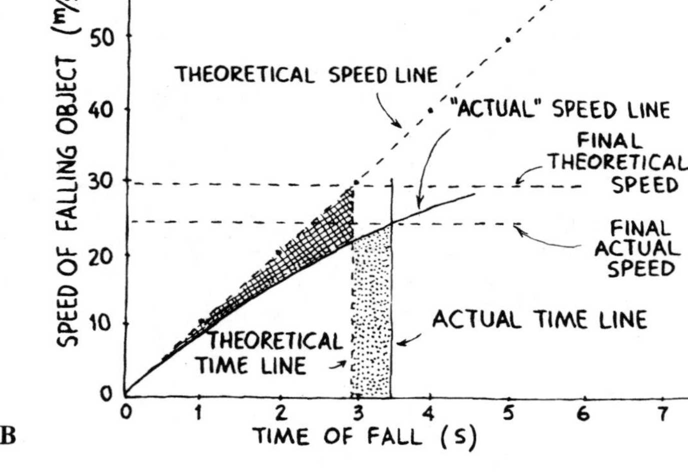

Fig. B

Since air resistance reduces the acceleration to below the theoretical value of 9.8 m/s^2, the falling speed is less than the theoretical speed. The difference is small at first, but grows as air resistance becomes greater and greater with the increasing speed. The graph of actual speed vs. time curves away from the theoretical straight line.

Is there a way to sketch the actual speed vs. time curve from only the distance fallen and the time of fall? Calculate the theoretical time of fall for no air resistance. The *height* from which the object is dropped is the same with or without air resistance. *The area under the actual speed vs. time curve must be the same as the area under the theoretical speed vs. time line.* It is the distance fallen. On a graph of theoretical speed vs. time, draw a vertical line from the theoretical time of fall on the horizontal axis up to the theoretical speed vs. time line. (In Figure B, this line is

labeled "theoretical time line.") Draw another vertical line upward from the *actual* time of fall on the horizontal axis. (This line is labeled "actual time line" in Figure B.) Sketch a curve of actual speed vs. time that crosses the second vertical line below the theoretical speed vs. time line. Sketch this curve so that the area added below it due to increased time of fall (stippled area) equals the area subtracted from below the theoretical speed vs. time line due to decreased speed (cross-hatched area). *The areas under the two graphs are then equal.* This is a fair approximation to the actual speed vs. time curve. The point where this curve crosses the vertical line of the actual time gives the probable impact speed of the tennis ball.

Procedure 

Devise dropping strategy.

Step 1: Your group should choose a strategy to drop a Ping-Pong ball or Styrofoam ball and clock its time of fall within 0.1 s or better. Consider a long-fall drop site, various releasing techniques, and reaction times associated with the timer you use.

Measure falling height.

Step 2: Devise a method that eliminates as much error as possible to measure the distance the object falls.

Submit proposal to teacher.

Step 3: Submit your plan to your teacher for approval.

Measure height and falling times.

Step 4: Measure the height and the falling times for your object using the approved methods of Steps 1 and 2.

height = ______________________

actual time of fall = ________________

Compute theoretical falling time.

Step 5: Using your measured value for the height, calculate the theoretical time of fall for your ball. Remember, this is the time it would take the ball to reach the ground if there were no air resistance.

theoretical time = ________________

Draw theoretical and actual speed lines.

Step 6: Using an overhead transparency or graph paper, trace Figure B, leaving out the actual speed vs. time curve. Draw one vertical line from the *theoretical* time of fall for your height up to the theoretical speed vs. time line. Draw the other vertical line from the *actual* time of fall up to the theoretical speed vs. time line.

Sketch curve on graph.

Step 7: Starting from the origin, sketch your approximation for the actual speed vs. time curve, out to the point where it crosses the actual time line, using the example mentioned in the discussion. The area of your stippled region should be the same as that of the cross-hatched region.

One possible way to do this is to tape a piece of cardboard to the wall and project your transparency onto it. Trace your two regions onto the cardboard and cut them out. Then measure the mass of the two regions. If their masses are not the same, adjust your actual speed curve and try again. Your approximation is done when the two regions have the same mass.

Predict probable impact speed.

Step 8: Draw a horizontal line from the upper right corner of your stippled region over to the speed axis. Where it intersects the speed axis is the object's probable impact speed.

Analysis

1. Have your teacher overlap your graph with those of others. How does your actual speed vs. time curve compare with theirs?

   Answers will vary depending on the object dropped.

2. What can you say about objects whose speed vs. time curves are close to the theoretical speed vs. time line?

   These objects encounter small air resistance.

3. What does the area under your speed vs. time graph represent?

   The height from which the object is dropped.

4. The equation for distance traveled is $d = \bar{v}t$. In this lab, the distance fallen is the same with or without air friction. How do the average speeds and times compare with and without air friction? Try to use different-sized symbols such as those used on page 66 of your text.

   $d = \bar{v}\mathbf{T} = \bar{\mathbf{V}}t$; **the average speed with air friction is less, but the time is greater.**

5. If you dropped a large leaf from the Empire State Building, what would its speed vs. time graph look like? How might it differ from that of a baseball?

   It would quickly become horizontal. The baseball's speed curve would take much longer and would arrive at a much higher value of speed before it became horizontal.

6. The *terminal* speed of a falling object is the speed at which it stops accelerating. How could you tell whether an object had reached its terminal speed by glancing at an actual speed vs. time graph?

   If the graph is horizontal, it's because the speed is no longer changing, so the object has reached its terminal speed.

15 Riding with the Wind

Purpose

To investigate the relationship between the components of the force that propels a sailboat.

Setup: 1
Lab Time: >1
Learning Cycle: concept development
Conceptual Level: difficult
Mathematical Level: moderate

Required Equipment/Supplies

dynamics cart fitted with an aluminum or cardboard sail
protractor
electric fan
ruler pulley
mass hangers
set of slotted masses
string

Adapted from PRISMS.

Pasco has an excellent fan cart that can be used as a sailboat. However, students must still use some kind of handheld fan, *not* the one on the cart!

Discussion

In this experiment, you will use a small model sailboat and an electric fan. The physics of sailing involves vectors—quantities that have both magnitude and direction. Such directed quantities include force, velocity, acceleration, and momentum. In this lab, you will be concerned with force vectors.

In Figure A, a crate is being pulled across a floor. The vector *F* represents the applied force. This force causes motion in the horizontal direction, and it also tends to lift the crate from the floor. The vector can be "resolved" into two vectors—one horizontal and the other vertical. The horizontal and vertical vectors are *components* of the original vector. The components form two sides of a rectangle. The vector *F* is the diagonal of this rectangle. When a vector is represented as the diagonal of a rectangle, its components are the two sides of the rectangle.

Whatever the direction of wind impact on a sail, the direction of the resulting force is always perpendicular to the sail surface. The magnitude of the force is smallest when the wind blows parallel to the sail—when it goes right on by without making any impact. The force is largest when the sail is perpendicular to the wind—when the wind makes full impact. Even if the wind hits the sail at another angle, the resulting force is always directed *perpendicular* to the sail.

The keel of a sailboat is a sort of fin on the bottom of the boat. It helps prevent the boat from moving sideways in the water. The wind force on the sail can be resolved into two components. One component, *K* (for *keel*), is parallel to the keel. The other component, *T* (for *tip*), is perpendicular to the keel, as shown in Figure B. Only the component *K* contributes to the motion of the boat. The component *T* tends to move the boat sideways and tip it over.

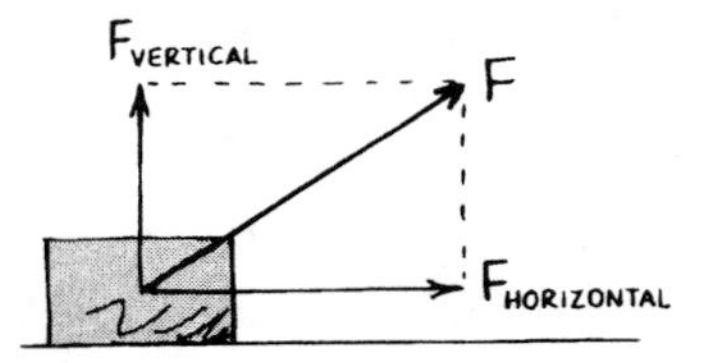

Fig. A

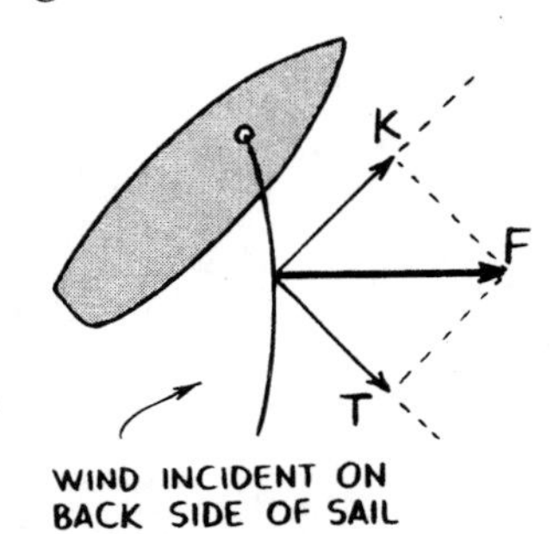

Fig. B

Any boat with a sail can sail downwind, that is, in the same direction as the wind. As the boat goes faster, the wind force on the sail decreases. If the boat is going as fast as the wind, then the sail simply sags. The wind force becomes zero. The fastest speed downwind is the speed of the wind.

A boat pointed directly into the wind with its sail at right angles to the keel is blown straight backward. No boat can sail *directly* into the wind. But a boat can *angle* into the wind so that a component of force *K* points in the forward direction. The procedure of going upwind is called *tacking*.

CAUTION: In the following procedure, you will be working with an electric fan. *Do not let hair or fingers get in the blades of the fan.*

Procedure  

Try to sail the boat directly upwind.

Step 1: Position the cart with its wheels (the keel) parallel to the wind from the fan. Position the sail perpendicular to the wheels. Start the fan blowing on the *front* of the cart with the wind parallel to the wheels.

1. What happens to the cart?

   The cart goes backward.

Sail the boat directly downwind.

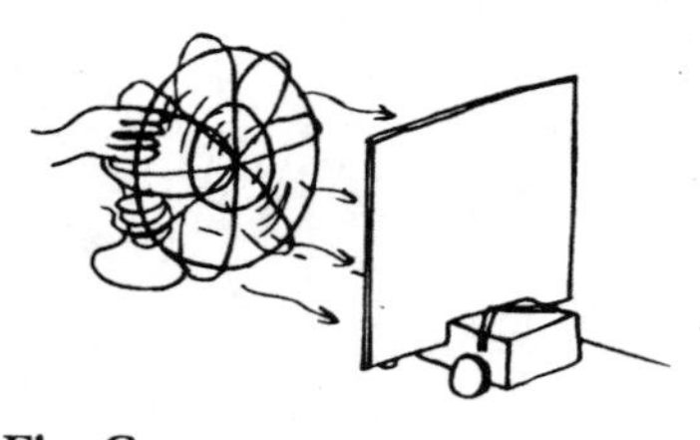

Fig. C

Step 2: Direct the wind from *behind* the cart and parallel to the wheels, keeping the sail perpendicular to the wheels, as shown in Figure C.

2. How does the cart move?

   The cart goes forward.

Sail the boat at an angle.

Step 3: Reposition the sail at a 45° angle to the wheels. Direct the wind behind, parallel to the wheels.

3. What happens to the cart?

   The cart goes forward although somewhat slower than in Step 2.

Step 4: Use a ruler to draw a vector that represents the wind force perpendicular to the sail. Remember that the *length* of the vector you draw

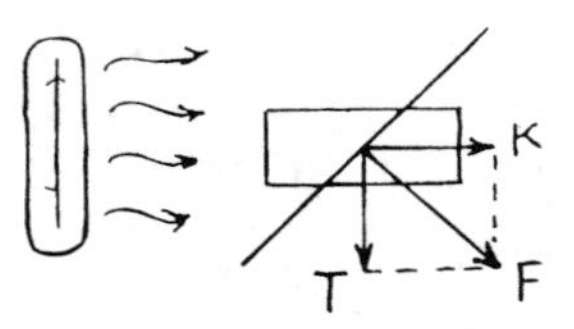

Name Period Date

represents the size of the force. Split the vector into its T and K components. Label the forces and each component on the diagram.

Change the wind direction.

The vector diagram should be identical to that in Step 4.

Step 5: Repeat Step 3 with the fan perpendicular to the wheels. Make a vector diagram for this case.

Change the wind direction.

Step 6: Repeat Step 3 but direct the wind at a 60° angle to the wheels from the front, as shown in Figure D.

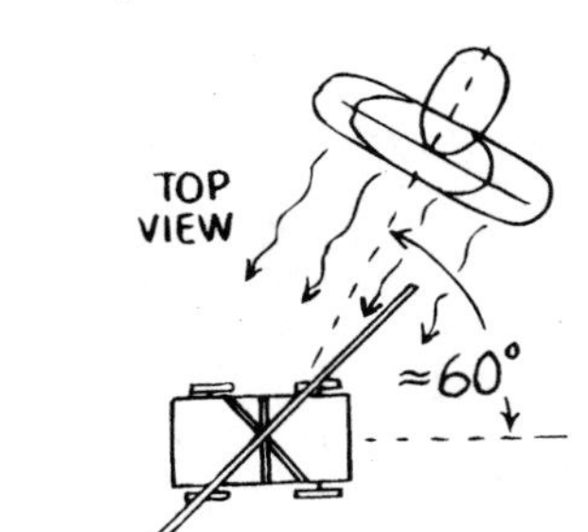

Fig. D

4. Does the cart sail against or with the wind?

It sails against the wind (at an angle).

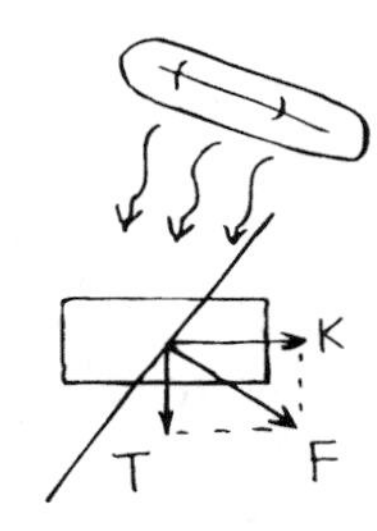

Draw a vector representing the wind force, and split it into its T and K components.

Step 7: Set up a pulley on the edge of the table, as shown in Figure E. Attach one end of a piece of lightweight string to the cart and thread the other end over the pulley. Attach a mass hanger to the string. Set the sail perpendicular to the direction of the cart. Place the fan directly between the pulley and the cart so that the full force of the wind strikes the sail at 90°. Holding the cart, turn the fan on its highest speed setting. The wind force depends on how close the fan is to the sail. Place the fan close to the sail. *Slowly* add very small masses to the mass hanger. Continue adding them until the weight of the masses (including the mass hanger) just balances the wind force on the cart. The wind force is balanced when adding the smallest mass you have causes the cart to move toward the pulley and removing the smallest mass causes the cart to move away from the pulley. Record the mass required to balance the cart.

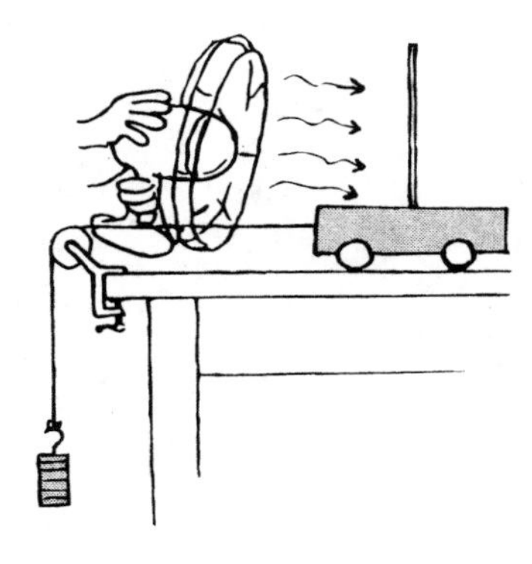

Fig. E

mass = ______________

Calculate the weight of this mass, using the approximation that a mass of 1 kg has a weight of 10 N.

weight = ______________

5. What is the amount of wind force acting on the sail in this configuration?

Answers will vary; a balancing mass of 100 g is typical, so the force would be about 1 N.

Step 8: Repeat Step 7, but this time orient the sail at a 45° angle to the wind. Record the mass and calculate the weight required to balance the cart.

mass = ______________

weight = ______________

6. What is the amount of wind force acting on the sail in this configuration?

Answers will vary; students may measure about half the answer of Step 7. The correct answer should be 0.707 times the answer of Step 6.

Analysis

7. Which orientation of the sail with respect to the wind provided the greatest wind force?

When the sail is perpendicular to the wind, the wind force is largest.

8. When the sail is oriented at 45° to the wind, is the wind force less than half, equal to half, or greater than half the wind force at 90°? Justify your answer with a vector diagram.

For a sail 45° to the wind, the wind force is greater than half the wind force when the wind is perpendicular to the sail. It takes two vector components, 0.707 (or about 0.7) units long and perpendicular to each other, to equal a resultant vector 1 unit long.

Name Period Date

16 Balloon Rockets

Purpose

To investigate action-reaction relationships.

Setup: <1
Lab Time: 1
Learning Cycle: exploratory
Conceptual Level: easy
Mathematical Level: none

Adapted from PRISMS.

Required Equipment/Supplies

balloons
paper clips
guide wire or string
tape
straws

Discussion

Early in this century, the physicist Robert Goddard proposed that rockets would someday be sent to the moon. He met strong opposition from people who thought that such a feat was impossible. They thought that a rocket could not work unless it had air to push against (although physics types knew better!). Your mission is a less ambitious one—to construct a balloon rocket system that will travel across the room!

Scientists look for the causes of observed effects. As you construct your balloon rockets and improve upon the design, think about how you would explain to a friend how and why your design works.

Devise a rocket system.

Activity

Procedure 

Step 1: String the guide wire or string across the room. Make a rocket system that will propel itself across the classroom on the wire or string. Use a balloon for the rocket power, and the straw, paper clips, and tape to make a guidance system. Draw a sketch or write a description of your design.

Modify system design.

Step 2: Getting there is only half the problem! Make a rocket system that will get across the room and then come *back* again. Draw a sketch or write a description of your design.

This activity is not unlike the work of the astronauts—challenging! Making two balloon "fuel tanks" propel across the room without snagging the guide wire requires both technique and ingenuity.

Analysis

1. Explain why an inflated balloon moves when the air escapes.

   The stretched balloon collapses and forces air out the nozzle (action) and the air that is squeezed out in turn pushes back on the balloon (reaction).

2. Would the rocket action still occur if there were no surrounding air?

   Yes, it is the air that is squeezed out of the balloon that pushes the balloon, not the surrounding air. The surrounding air with its air resistance actually hinders acceleration of the balloon, and without it, the balloon would have a greater acceleration. Rockets would have a more difficult time getting to the moon if there were air in outer space.

17 Tension

Purpose

To introduce the concept of tension in a string.

Setup: <1

Lab Time: >1

Learning Cycle: concept development

Conceptual Level: moderate

Mathematical Level: none

Required Equipment/Supplies

3 large paper clips
3 large identical rubber bands
1 m of strong string
500-g hook mass
ring stand
spring scale
ruler
marking pen

Rubber bands should be thick enough so that the percentage stretching is relatively small for the forces applied in the experiment.

Discussion

When you put bananas on a hanging scale at the supermarket, a spring is stretched. The greater the weight of the bananas (the force that is exerted), the more the spring is stretched. Rubber bands act in a similar way when stretched. For stretches that are not too great, the amount of the stretch is directly proportional to the force.

If you hang a load from a string, you will produce a tension in the string and stretch it a small amount. The amount of tension can be measured by attaching the top end of the string to the hook of a spring scale hanging from a support. The scale will register the tension as the sum of the weights of the load and the string. The weight of the string is usually small enough to be neglected, and the tension is simply the weight of the load.

Is this tension the same all along the string? To find the tension at the lower end of the string you could place a second scale there, but the weight of the added scale would increase the tension in the top scale. You need a scale with a tiny mass. Since a rubber band stretches proportionally with the force, you could use it to measure the tension at the lower end of the string. Its tiny weight will not noticeably affect the reading in the top scale. In this activity, you will use rubber bands to investigate the tensions at different places along short and long strings supporting the same load.

Procedure 

Step 1: Without stretching them, place three identical large rubber bands flat on a table. Carefully mark two ink dots 1 cm apart on each of the rubber bands.

Mark rubber bands.

Cut string.

Step 2: Cut a 1-meter string into 2 lengths of 25 cm each, and 1 length of 50 cm. Tie each end of each string into a loop.

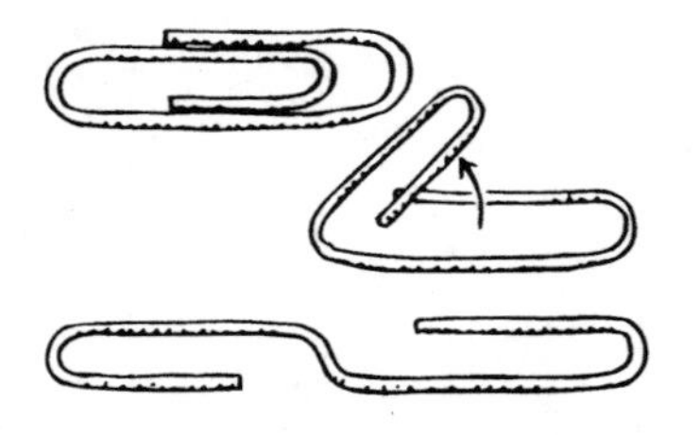

Fig. A

Step 3: Bend the paper clips into double hooks, as shown in Figure A.

Step 4: Suspend the spring scale from the top of a ring stand such that it extends over the edge of the table. Place one end of a rubber band over the scale hook. Suspend the 500-g load from the lower end of the rubber band as shown in Figure B. Note that the weight of the load stretches both the rubber band and the spring inside the spring scale.

While the load is suspended, measure and record the distance between the ink marks on the rubber band. Also record the reading on the spring scale.

distance between marks = ____________

tension (scale reading) = ____________

3-rubber-band stretch.

Step 5: With the use of paper-clip hooks, repeat the previous step using three rubber bands connected as shown in Figure C. Measure the stretch of the three bands. Record your results.

distance between marks of top band = ____________

distance between marks of middle band = ____________

distance between marks of bottom band = ____________

tension (spring-scale reading) = ____________

Hang load from one band and string.

Step 6: Disconnect the load and two of the rubber bands. Using a paper-clip hook, connect one end of a 25-cm length of string to the remaining rubber band. Suspend the load on the other end of the string, as shown in Figure D. Record the spring-scale reading and the distance between the marks of the rubber band.

tension (spring-scale reading) = ____________

distance between marks of band = ____________

Hang load from string and two bands.

Step 7: Using a paper-clip hook, insert a rubber band between the lower end of the string and the load, as shown in Figure E. While you are attaching the bottom rubber band, mentally predict the amount of stretch of the added band. Compare the tensions at the top and the bottom of the string by measuring the stretches of the two rubber bands. Record your findings.

distance between marks of top band = ____________

distance between marks of bottom band = ____________

Name Period Date

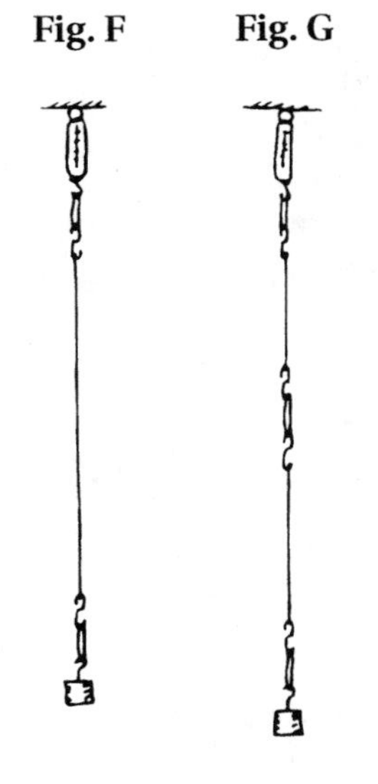

Step 8: Repeat Step 7 using the 50-cm piece of string, as shown in Figure F, to see whether the tension in the string depends on the length of the string. Make a mental prediction before you measure and record your findings.

Repeat using longer string.

distance between marks of top band = ______________

distance between marks of bottom band = ______________

Step 9: Repeat the previous step, substituting two 25-cm pieces of string joined by paper-clip hooks with a rubber band in the middle (as shown in Figure G) for the 50-cm string. This arrangement is to compare the tension at the middle of a string with the tension at its ends. While you are substituting, mentally predict how far apart the ink marks will be. After your prediction, measure and record your findings.

Add rubber band between two strings.

The distance between the marks should be the same for all the rubber bands. The tension as recorded on the spring scale should always be the same. For a 500-g load the scale should read 500-g or 4.9 N.

distance between marks of top band = ______________

distance between marks of middle band = ______________

distance between marks of bottom band = ______________

Analysis

1. Does the length of the string have any effect on the reading of the spring scale? What evidence can you cite?

The length of the string that transmits the weight of the load to the spring scale has no effect on the spring-scale reading (because the string's weight is negligible). Evidence for this is the same scale readings of Steps 5 and 7.

2. Does the tension in the string depend on the length of the string? What evidence can you cite?

The length of the string has no effect on the magnitude of the force transmitted through it as tension. Evidence for this is the same reading on the scale throughout the different steps. (The weights of the rubber bands, paper clips, and string are negligible in comparison to the 500-g load.)

3. How does the tension at different distances along a stretched string compare? What evidence can you cite?

The tension is the same all along a stretched string. This is evidenced by the same amount of the stretch of the rubber bands in Step 8.

4. How much would each rubber band stretch if the 500-g load were suspended by two side-by-side bands as shown in Figure H? Make your prediction, then set up the experiment. Explain your result.

predicted stretch = ____________

actual stretch = ____________

When two side-by-side rubber bands support the load, the tension in each is reduced by half. Each band supports half the load and the amount of stretch is therefore half as much as it was before. (The distance between the ink marks is greater than 1 cm by an amount that is half the normal stretching distance when a single rubber band supports the load.)

Name Period Date

18 Tug-of-War

Experiment

Purpose

To investigate the tension in a string, the function of a simple pulley, and a simple "tug-of-war."

Setup: <1

Lab Time: 1

Learning Cycle: concept development

Conceptual Level: moderate

Mathematical Level: none

Required Equipment/Supplies

5 large paper clips
2 large identical rubber bands
2 m of strong string
2 500-g hook masses
2 low-friction pulleys
2 ring stands
spring scale
measuring rule

Discussion

Suppose you push on the back of a stalled car. You are certainly aware that you are exerting a force on the car. Are you equally aware that the car is exerting a force on you? And that the magnitudes of the car's force on you and your force on the car are the same? A force cannot exist alone. Forces are always the result of interactions between two things, and they come in balanced pairs.

Now suppose you get a friend to tie a rope to the front of the car and pull on it. The rope will be pulling back on your friend with exactly the same magnitude of force that she is exerting on the rope. The other end of the rope will be pulling on the car and the car will be pulling equally back on it. There are two interaction pairs, one where your friend grasps the rope and one where the rope is attached to the car.

A rope or string is a transmitter of force. If it is not moving or it is moving but has a negligible mass, the forces at its two ends will also be equal. In this activity, you will learn about balanced pairs of interaction forces and about the way a string transmits forces.

Procedure

Step 1: Suspend a 500-g load from a string that is held by a spring scale as shown in Figure A.

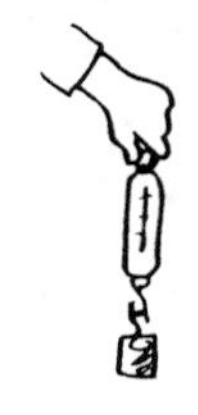

Fig. A

Put tension on a string.

1. What does the scale reading tell you about the tension in the string?

   The tension in the string is equal to the weight of the load

Hang weight over pulley.

Step 2: Drape the string over a pulley such that both ends of the string hang vertically, as shown in Figure B. Hold the scale steady so that it supports the hanging load.

2. What does the scale read, and how does this force compare with the weight of the load?

The tension in the string is still equal to the weight of the load, as evidenced by the same scale reading.

3. How does it compare with the tension in the string?

The scale reading is equal to the tension in the string.

Move spring scale to different positions along vertical.

Step 3: Move the spring scale first to a higher, then to a lower position, keeping the strings on each side of the pulley vertical.

4. Does the reading at the higher position change?

No; the reading does not change.

5. Move the scale to a lower position. Does the reading at the lower position change? Briefly explain these results.

The scale reading does not change with different positions of the string. The tension does not depend on the length of the strings, but rather on the force transmitted through the string.

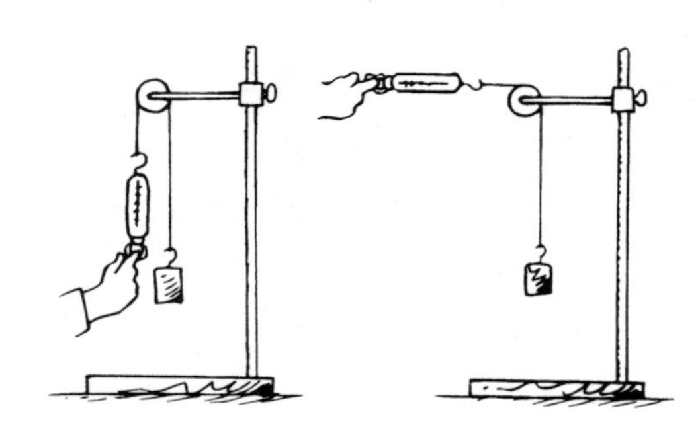

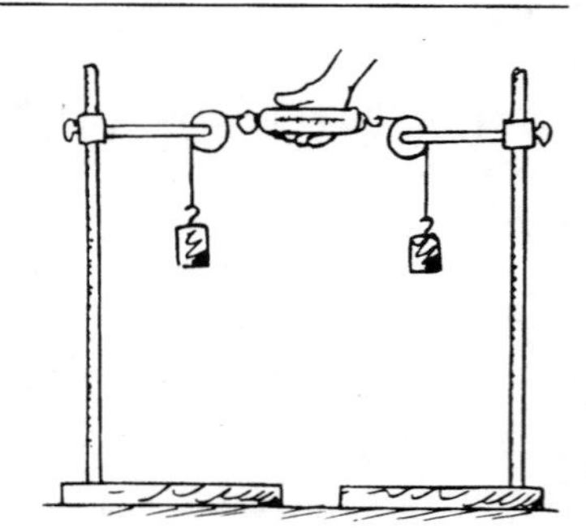

Fig. B Fig. C Fig. D Fig. E Fig. F

Name Period Date

Step 4: Move the spring scale to various angles to the vertical, until the scale is horizontal, as shown in Figure C.

Move spring scale to different positions away from vertical.

6. Does the reading on the scale ever deviate from what you measured in the previous steps? Briefly explain your result.

The reading does not change. The pulley serves only to change the direction of the force that is transmitted through the string.

Step 5: Remove the string from the pulley and drape it over a horizontal rod. Repeat Step 4, as shown in Figure D.

Hang weight over rod.

7. Do you find a difference between the results of Steps 4 and 5? Explain.

In this case friction on the string tends to prevent movement of the load, so the tension is *not* the same on the two sides of the rod. The force of friction acts in a direction to oppose sliding of the string. It provides part of the force needed to support the load, so the tension on the side of the string held by the hand is less. If the hand is pulled down harder, to make the load rise, the tension in the hand-held side is greater than the tension in the side that supports the load. It is greater by the amount of friction that acts between the rod and the string.

Step 6: Attach a spring scale to each end of the string. Drape the string over the pulley and attach equal masses to each end, as shown in Figure E.

Pull on both ends of string over pulley.

8. What do the scales read?

The scales will read the same if the pulls are equal. If the pull is harder by an amount equal to the friction between the rod and string (without string slippage), then the scale reading is greater than that of the partner's. If the pull is harder than the friction that is developed, the string slides over the rod.

9. What role does friction play in the function of a pulley?

If the bearings in the pulley are good, the friction is small and its effects are also small.

Have mini tug-of-war with two spring scales.

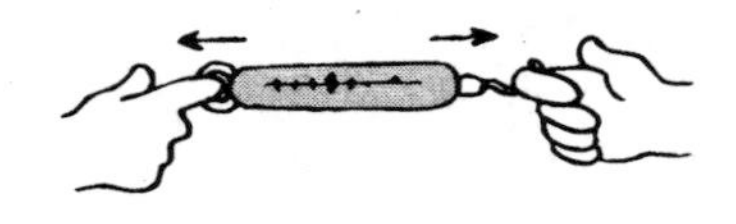

Step 7: Have your partner hold one end of a spring scale stationary while you pull horizontally on the other end. Pull until the scale reads the same force as it did when suspending the mass. Record the following observations.

All answers will be the same, 4.9 N.

force you exert on the scale = ____________

force the scale exerts on you = ____________

force your partner exerts on the scale = ____________

force the scale exerts on your partner = ____________

Attach string to wall and tug.

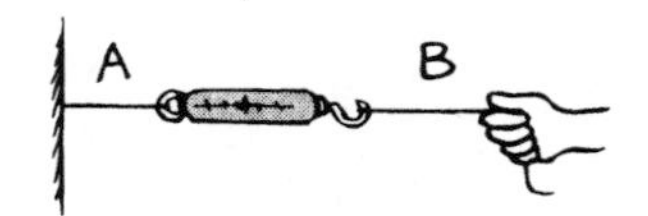

Step 8: Attach strings on both ends of the spring scale. Fasten one end to the wall or a steady support. Call this String A. Pull horizontally on the other string, String B, until the scale reads the same as in the previous step. Record the following observations.

force you exert on String A = ____________

force String A exerts on scale = ____________

force the scale exerts on String B = ____________

force String B exerts on the wall = ____________

force the wall exerts on String B = ____________

10. What is the essential difference between the situations in Step 7 and Step 8?

There is no essential difference between these steps. In Step 7 the force is transmitted through the scale to the partner, and in Step 8 the force is transmitted to the wall.

Think and Explain

11. From a microscopic point of view, how does the spring or string transmit the force you are exerting on your partner or the wall?

The force is transmitted undiminished between molecules throughout the material. From a microscopic point of view, forces between molecules in the string react to the pull.

Step 9: Study Figure F and predict the reading on the scale when two 500-g loads are supported at each end of the strings. Then assemble the apparatus and check your prediction.

predicted scale reading = ____________

actual scale reading = Same as in Steps 7 and 8.

19 Go Cart

Purpose

To investigate the momentum imparted during elastic and inelastic collisions.

Setup: 1
Lab Time: >1
Learning Cycle: exploratory
Conceptual Level: moderate
Mathematical Level: moderate

Required Equipment/Supplies

"bouncing dart" from Arbor Scientific
ring stand with ring
dynamics cart (with a mass of 1 kg or more)
string
pendulum clamp
C clamp
meterstick
brick or heavy weight

Thanks to Evan Jones for his ideas and suggestions for this lab.

Activity

Discussion

If you fell from a tree limb onto a trampoline, you'd bounce. If you fell into a large pile of leaves, you'd come to rest without bouncing. In which case, if either, is the change in your momentum greater? This activity will help you answer that question. You'll compare the changes in momentum in the collision of a "bouncing dart" where bouncing does take place and where it doesn't.

The dart consists of a thick wooden dowel with a rubber tip on *each* end. Although the tips look and feel the same, the tips are made of different kinds of rubber. One end acts somewhat like a very bouncy ball. The other end acts somewhat like a lump of clay. They have different *elasticities.* Bounce each end of the dart on the table and you'll easily see which end is more elastic. In the activity, you'll do the same against the dynamics cart using the dart as a pendulum.

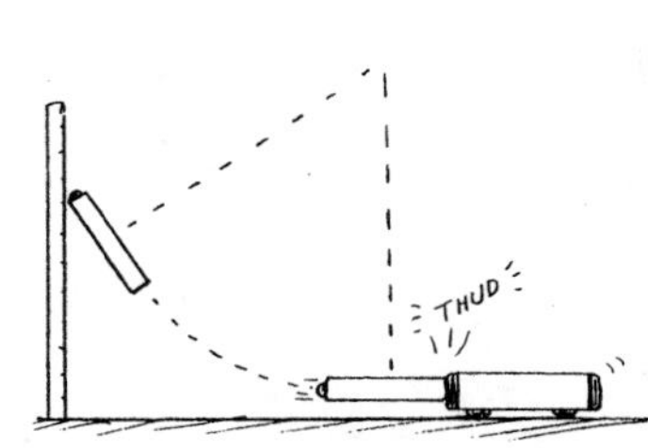

Fig. A

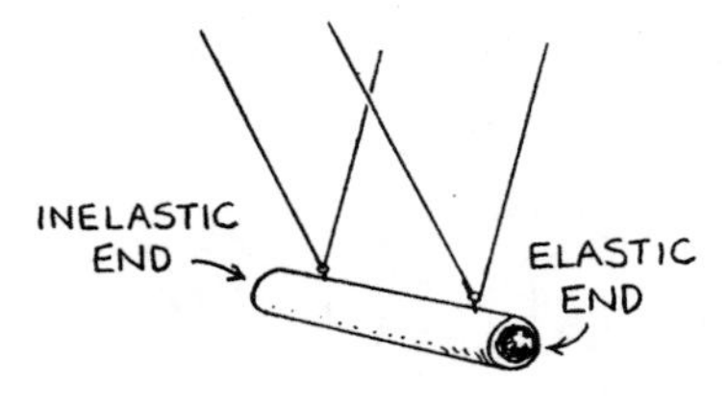

Procedure

Step 1: Attach the dart to the ring stand as a pendulum, using a heavy weight to secure the base of the ring stand. To prevent the dart from swinging into the weight, position the ring on the stand so that it faces the opposite direction. Adjust the string so that the dart strikes the middle of one end of the cart when the dart is at the lowest point of its swing.

For best results, the dart should be mounted in such a way that it can't vibrate from side to side as it is swinging. One way is to mount the dart on the end of a dowel with a PVC "T" on the pivot end.

Step 2: Elevate the dart so that when impact is made, the cart will roll forward a foot or so on a level table or floor when struck by the inelastic end of the dart. Use a meterstick to measure the vertical distance

Observe collision without bouncing.

between the release point of the dart and the bottom of its swing. Repeat several times. Record the average stopping distance of the cart.

vertical distance = ______________

stopping distance (no bouncing) = ______________

Observe collision with bouncing.

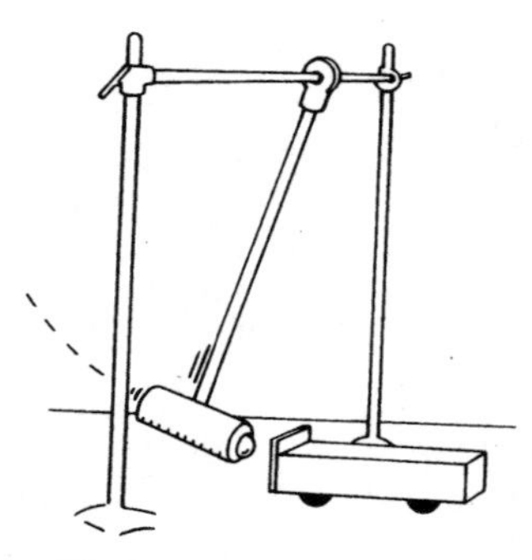

Alternate setup

Step 3: Repeat using the elastic end of the dart. Be sure to release the dart from the *same position* as in Step 2. Note what happens to the dart after it hits the cart. Make sure to release the dart from the same height each time. Repeat several more times to see whether your results are consistent. Record the average stopping distance of the cart.

stopping distance (with bouncing) = ______________

Write down your observations.

The dart bounces back after hitting the block.

Analysis

1. Define the momentum of the swinging dart before it hits the cart to be positive, so that momentum in the opposite direction is negative. After the dart bounces off the cart, is its momentum negative or positive?

   The dart's momentum after the collision is negative because it is in the opposite direction from the original momentum. (Even in a perfectly elastic collision, the dart does *not* rebound with the same speed it had just before hitting the cart, so the magnitude of the momentum change is less than $2mv$.)

2. When does the dart acquire the greater momentum—when it bounces off the cart or when it doesn't? Explain.

   When the dart strikes the block without bouncing, its momentum goes to nearly zero and its change in momentum is approximately $-mv$ (magnitude mv). When it bounces back in the other direction, its change in momentum has a magnitude that is greater than mv.

Name Period Date

3. When does the *cart* undergo the greater change in momentum—when struck by the end of the dart that bounces or by the end of the dart that doesn't bounce? Explain.

Since momentum is conserved, the change in momentum of the cart is the same in magnitude as the change of momentum of the dart. It is greater when the dart bounces.

4. How do the stopping distances of the cart compare?

The cart rolls farther because of the greater impulse produced when the dart bounces. How much farther depends on the mass of the cart and the dart, and how accurately the dart strikes the back surface of the cart at 90°, etc.

5. How would you account for the difference in stopping distances?

The greater change in momentum for bouncing, and the corresponding greater impulse, gives the cart a correspondingly greater *momentum* and *speed*.

Name Period Date

20 Tailgated by a Dart

Experiment

Purpose

To estimate the speed of an object by applying conservation of momentum to an inelastic collision.

Setup: 1
Lab Time: 1
Learning Cycle: application
Conceptual Level: moderate
Mathematical Level: moderate

Required Equipment/Supplies

opposite types of Velcro® hook and loop fastener tape
toy car
toy dart gun using rubber darts
stopwatch
meterstick
balance

Optional: photogate timer or computer, light probe with interface, and light source.

Thanks to Roy Unruh for his ideas and suggestions for this lab.

This lab can be simulated nicely on the computer using the program of the same name on *Laboratory Simulations-II.*

Discussion

If you catch a heavy ball while standing motionless on a skateboard, the momentum of the ball is transferred to you and sets you in motion. If you measure your speed and the masses of the ball, the skateboard, and yourself, then you know the momentum of everything after the ball is caught. But this is equal to the momentum of the ball just before you caught it, from the law of conservation of momentum. To find the speed of the ball just before you caught it, divide the momentum by the mass of the ball. In this experiment, you will find the speeds of a toy dart before and after it collides with a toy car.

I HEAR, AND I FORGET.
I SEE, AND I REMEMBER.
I DO, AND I UNDERSTAND.
LAB EXPERIMENTS ARE DOING PHYSICS -
SIGH

Procedure  

Step 1: Fasten one type of Velcro tape to the back end of a toy car of small mass and low wheel friction. Fasten the opposite type of Velcro to the suction-cup end of a rubber dart. When the toy car is hit, it must be free to coast in a straight line on a level table or the floor until it comes to a stop.

Practice shooting the dart onto the back end of the car. The dart should stick to the car and cause it to coast.

Tailgate car with dart.

A track helps constrain the car's motion to a straight line but increases the friction.

1. What is the relationship between the momentum of the dart before the impact and the combined momentum of the dart *and* car just immediately after the impact?

The momentum of the dart before the collision is equal to the combined momentum of the dart-and-car combination after the collision.

Step 2: Measure the distance and time that the car coasts after it is hit by the dart, until it comes to a stop. Record your data in Data Table A. Repeat for two more trials.

Step 3: Calculate the average speed of the car after impact for the three trials, and record in Data Table A.

TRIAL	COASTING DISTANCE	COASTING TIME	AVERAGE SPEED OF CAR AFTER IMPACT	SPEED OF CAR UPON IMPACT	INITIAL SPEED OF DART
1					
2					
3					

Data Table A

2. Was the speed of the car constant as it coasted? Explain.

No, the speed was not constant. If it had been, the car would not have coasted to a stop.

Determine speed upon impact.

3. If the retarding force on the car is assumed to be nearly constant, how does the speed of the car immediately after impact compare with the average speed?

The speed immediately after impact is twice the average speed.

Enter values for the speed of the car upon impact in Data Table A.

Step 4: Find the masses needed to compute the momenta.

mass of car = ______________

mass of dart = ______________

Name Period Date

Step 5: Write an equation showing the momenta before and after the collision.

(mass of dart) × (initial speed of dart) =

(combined mass of dart and car) × (speed of car upon impact)

Compute the initial speed of the dart before impact for each of the three trials. Record your values for the initial speed of the dart in Data Table A.

Step 6: Use a photogate timer or a computer with a light probe to measure the speed of the toy car just after it collides with the dart. Calculate the initial speed of the dart before impact from this measurement.

speed of car upon impact = ____________________

initial speed of dart before impact = ____________

4. How does the speed of the car upon impact, as measured by the light probe or photogate timer, compare with the value you obtained in Step 3?

The speed of the car measured by the light probe or photogate timer should agree fairly well with the computed value.

Analysis

5. Is the momentum of the tailgated car constant the whole time it is moving? Explain.

No, once the tailgated car is moving, there is an external force on it—friction—that causes it to lose its momentum and come to a stop.

Name Period Date

21 Making the Grade

Purpose

To investigate the force and the distance involved in moving an object up an incline.

Setup: <1
Lab Time: 1
Learning Cycle: exploratory
Conceptual Level: easy
Mathematical Level: easy

Adapted from PRISMS.

Required Equipment/Supplies

board for inclined plane
spring scale
meterstick
ring stand
clamp
cart

Activity

Discussion

One of the simplest machines that makes doing work easier is the inclined plane, or ramp. It is much easier to push a heavy load up a ramp than it is to lift it vertically to the same height. When it is lifted vertically, a greater lifting force is required but the distance moved is less. When it is pushed up a ramp, the distance moved is greater but the force required is less. This fact illustrates one of the most powerful laws of physics, the law of energy conservation.

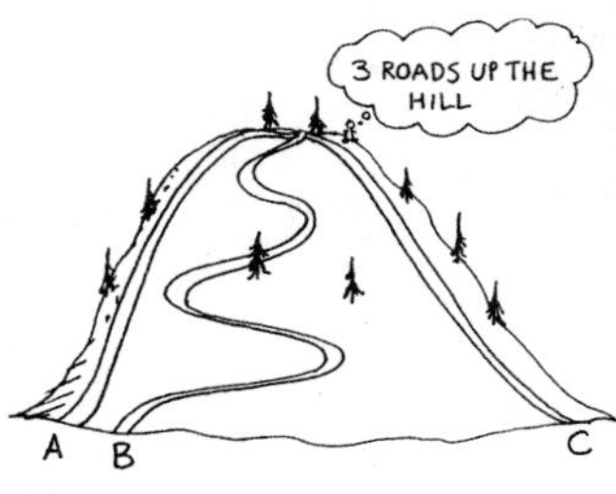

Fig. A

1. A hill has three paths up its sides to a flat summit area, D, as shown in Figure A. The three path lengths AD, BD, and CD are all different, but the vertical height is the same. Not including the energy used to overcome the internal friction of a car, which path requires the most energy (gasoline) for a car driving up it? Explain your answer.

The amount of work done is independent of the path taken. Therefore, all the paths would require the same amount of energy (gasoline) since the work done against the pull of gravity is the same for each path.

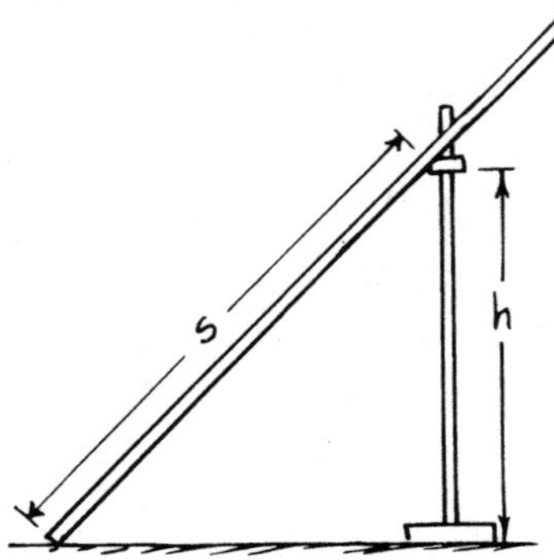

Fig. B

Raise the cart.

Procedure

Step 1: Place a clamp on a ring stand. Clamp the board in place at an angle of 45°, as shown in Figure B. Pull the cart up the inclined plane with a spring scale kept parallel to the plane, to measure the force. Measure the distance *s* from the bottom of the incline to the ring stand clamp. Record the force and distance in Data Table A.

Data Table A

	10°	30°	45°	60°
FORCE (N)				
DISTANCE (cm)				

Change the angle of the incline.

Step 2: Vary the angle while keeping the height *h* the same by sliding the board up or down inside the clamp to make angles of 10°, 30°, and 60°. For each of the different angles (and distances), pull the cart parallel to the board. Record your force and distance data in Data Table A.

Analysis

2. What pattern or relationship do you find between the forces and the distances?

The longer paths have smaller slopes, and thus less force is required to move the car up them.

22 Muscle Up!

Purpose

To determine the power that can be produced by various muscles of the human body.

Setup: <1
Lab Time: >1
Learning Cycle: exploratory
Conceptual Level: easy
Mathematical Level: easy

Required Equipment/Supplies

bleachers and/or stairs
stopwatch
meterstick
weights
rope

Discussion

Power is usually associated with mechanical engines or electric motors. Many other devices also consume power to make light or heat. A lighted incandescent bulb may dissipate 100 watts of power. The human body also dissipates power as it converts the energy of food to heat and work. The human body is subject to the same laws of physics that govern mechanical and electrical devices.

The different muscle groups of the body are capable of producing forces that can act through distances. Work is the product of the force and the distance, provided they both act in the same direction. When a person runs up stairs, the force lifted is the person's weight, and the distance is the vertical distance moved—not the distance along the stairs. If the time it takes to do work is measured, the power output of the body, which is the work divided by the time, can be determined in watts.

Activity

Procedure 

Step 1: Select five different activities from the following list:

Measure force and distance.

You might suggest your students do each activity for 100 s (that is, one minute and 40 s).

Possible Activities
Lift a mass with your wrist only, forearm only, arm only, foot only, or leg only.
Do push-ups, sit-ups, or some other exercise.
Run up stairs or bleachers.
Pull a weight with a rope.
Jump with or without weights attached.

Perform these activities, and record in Data Table A the *force* in newtons that acted, the *distance* in meters moved against the force, the number of repetitions (or "reps"), and the *time* in seconds required. Then calculate the *power* in watts. (One hundred seconds is a convenient time interval.)

Count the number of reps and measure the time.

	1	2	3	4	5	6	7	8	9	10
FORCE										
DISTANCE										
# REPS										
WORK										
TIME										
POWER										

Data Table A

Point out to your students that they do positive work on a weight when they lift it (force and displacement in same direction) and negative work on a weight when they lower it (force and displacement in opposite directions). A weight being lowered does work on the person (even if the person is getting tired)! A similar analysis applies to running.

Step 2: Complete the table by recording the results of four other activities performed by other class members.

Analysis

1. What name is given to the rate at which work is done? What are the units of this rate?

   The rate at which work is done is the power. It is measured in watts.

2. In which activity done by your class was the largest power produced? Which muscle groups were used in this activity?

   Answers will vary. Usually the leg muscles are the most powerful muscle group.

3. Did the activity that used the largest force result in the largest power produced? Explain how a large force can result in a relatively small power.

   The largest force does not necessarily result in the most power if it acts for a longer time.

4. Can a pulley, winch, or lever increase the rate at which a person can do work? Pay careful attention to the wording of this question, and explain your answer.

   A pulley, winch, or lever can't increase the rate at which a person can do work. It can allow the person to work at a lower rate by lengthening the time, thus making a job easier. Usually a pulley, winch, or lever decreases the required power input, so the same work is accomplished in more time.

23 Cut Short

Purpose

To illustrate the principle of conservation of energy with a pendulum.

Setup: <1
Lab Time: <1
Learning Cycle: exploratory
Conceptual Level: easy
Mathematical Level: none

Required Equipment/Supplies

3 ring stands
pendulum clamp
string
steel ball
rod
clamp

If time is short, you might consider doing this activity as a demonstration and have your students proceed to "Conserving Your Energy."

Activity

Discussion

A pendulum swinging to and fro illustrates the conservation of energy. Raise the pendulum bob to give it potential energy. Release it and the potential energy is converted to kinetic energy as the bob approaches its lowest point. Then, as the bob swings up on the other side, kinetic energy is converted to potential energy. Back and forth, the forms of energy change while their sum is constant. Energy is conserved. What happens if the length of the pendulum is suddenly changed? How does the resulting motion illustrate energy conservation?

Procedure

Step 1: Attach a pendulum clamp to the top of a ring stand set between two other ring stands, as shown in Figure A. Attach a steel ball to a piece of string that is nearly as long as the ring stand is tall.

Make pendulum.

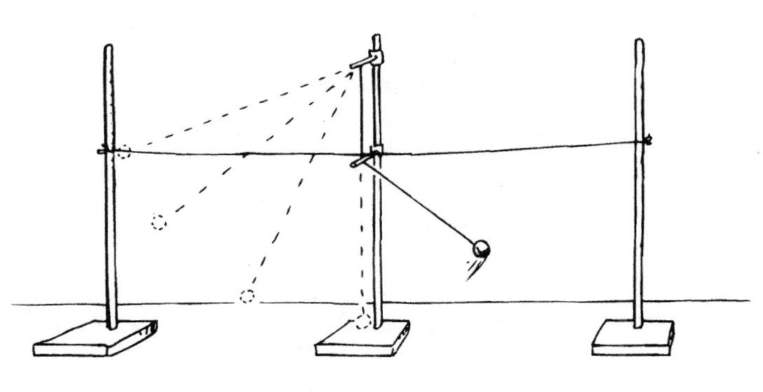

Fig. A

Step 2: Tie a string horizontally from one empty ring stand to the other, as shown in Figure A. The string should be about two-thirds as high as the pendulum clamp.

Set up level string.

Attach crossbar.

Step 3: Attach a rod to the central ring stand at the same height as the horizontal string (Figure A). The rod should touch the pendulum string when the string is vertical.

Predict height.

Step 4: Predict what height the ball will reach if the ball is released at the same height as the horizontal string and the pendulum string is stopped by the rod. Check one:

Prediction	Observation	
☐	☐	a. The ball will go higher than the horizontal string.
☐	☒	b. The ball will go just as high as the horizontal string.
☐	☐	c. The ball will not go as high as the horizontal string.

Release pendulum.

Step 5: Release the pendulum! Record whether you observe a, b, or c.

Raise rod.

Step 6: Predict what would happen if the rod were attached higher than the string. Perform the experiment to confirm or deny your prediction.

Prediction: ______________________________

Observation: **The ball will go just as high as the horizontal string.**

Lower rod.

Step 7: Predict what would happen if the rod were attached lower than the string. Perform the experiment to confirm or deny your prediction.

Prediction: ______________________________

Observation: **The ball will go just as high as the horizontal string.**

Analysis

1. Explain your observations in terms of potential and kinetic energy and the conservation of energy.

   Explanations should state that the sum of the kinetic energy (KE) and potential energy (PE) is conserved.

2. Is there an upper limit on how high the rod can be? If so, explain why you think there are limits.

   No; the only practical limit on the height of the rod is the height of the pendulum clamp.

3. Is there a lower limit on how low the rod can be? If so, explain why you think there are limits.

   Yes; if the rod is lower than two-fifths the distance between the lowest position of the ball and the height of the string, the ball does a loop-the-loop.

Conserving Your Energy

Experiment

Purpose

To measure the potential and kinetic energies of a pendulum in order to see whether energy is conserved.

Setup: 1
Lab Time: 1
Learning Cycle: concept development
Conceptual Level: moderate
Mathematical Level: moderate

This lab makes a great follow-up to the previous activity, "Cut Short."

Required Equipment/Supplies

ring stand
pendulum clamp
pendulum bob
balance
string
meterstick
photogate timer or
 computer
 light probe with interface
 light source

Discussion

In Activity 23, "Cut Short," you saw that the height to which a pendulum swings is related to its initial height. The work done to elevate it to its initial height (the force times the distance) becomes stored as potential energy with respect to the bottom of the swing. At the top of the swing, all the energy of the pendulum is in the form of potential energy. At the bottom of the swing, all the energy of the pendulum is in the form of kinetic energy.

The *total energy* of a system is the sum of its kinetic and potential energies. If energy is conserved, the sum of the kinetic energy and potential energy at one moment will equal their sum at any other moment. For a pendulum, the kinetic energy is zero at the top, and the potential energy is minimum at the bottom. Thus, if the energy of a pendulum is conserved, the extra potential energy at the top must equal the kinetic energy at the bottom. For convenience, potential energy at the bottom can be defined to be zero. In this experiment, you will measure kinetic and potential energy and see if their sum is conserved.

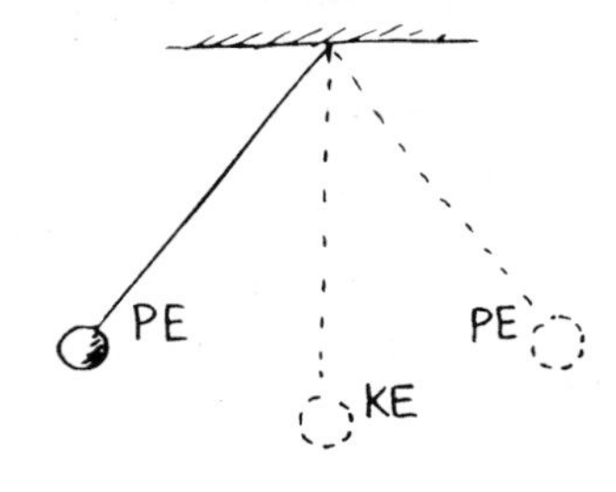

Procedure

Step 1: Devise an experiment with the equipment listed to test the conservation of energy. Write down your procedure in the space following. Include a diagram of your pendulum and label it with all the quantities, such as the height, potential energy, kinetic energy, speed, and so on.

Step 2: Perform your experiment. Record your data below in the form of a table.

Analysis

1. What units of potential energy did you use for the pendulum bob?

   The units of PE will be joules if the mass is measured in kilograms and the height in meters. Measurements in grams and centimeters must be converted to kilograms and meters to get joules.

Name Period Date

2. What units of kinetic energy did you use for the pendulum bob?

The units of KE will be joules if the mass is measured in kilograms, and the speed in meters per second. Conversion to these units is necessary to get the energy in joules.

3. List the sources of error in your experiment. Which one do you think is the most significant?

Sources of error include the measurement of the mass and the height of the elevated pendulum bob, the determination of the speed of the bob at the bottom of its swing, and the air resistance on the bob and the string. The source of largest error is probably the determination of the speed even when using a light probe or photogate. Your students probably correctly assume that the speed of the bob is the width of the bob divided by the eclipse time as the bob passes through the beam of the light probe. The uncertainties depend mainly on the spreading effects of the light beam and whether or not the beam is eclipsed by a full diameter of the ball.

4. Based on your data, does the total energy of the pendulum remain the same throughout its swing?

You can calculate the percentage by which energy appears not to be conserved using the equation

$$\% \text{ difference} = \frac{PE - KE}{PE} \times 100\%$$

where PE is the measured change of potential energy during the downward swing and KE is the measured gain of kinetic energy during this swing. Either PE or KE could be used in the denominator, since they are nearly equal.

Careful measurements of the bob's PE (mgh) and its KE ($\frac{1}{2}mv^2$) often yield results within a 3% difference or less.

25 How Hot Are Your Hot Wheels?

Experiment

Purpose

To measure the efficiency of a toy car on an inclined track.

Setup: 1
Lab Time: 1
Learning Cycle: application
Conceptual Level: easy
Mathematical Level: moderate

Adapted from PRISMS.

Required Equipment/Supplies

toy car
3 m of toy car track
meterstick
tape
2 ring stands
2 clamps

Discussion

According to the law of energy conservation, energy is neither created nor destroyed. Instead, it *transforms* from one kind to another, finally ending up as heat energy. The potential energy of an elevated toy car on a track transforms into kinetic energy as the car rolls to the bottom of the track, but some energy becomes heat because of friction. The kinetic energy of the car at the bottom of the track is transformed back into potential energy as the car rolls to higher elevation, although again some of the energy becomes heat. The car does not reach its initial height when it moves back up the incline, because some of its energy has been transformed into heat.

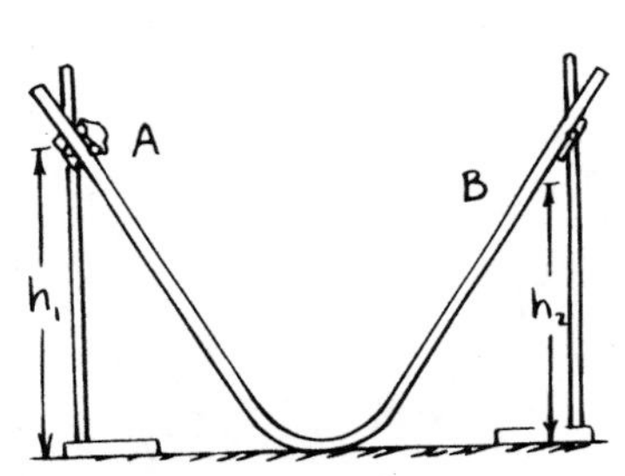

Fig. A

Procedure

Set up track.

Step 1: Set up a toy car track as shown in Figure A. Both ends of the track should be elevated to a height of 1 meter above the table or floor. Secure the track to the table or floor and supporting ring stands with tape to eliminate motion of the track.

Measure the initial and final height of the car.

Step 2: Mark starting point A with a piece of masking tape and record its height h_1. Release the car from starting point A. Record the height h_2 of point B to which the car rises on the other end of the track.

h_1 = _______________

h_2 = _______________

Compute efficiency.

Step 3: Efficiency is defined as the useful energy out divided by the total energy in. It is a ratio or a percentage. In this activity, we can define the total energy *in* as the change in potential energy as the car rolls from its highest to its lowest point. This is energy supplied by earth's gravity. The useful energy out we can take to be the potential energy change as the car rolls from its lowest point until it stops at point B. This is energy that the car now possesses relative to the lowest point of its travel. Since the potential energy at any height h above a reference level is mgh, the ratio of the potential energy transferred back into the car at point B (the output energy) to the potential energy lost by the car in rolling down to point A (the input energy) is equal to the ratio of the final height h_2 to the initial height h_1. This ratio of h_2 to h_1 can be called the "efficiency" of the car-and-track system from point A to point B. It shows the fraction of the energy supplied by gravity in rolling down the track that is retained after it rolls up the track. Compute this efficiency.

$$\text{efficiency} = \frac{PE_{\text{point B}}}{PE_{\text{point A}}} = \frac{h_2}{h_1}$$

Analysis

1. Is the efficiency of the car-and-track system changed if the track is not taped?

   The efficiency of the car decreases if the track is not secure.

2. In what units is efficiency measured?

   Since the efficiency is a ratio of two values of energy, it has no units—it is dimensionless. When the decimal ratio is multiplied by 100%, then the efficiency is expressed as a percent.

3. Is the efficiency of the car-and-track system changed if the height of the track is altered?

   Answers may vary. If the height difference is small, little difference in efficiency will be noticed. If the height difference is large, then the efficiencies may differ slightly because frictional effects vary at different speeds.

Name Period Date

26 Wrap Your Energy in a Bow

Purpose

To determine the energy transferred into an archer's bow as the string is pulled back.

Setup: 1
Lab Time: >1
Learning Cycle: application
Conceptual Level: moderate
Mathematical Level: moderate

Adapted from PRISMS.

This lab can be simulated on the computer using the program of the same name on *Laboratory Simulations-III.* The "Sample Calculation" feature makes an excellent class demonstration of the conservation principle in a very visual manner.

Required Equipment/Supplies

archer's recurve or compound bow
large-capacity spring scale
meterstick
clamp
graph paper

Discussion

The kinetic energy of an arrow is obtained from the potential energy of the drawn bow, which in turn is obtained from the work done in drawing the bow. This work is equal to the average force acting on the bowstring multiplied by the distance it is drawn.

In this experiment, you will measure the amounts of force required to hold the center of a bowstring at various distances from its position of rest, and plot these data on a force vs. distance graph. The force is relatively small for small deflections, and becomes progressively larger as the bow is bent further. The area under the force vs. distance curve out to some final deflection is equal to the average force multiplied by the total distance. This equals the work done in drawing the bow to that distance. Therefore, your graph will show not only the relationship of the force to the distance stretched, but also the potential energy possessed by the fully drawn bow.

The effect of a constant force of 10 N acting over a distance of 2 m is represented in the graph of Figure A. The work done equals the area of the rectangle.

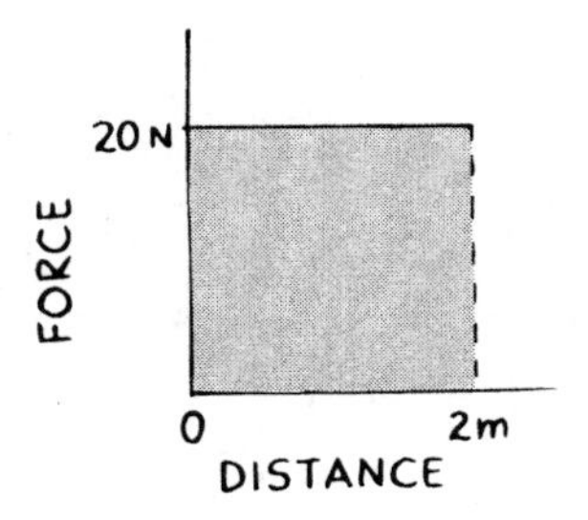

Fig. A

$$\text{work} = F \times d = (20\text{ N}) \times (2\text{ m}) = 40\text{ N}\cdot\text{m} = 40\text{ J}$$

When the force is not constant, as in Figure B, the work done on the system still equals the area under the graph (between the graph and the horizontal axis). In this case, the total area under the graph equals the area of the triangle plus the area of the rectangle.

$$\begin{aligned}\text{work} &= \text{total area} = \text{area of triangle} + \text{area of rectangle}\\ &= [(1/2)\text{ (base)} \times \text{(height)}] + [\text{(base)} \times \text{(height)}]\\ &= [(1/2)\text{ (2 m)} \times \text{(20 N)}] + [\text{(3 m)} \times \text{(20 N)}]\\ &= (20\text{ N}\cdot\text{m}) + (60\text{ N}\cdot\text{m})\\ &= 80\text{ N}\cdot\text{m}\\ &= 80\text{ J}\end{aligned}$$

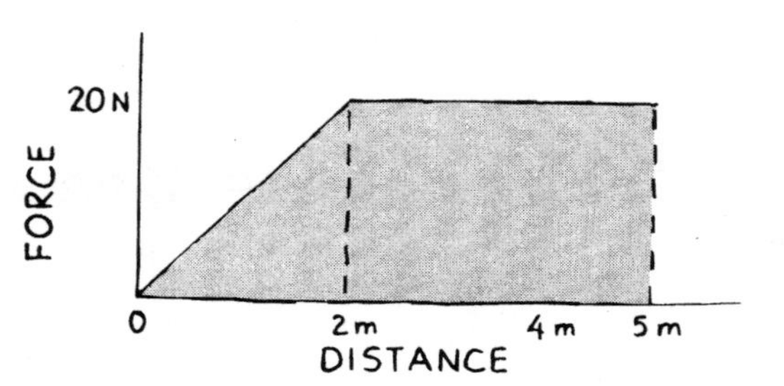

Fig. B

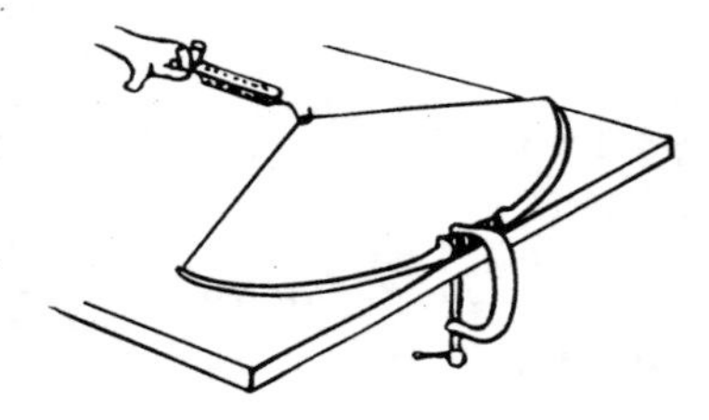

Fig. C

Don't be discouraged by the large discrepancy between the predicted height and the actual height. Values between 50 and 75% are not unusual. This means a significant amount of energy goes into friction of the air on the arrow, heating of the bow, and vibration.

Procedure

Step 1: Fasten the bow at its handle with a clamp in a vertical position, as shown in Figure C. You will pull horizontally on the bowstring with a spring scale. You will measure the distance the bowstring is stretched from its original position and the force required to hold the bowstring that far out. Prepare a table in which to record your data. Show the stretch distances in centimeters in the first column, and their equivalent values in meters in the second column. Show the force readings in the third column. If they are not in newtons, show the equivalent force values in newtons in a fourth column. Leave 10 rows for data in your table.

Stretch bowstring and measure forces.

Step 2: Stretch the bowstring by 1.0 cm, and record the stretch distance and force reading. Continue to stretch the bowstring in 1.0-cm increments, and record your data in the table. Compute the stretch distances in meters and the equivalent force values in newtons.

Make a graph.

Step 3: Plot a graph of the force (vertical axis) vs. distance (horizontal axis). Use the units newtons and meters.

Compute area.

Step 4: Estimate the area under the graph in units of newtons times meters, N•m. Since 1 N•m = 1 J, this area is the total energy transferred to the bow. When the bow is drawn, this energy is in the form of elastic potential energy.

area = ________________

Name Period Date

Analysis

1. If a 50-g arrow were shot straight up with the bow stretched to the maximum displacement of your data, how high would it go? (To find the answer in meters, express 50 g as 0.05 kg.)

To find the height the arrow will go, equate the PE of the drawn bow to the KE of the fired arrow. Then equate the KE of the fired arrow to the gravitational PE at the top of its flight.

$$PE_{bow} = KE_{arrow} = PE_{arrow}$$

$$PE_{bow} = PE_{arrow}$$

$$PE_{bow} = mgh$$

The height h of the arrow is equal to the PE in the drawn bow divided by the weight mg of the arrow, or PE/[(0.50 kg) × (9.8 m/s^2)]. The PE of the drawn bow is the work done drawing it back, as calculated in Step 4.

2. How high would a 75-g arrow go?

To calculate the height of the arrow, use the same computation as in Question 1 except use 0.075 kg for the mass of the arrow. The arrow goes only two-thirds as high.

3. At what speed would the 50-g arrow leave the bow?

The speed of the arrow is found by equating the PE of the drawn bow with the KE of the fired arrow and solving this equation for v.

$$v = \sqrt{2PE/m}$$

4. List three other devices that transform potential energy into work on an object.

Answers will vary among devices using stretchy objects: springs, rubber bands, bent wood, anything flexible.

Name Period Date

27 On a Roll

Experiment

Purpose

To investigate the relationship between the stopping distance and the height from which a ball rolls down an incline.

Setup: <1

Lab Time: 1

Learning Cycle: concept development

Conceptual Level: moderate

Mathematical Level: moderate

Thanks to John Hubisz for his ideas and suggestions for this lab.

Required Equipment/Supplies

6-ft length of 5/8-inch aluminum channel
support about 30 cm high
marble
wood ball
steel ball
tennis ball
meterstick
piece of carpet (10 to 20 feet long and a few feet wide)
graph paper or overhead transparency

Optional Equipment/Supplies

computer
light probe with interface
light source
data plotting software

Discussion

When a moving object encounters friction, its speed decreases unless a force is applied to overcome the friction. The greater the initial speed, the more work by the friction force is necessary to reduce the speed to zero. The work done by friction is the force of friction multiplied by the distance the object moves. The friction force remains more or less constant for different speeds as long as the object and the surface stay the same.

In this experiment, you will investigate the relationship between the initial height of a rolling ball and the distance it takes to roll to a stop.

Procedure

Step 1: Assemble the apparatus shown in Figure A. Elevate the ramp to a height that keeps the marble on the carpet when started at the top of the ramp.

Assemble apparatus.

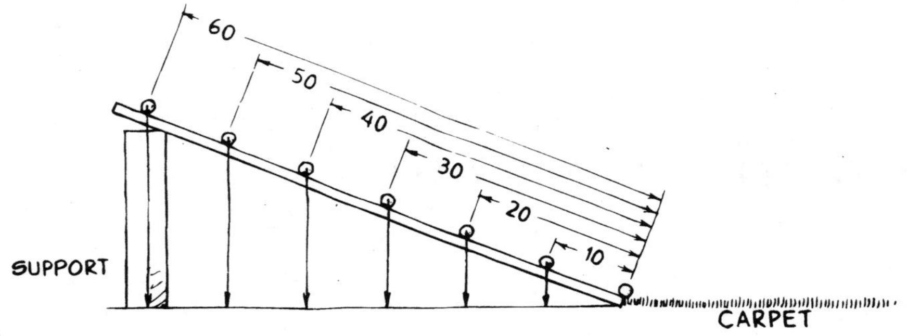

Fig. A

Roll marble onto carpet.

Step 2: Release the marble at intervals of 10 cm along the ramp as shown in Figure A. Measure the vertical height from the floor or table to the release point on the ramp. Also measure the distance required for the marble to roll to a complete stop on the carpet. Roll the marble three times from each height and record the stopping distances in Data Table A.

Repeat using different balls.

Step 3: Repeat Step 2 using a wood ball, a steel ball, and a tennis ball. Record the heights and distances in Data Table A.

Graph data.

Step 4: On the graph paper provided by your teacher, construct a graph of average stopping distance (vertical axis) vs. height (horizontal axis) for each of the four balls.

1. Describe the shapes of the four graphs you made.

They are all straight lines. The stopping distance is directly proportional to the height of the release point.

2. How did the stopping distances of the different types of balls compare?

The different kinds of balls have different stopping distances for releases from the same height, due to the different amounts of rolling friction that act on them.

3. Compare your results with those of the rest of the class. Does the mass of the ball affect the stopping distance?

The same kinds of balls released from the same height roll to a stop in more or less the same distance no matter what the mass.

Name Period Date

Data Table A

OBJECT	POSITION ON RAMP	HEIGHT	STOPPING DISTANCE			AVERAGE STOPPING DISTANCE	SPEED AT BOTTOM	SPEED SQUARED
			TRIAL 1	TRIAL 2	TRIAL 3			
MARBLE								
STEEL BALL								
TENNIS BALL								
WOOD BALL								

4. Based on your data, what factors would seem to determine the stopping distance of automobiles?

Some factors are the initial speed and the force of friction.

Going Further (Optional)

Graph speed vs. height.

Step 5: Use a light probe to time the marble at the bottom of the incline. Repeat for each of the release heights. Compute the speeds by dividing the diameter of the marble by the measured times. Record the speeds in Data Table A. Make a graph of speed (vertical axis) vs. height (horizontal axis) on graph paper.

Repeat using different balls.

Step 6: Repeat Step 5 for the steel ball, wood ball, and tennis ball.

5. Describe the shape of the four graphs you made.

The graphs of speed vs. height are curves.

Graph square of speed vs. height.

Using a computer to generate the graphs greatly reduces drudge work and saves time that can be used to learn more physics.

Step 7: Compute the square of the speed and record in Data Table A or use data plotting software to plot the square of the speed (vertical axis) vs. height (horizontal axis) for the marble, steel ball, wood ball, and tennis ball.

6. Describe the shape of the four graphs you made.

The graphs of the square of the speed vs. height are straight lines. The kinetic energy at the bottom ($\frac{1}{2}mv^2$) is directly proportional to the height of the release point.

Name Period Date

28 Releasing Your Potential

Experiment

Purpose

To find relationships among the height, speed, mass, kinetic energy, and potential energy.

Setup: <1
Lab Time: >1
Learning Cycle: exploratory
Conceptual Level: difficult
Mathematical Level: difficult

Thanks to Rex Rice for his ideas and suggestions for this lab.

Required Equipment/Supplies

pendulum apparatus as in Figure A
2 steel balls of different mass
graph paper or overhead transparencies
meterstick

Optional Equipment/Supplies

computer
light probe with interface
light source

Discussion

Drop two balls of different mass and they fall together. Tie them separately to two strings of the same length and they will swing together as pendulums. The speeds they achieve in falling or in swinging do not depend on their mass, but only on the vertical distance they have moved downward from rest. In this experiment, you will use a rigid pendulum (see Figure A) raised to a certain height. At the bottom of the pendulum's swing, a crossbar stops the pendulum, but the ball leaves the holder and keeps going.

How far downrange does the ball travel? The horizontal distance from the crossbar depends on *how fast* the ball is going and *how long* it remains in the air. How fast it is launched depends on *the launcher*. How long it remains in the air depends on how high it is above the floor or table.

Although you can make this apparatus yourself, a ready-to-go kit is available from Arbor Scientific Co.

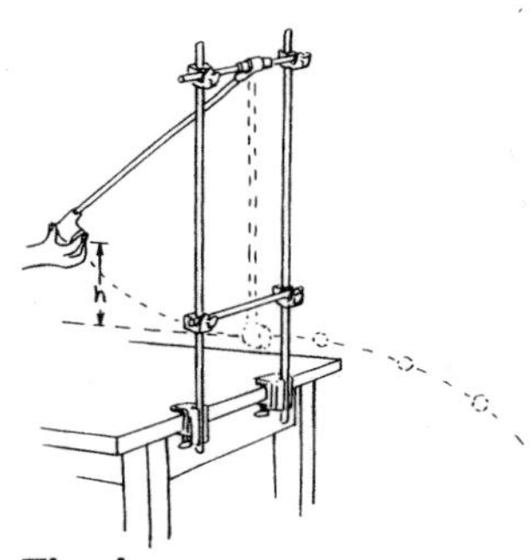

Fig. A

Measure vertical height.

Procedure 

Step 1: Devise an appropriate method for measuring the vertical height h the pendulum ball falls. Record your method in the following space.

___

___

___

Launch ball with pendulum.

Step 2: Raise the pendulum to the desired vertical height, using your finger to hold the ball in place. Take your finger away in such a manner that you do not push the pendulum up or down. Both the ball and pendulum swing down together, and the ball is launched upon impact with the crossbar. Practice your technique until you get consistent landings of the ball downrange.

Measure the range.

Step 3: When your results have become consistent, release the ball three times from the same height. Use a meterstick to measure the downrange distance for each trial. Repeat the experiment for six different heights. Record each average distance and height in Data Table A.

HEIGHT	DISTANCE			
	TRIAL 1	TRIAL 2	TRIAL 3	AVERAGE

Data Table A

Calculate minimum launch speed.

Step 4: Suppose the ball were attached to a lightweight string (as in a simple pendulum) that struck a razor mounted on the crossbar as shown in Figure A. If the ball is released from a sufficient height, its inertia will cause the string to be cut as it strikes the crossbar, projecting the ball horizontally. From the law of conservation of energy, the kinetic energy of the ball as it is launched from the low point of its swing is equal to the potential energy that it lost in swinging down, so $KE_{gained} = PE_{lost}$ or $\frac{1}{2}mv^2 = mgh$.

The launcher pictured in Figure A has its mass distributed along its length, so strictly speaking it isn't a *simple* pendulum. We'll see in Chapter 11 that it has less "rotational inertia" and swings a bit faster than if all its mass were concentrated at its bottom. For simplicity, we won't treat this complication here, and acknowledge that the speed calculated for a simple pendulum, $v = \sqrt{2gh}$, is a *lower limit*—the *minimum* launch speed. Record your computation of the minimum launch speed for each height from which the pendulum was released, in the second column of Data Table B.

Name Period Date

HEIGHT	LAUNCH SPEED COMPUTED	LAUNCH SPEED AS MEASURED BY PHOTOGATE

Data Table B

Step 5: (Optional) Use a single light probe to measure the launch speed of the ball at each of the six heights. Record your results in the third column of Data Table B.

1. How do the minimum and measured speeds of the ball compare?

The closer the launcher approximates a simple pendulum, the closer the calculated and measured speeds should compare. If a physical pendulum is used, such as the one from Arbor Scientific Co. depicted in Figure A, disagreement may vary anywhere from 5 to 25%—depending on the particular launcher.

Step 6: In this step, you will investigate the relationship between the mass of the ball and its launch speed. Use a ball with a different mass, and release it from the same six heights as before. Record the downrange distances in Data Table C.

Use balls of different mass.

HEIGHT	DISTANCE			
	TRIAL 1	TRIAL 2	TRIAL 3	AVERAGE

Data Table C

If a computer and data plotting software are available, students can use them to make their graphs.

Graph data.

Step 7: You now have a tremendous amount of data. What does it mean? What is the pattern? You can often visualize a pattern by making a graph. Graph the pairs of variables suggested. You may want to graph other quantities instead. Use overhead transparencies or graph paper. Each member of your team should make one graph, then all of you can pool your results. If a computer and data plotting software are available, use them to make your graphs.

If students are using data plotting software, have them vary the power of the *x* and *y* values until the graph is a straight line.

Suggested graphs

(a) Distance or range (vertical axis) vs. height from which the pendulum is released (horizontal axis) for the same mass.

(b) Launch speed (vertical axis) vs. release height (horizontal axis) for the same mass.

(c) Mass (vertical axis) vs. distance (horizontal axis) for the same height. If you are using data plotting software, vary the powers of the *x* and *y* values until the graph is a straight line.

Analysis

2. Describe what happens to the kinetic energy of the ball as it swings from the release height to the launch position.

The kinetic energy increases.

3. Describe what happens to the potential energy of the ball as it swings from the release height to the launch position.

The potential energy decreases.

4. Describe what happens to the total energy of the ball as it swings from the release height to the launch position.

The total energy, KE + PE, increases! This is because the work being done on the ball to increase its kinetic energy comes only partly from gravity. Some of it comes from the push of the launcher.

Going Further

Have your teacher assign you a specified distance downrange. Try to predict the angle from the vertical or height at which the pendulum must be released in order to score a bull's-eye.

29 Slip-Stick

Purpose

To investigate three types of friction and to measure the coefficient of friction for each type.

Setup: <1

Lab Time: 1

Learning Cycle: concept development

Conceptual Level: moderate

Mathematical Level: moderate

Experiment

Required Equipment/Supplies

friction block (foot-long 2 × 4 with an eye hook)
spring scale with maximum capacity greater than the weight of the friction block
set of slotted masses
flat board
meterstick
shoe

This lab avoids referencing that $\mu = \tan\theta$ at the angle of repose. You may wish to augment the Discussion of the lab by doing so, depending on the mathematical level of your students.

Discussion

The force that presses an object against a surface is the entire weight of the object only when the supporting surface is horizontal. When the object is on an incline, the component of gravitational force pressing the object against the surface is *less* than the object's weight. This component that is perpendicular (normal) to the surface is the *normal force.* For a block on an incline, the normal force varies with the angle. Although an object presses with its full weight against a horizontal surface, it presses with only half its weight against a 60° incline. So the normal force is half the weight at this angle. The normal force is zero when the incline is vertical because then the surfaces do not press against each other at all.

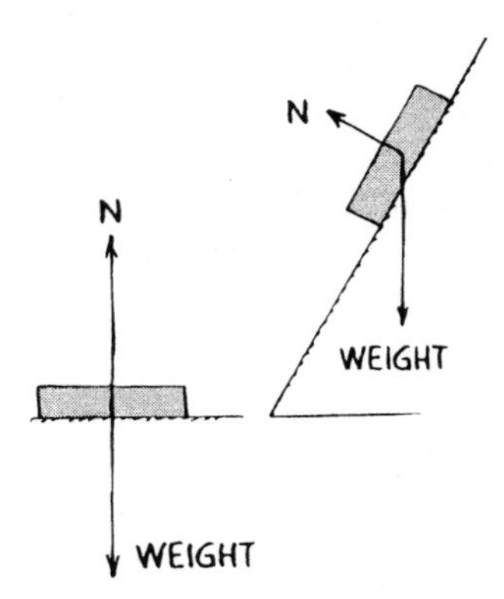

The normal force can be greater than the object's weight if you press down on the object. In general, the coefficient of friction is defined by replacing weight in the formula above by the normal force, whatever the source of the force. So, in general, the force of friction F_f depends on the coefficient of friction μ and the normal force N:

$$F_f = \mu N$$

so that the coefficient of friction, μ, equals

$$\mu = \frac{\text{friction force}}{\text{weight}}$$

The coefficient of friction μ is greatest when the two surfaces are at rest, just before motion starts. (Then the ridges and valleys have had time to sink into each other and are meshed.) Once sliding begins, μ is slightly less. The coefficient of friction for sliding objects is called the *coefficient of sliding friction* (or coefficient of kinetic friction). When friction holds an object at rest, we define the *coefficient of static friction* as the greatest friction force than can act without motion divided by the normal force. A partial list of coefficients of both sliding and static friction is shown in Figure A.

SURFACES	μ_s (STATIC)	μ_k (KINETIC)
STEEL ON STEEL, DRY	0.6	0.3
STEEL ON WOOD, DRY	0.4	0.2
STEEL ON ICE	0.1	0.06
WOOD ON WOOD, DRY	0.35	0.15
METAL ON METAL, GREASED	0.15	0.08

Fig. A

Friction also occurs for objects moving through fluids. This friction, known as *fluid friction*, does not follow laws as simple as those that govern sliding friction for solids. Air is a fluid, and the motion of a leaf falling to the ground is quite complicated! In this experiment, you will be concerned only with the friction between two solid surfaces in contact.

Friction always acts in a direction to oppose motion. For a ball moving upward in the air, the friction force is downward. When the ball moves downward, the friction force is upward. For a block sliding along a surface to the right, the friction force is to the left. Friction forces are always opposite to the direction of motion.

Part A: Computing the Coefficients of Static and Sliding Friction

Procedure

Drag friction block.

Step 1: Weigh the friction block by suspending it from the spring scale. Record the weight in Data Table A. Determine the coefficients of static and sliding friction by dragging the friction block horizontally with a spring scale. Be sure to hold the scale horizontally. The static friction force F_f is the maximum force that acts just before the block starts moving. The sliding friction force $F_{f'}$ is the force it takes to keep the block moving at *constant velocity*. Your scale will vibrate around some average

value; make the best judgment you can of the values of the static and sliding friction forces. Record your data in Data Table A.

F_f FORCE TO JUST GET GOING	F_f' DRAG FORCE AT CONSTANT VELOCITY	W WEIGHT OF CART	$\mu_{STATIC} = \frac{F_f}{W}$	$\mu_{SLIDING} = \frac{F_f'}{W}$

Data Table A

Step 2: Drag the block at different speeds. Note any changes in the sliding friction force.

Change dragging speed.

1. Does the dragging speed have any effect on the coefficient of sliding friction, $\mu_{sliding}$? Explain.

   The dragging speed should have very little effect on the coefficient of sliding friction.

Step 3: Increase the force pressing the surfaces together by adding slotted masses to the friction block. Record the weight of both the block and the added masses in Data Table A. Find both friction forces and coefficients of friction for at least six different weights and record in Data Table A.

Add weights to block.

Analysis

2. At each weight, how does μ_{static} compare with $\mu_{sliding}$?

   It requires appreciably greater force to overcome the force of static friction than to keep the object moving. Once it is moving, a smaller force is required. Thus μ_{static} is greater than $\mu_{sliding}$.

3. Does $\mu_{sliding}$ depend on the weight of the friction block? Explain.

No, $\mu_{sliding}$ should not depend appreciably on the weight. As the weight increases, the friction force increases in direct proportion, and the *ratio* of forces does not change significantly.

4. Tables in physics books rarely list coefficients of friction with more than two significant figures. From your experience, why are more than two significant figures not listed?

There is sufficient variation in surfaces of the same material to prevent precise, reproducible results.

5. If you press down upon a sliding block, the force of friction increases but μ does not. Explain.

You increase the normal force as well as the force of friction, and the ratio of the two does not change appreciably.

6. Why are there no units for μ?

The coefficient of friction has no units because it is the ratio of two forces.

Part B: The Effect of Surface Area on Friction

Procedure

Drag friction block.

Step 4: Drag a friction block of known weight at constant speed by means of a horizontal spring scale. Record the friction force and the weight of the block in Data Table B.

CONFIGURATION	F_f FORCE OF FRICTION	W NORMAL FORCE	A AREA OF CONTACT	μ COEFFICIENT OF FRICTION
1				
2				
AVERAGE				

Data Table B

Step 5: Repeat Step 4, but use a different side of the block (with a different area). Record the friction force in Data Table B.

Use different side of friction block.

Step 6: Compute μ_{sliding} for both steps and list in Data Table B.

Compute μ.

7. Does the area make a difference in the coefficient of friction?

No, the area does not affect the coefficient of friction.

Going Further: Friction on an Incline

Place an object on an inclined plane and it may or may not slide. If friction is enough to hold it still, then tip the incline at a steeper angle until the object just begins to slide.

The coefficient of friction of a shoe is critical to its function. When will a shoe on an incline start to slip? Study Figure B. To make the geometry clearer, a cube can represent the shoe on the incline, as in Figure C. Triangle B shows the vector components of the shoe's weight. The component perpendicular to the incline is the normal force N; it acts to press the shoe to the surface. The component parallel to the incline, which points downward, tends to produce sliding. Before sliding starts, the friction force F_f is equal in magnitude but opposite in direction to this component. By tilting the incline, we can vary the normal force and the friction force on the shoe.

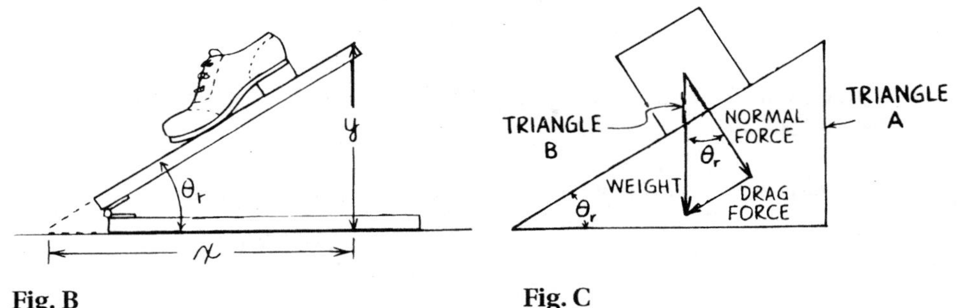

Fig. B **Fig. C**

At the angle at which the shoe starts to slip (the *angle of repose*) θ_r, the component of weight parallel to the surface is just enough to overcome friction and the shoe breaks free. At that angle, the parallel weight component and the friction force are at their maximum. The ratio of this friction force to the normal force gives the coefficient of static friction.

In Figures B and C, the angle of incline has been set to be the angle of repose, θ_r. With the help of geometry, it can be proven that triangles A and B in Figure C are *similar triangles*—that is, they have the same angles—and triangle B is a shrunken version of triangle A. The importance of similar triangles is that the *ratios* of corresponding parts of the two triangles are *equal.* Thus, the ratio of the parallel component to the normal force equals the ratio of the height y to the horizontal distance x. The coefficient of static friction is the ratio of those sides.

$$\mu_{\text{static}} = \frac{F_f}{N} = \frac{y}{x}$$

Procedure

Tilt incline until shoe slips.

Step 7: Put your shoe on the board, and slowly tilt the board up until the shoe just begins to slip. Using a meterstick, measure the horizontal distance x and the vertical distance y. Compute the coefficient of static friction μ_{static} from the equation in the preceding paragraph.

μ_{static} = ____________

Tap the incline.

Step 8: Repeat Step 7, except have a partner tap the board constantly as you approach the angle of repose. Adjust the board angle so that the shoe slides at constant speed. Compute the coefficient of sliding (kinetic) friction, $\mu_{sliding}$, which equals the ratio of y to x at that board angle.

$\mu_{sliding}$ = ____________

8. Did you measure any difference between μ_{static} and $\mu_{sliding}$?

Yes, μ_{static} should be greater than $\mu_{sliding}$.

Drag shoe on level.

Step 9: With a spring scale, drag your shoe along the same board when it is level. Compute $\mu_{sliding}$ by dividing the force of friction (the scale reading) by the weight of the shoe. Compare your result with that of Step 8.

$\mu_{sliding}$ = ____________

9. Are your values for $\mu_{sliding}$ from Steps 8 and 9 equal? Explain any differences.

They should be the same, within the uncertainties of the measurements.

10. Does the force of sliding friction between two surfaces depend on whether the supporting surface is inclined or horizontal?

Yes, the force of friction is greater when the surface is horizontal.

11. Does the coefficient of sliding friction between two surfaces depend on whether the supporting surface is inclined or horizontal?

No, the coefficient of friction is the same.

12. Explain why Questions 10 and 11 are different from each other.

Question 10 asks about a force, whereas Question 11 asks about a ratio of forces.

30 Going in Circles

Purpose

To determine the relative magnitude and direction of acceleration of an object at different positions on a rotating turntable.

Setup: <1
Lab Time: 1
Learning Cycle: exploratory
Conceptual Level: difficult
Mathematical Level: none

Adapted from PRISMS.

Required Equipment/Supplies

dynamics cart
accelerometer
mass
pulley
string
support for pulley and mass
phonograph turntable
masking tape

Activity

Discussion

How can a passenger in a smooth-riding train tell when the train undergoes an acceleration? The answer is easy: He or she can simply watch to see whether the surface of the liquid in a container is disturbed, or look for motion at the bottom of the hanging curtains. These are two simple accelerometers.

Procedure

Step 1: Attach the accelerometer securely to a cart, and place the cart on a smooth surface such as a table.

Observe accelerometer moving in straight line.

Step 2: Attach the cart to the mass with a string. Attach a pulley to the edge of the table and place the string on the pulley.

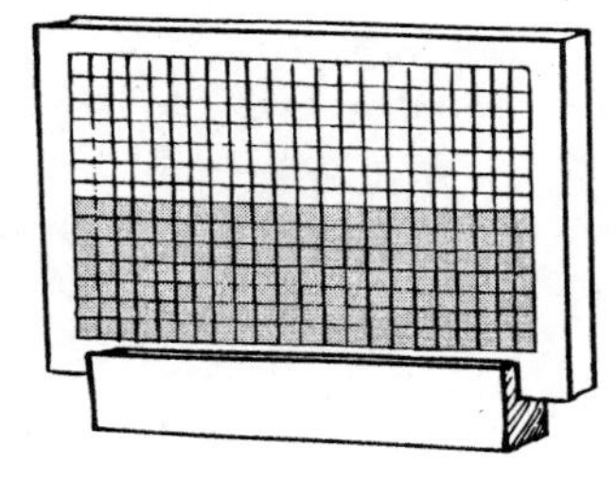

Step 3: Allow the liquid in the accelerometer to settle down (level) while holding it in position ready to release. Allow the mass to fall and cause the cart to accelerate. Observe and sketch the surface of the liquid in the accelerometer. Label the direction of the acceleration of the cart.

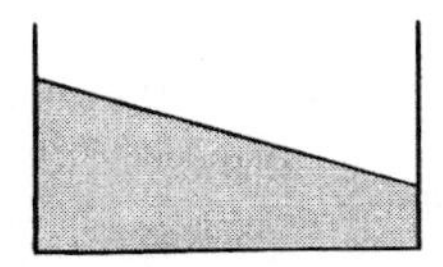

1. What happens to the surface as the cart continues to accelerate?

The liquid surface remains tipped at the same angle for constant acceleration.

Repeat the experiment several times.

2. Does the surface behave the same way each time?

Yes, the angle of the surface should be the same if the same falling mass is used each time. (At greater acceleration, produced by a greater falling mass, the angle is greater.)

Observe accelerometer on turntable.

Step 4: Remove the accelerometer from the cart, and securely attach the accelerometer to the turntable. Make sure that the accelerometer will not fly off.

Step 5: Place the center of the accelerometer at the center of the turntable, and rotate the turntable at various speeds. Observe and sketch the surface of the liquid each time.

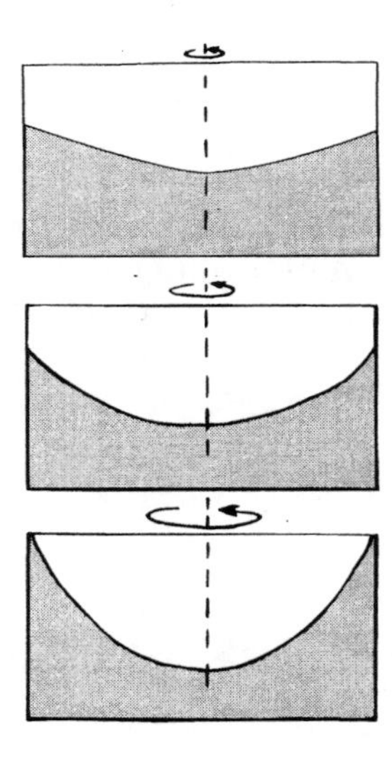

3. What do you conclude about the magnitude and direction of the acceleration on the rotating turntable?

The magnitude of the acceleration is greater at greater distances from the center and at greater rotational speeds, as evidenced by the steeper slope of the liquid. The direction of the acceleration is toward the center of the turntable.

Name Period Date

Analysis

4. Figures A, B, C, and D show the surface of the liquid for a particular situation. What situation of the ones you observed most closely matches these figures?

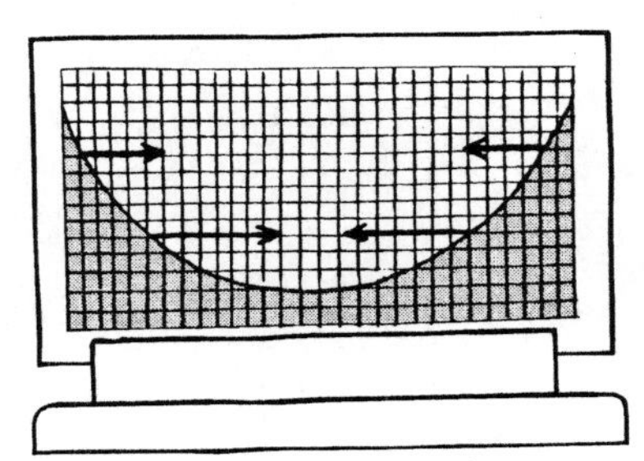
Fig. A

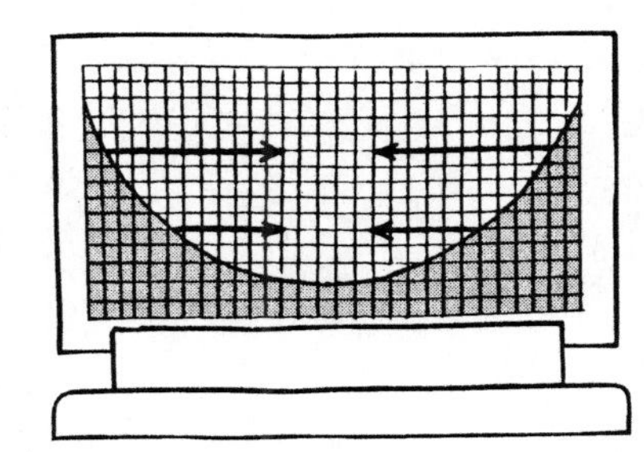
Fig. B

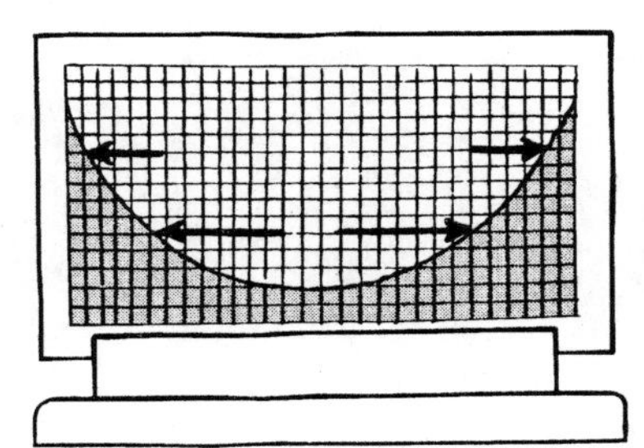
Fig. C

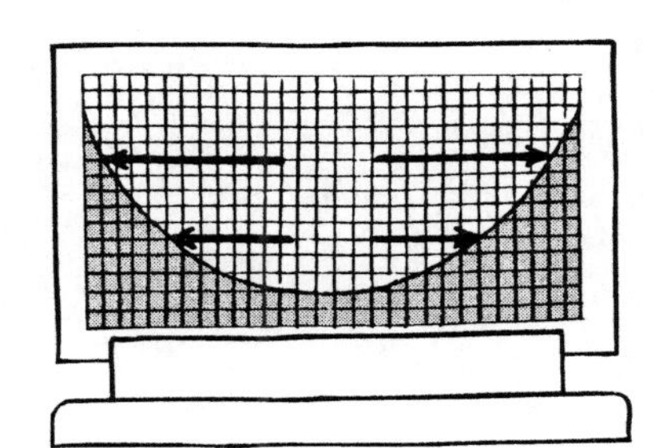
Fig. D

The situation shown in the figures is that of an accelerometer in the center of a rotating system such as a turntable.

5. The four figures show four possible sets of arrows representing the magnitude and direction of the acceleration at different points. Which figure is most correct? Explain the reason(s) for your choice.

Figure B is most correct, as the accelerations are toward the center and are greater closer to the edge.

Name Period Date

31 Where's Your CG?

Purpose

To locate your center of gravity.

Setup: <1
Lab Time: 1
Learning Cycle: exploratory
Conceptual Level: difficult
Mathematical Level: none

Required Equipment/Supplies

2 bathroom scales
8' × 2" × 12" "reaction" board
meterstick
bricks or old books
2 large triangular supports (such as 1-inch angle iron or machine-shop files)

Discussion

Activity

For a symmetrical object such as a ball or donut, the center of gravity, or CG, is located in the geometric center of the object. For asymmetrical objects, such as a baseball bat or a person, the CG is located closer to the heavier end. When your arms are at your side, your center of gravity is near your navel, where your umbilical cord was attached. This is no coincidence—unborns sometimes rotate about their CG.

Your CG is a point that moves when you move. When you raise your hands above your head, your CG is a little higher than when your hands are by your sides. In this activity, you will locate your CG when your hands are by your sides.

Procedure 

Weigh yourself.

Step 1: Using a bathroom scale, weigh yourself and then a reaction board. Record these weights.

weight of self = ________ weight of board = ________

Measure your height.

Step 2: Measure your height in centimeters.

height = ________

Position reaction board.

Step 3: Place one triangular support on each bathroom scale. You may need to add bricks or books between the scale and the support if the bubble on the scale causes the support to rock. Position the scales so that the tops of the triangular supports are separated by a distance equal to your height. Place the reaction board on the two supports. *The overhangs of the ends of the reaction board on each support should be equal.*

Each overhang should be half the difference between the board length and the student's height.

Read the weight reading on each bathroom scale. These readings should each be half the weight of the reaction board (plus the books or bricks, if any). If they are not, either adjust the calibration knob on one scale until the readings are equal or record the readings for future calculations.

Position yourself on reaction board.

It may be necessary to utilize a small plane mirror to facilitate reading the scale.

Step 4: Lie down on the reaction board so that the tip of your head is over one support and the bottoms of your feet are over the other support, as in Figure A. Remain flat on the board with your hands by your sides. Have someone record the reading on each bathroom scale.

weight at head = _______ weight at feet = ________

Adjust your position.

A thick block of wood, a brick, or a few books help make the scale reading more readable. Books that compress will decrease the scale reading somewhat, but not appreciably.

Step 5: After you are informed of the weight readings, adjust your position along the length of the board until the two readings become equal. Have someone measure how far the bottoms of your feet are from the support near your feet.

distance from support to feet = _________

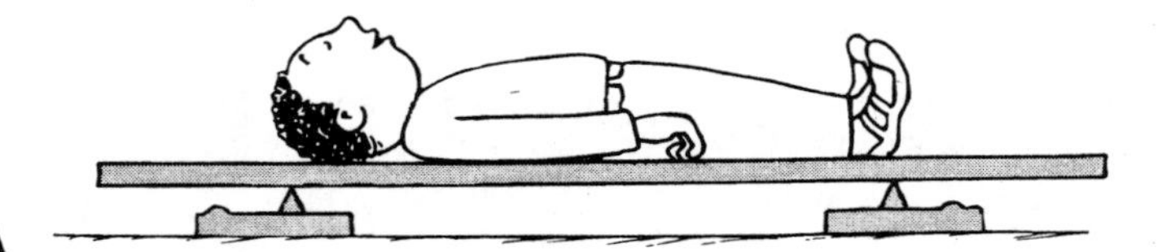

Fig. A

1. Did you have to move toward the foot end or toward the head end?

Students will have to move toward the foot end. (Typically, the distance between a person's CG and the feet is about 0.6 of the total height.)

2. When the readings on the scales are equal, where is your CG in relation to the supports?

When the two weight readings of the student are equal, the student's CG is midway between the two supports.

Determine your CG.

Step 6: Determine the location of your CG, in relation to the bottoms of your feet.

location of CG = _________ cm from the feet

Place your finger on your navel and have someone measure the distance from the bottom of your feet to your navel.

location of navel = _______ cm from the feet

Analysis

3. How close is your CG to your navel?

The CG, as determined by this activity, should be within a few centimeters of the navel.

4. What would happen if the two weight readings of the board were not equal?

If the weight of the board were not evenly distributed on each support, the center of gravity of the board would not be midway between the two supports as assumed.

5. When an astronaut spins when doing acrobatics aboard an orbiting space vehicle, what point does the body spin about?

Astronauts spin about their center of gravity (mass).

Name Period Date

32 Torque Feeler

Purpose

To illustrate the qualitative differences between torque and force.

Setup: <1
Lab Time: <1
Learning Cycle: exploratory
Conceptual Level: easy
Mathematical Level: none

Required Equipment/Supplies

meterstick
meterstick clamp
1-kg mass
mass hanger

Like many of the activities in this manual, *Torque Feeler* should not be skipped because it is so short and simple. Proceeding to concept development *after* students wrestle with these concepts for themselves will increase your effectiveness dramatically.

Discussion

Torque and force are sometimes confused because of their similarities. Their differences should be evident in this activity.

Activity

Procedure

Step 1: Hold the end of a meterstick in your hand so that your index finger is at the 5-cm mark. With the stick held horizontally, position the mass hanger at the 10-cm mark, and suspend the 1-kg mass from it. Rotate the stick to raise and lower the free end of the stick. Note how hard or easy it is to raise and lower the free end.

Rotate meterstick.

Step 2: Move the mass hanger to the 20-cm mark. Rotate the stick up and down about the pivot point (your index finger) as before. Repeat this procedure with the mass at the 40-cm, 60-cm, 80-cm, and 95-cm marks.

Move mass farther from pivot point.

1. Does it get easier or harder to rotate the stick as the mass gets farther from the pivot point?

It gets harder to rotate the stick the farther away the mass gets from the pivot point.

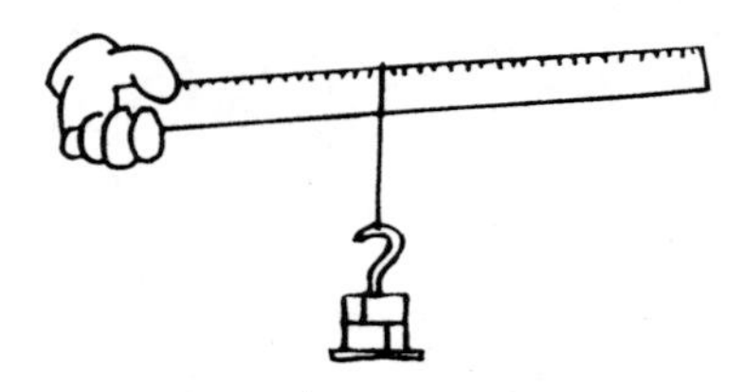

Analysis

2. Does the weight of the mass increase as you move the mass away from the pivot point (your index finger)?

The weight of the mass does not increase as you move it away from the pivot point.

3. If the weight of the mass is not getting any larger, why does the difficulty in rotating the stick increase in Step 2?

The pull of the earth on the mass at the larger distance exerts a larger *torque* on the meterstick. Torque depends on distance from the pivot point as well as on the force.

33 Weighing an Elephant

Experiment

Purpose

To determine the relationship between masses and distances from the fulcrum for a balanced seesaw.

Setup: <1

Lab Time: >1

Learning Cycle: concept development

Conceptual Level: moderate

Mathematical Level: moderate

Thanks to Art Farmer for his ideas and suggestions for this lab.

Required Equipment/Supplies

meterstick
wedge or knife-edge
2 50-g mass hangers
slotted masses or set of hook masses
knife-edge level clamps

Discussion

An object at rest is in equilibrium (review Section 4.7 of the text). The sum of the forces exerted on it is zero. The resting object also shows another aspect of equilibrium. Because the object has no rotation, the sum of the torques exerted on it is zero. When a force causes an object to start turning or rotating (or changes its rotation), a nonzero net torque is present.

The seesaw is a simple mechanical device that rotates about a pivot or fulcrum. It is a type of lever. Although the work done by a device can never be more than the work or energy invested in it, levers make work *easier* to accomplish for a variety of tasks.

Suppose you are an animal trainer at the circus. You have a very strong, very light, wooden plank. You want to balance a 600-kg baby elephant on a seesaw using only your own body weight. Suppose your body has a mass of 50 kg. The elephant is to stand 2 m from the fulcrum. How far from the fulcrum must you stand on the other side in order for you to balance the elephant?

Laboratories do not have elephants or masses of that size. They do have a variety of smaller masses, metersticks, and fulcrums that enable you to discover how levers work, describe their forces and torques mathematically, and finally solve the elephant problem.

Procedure

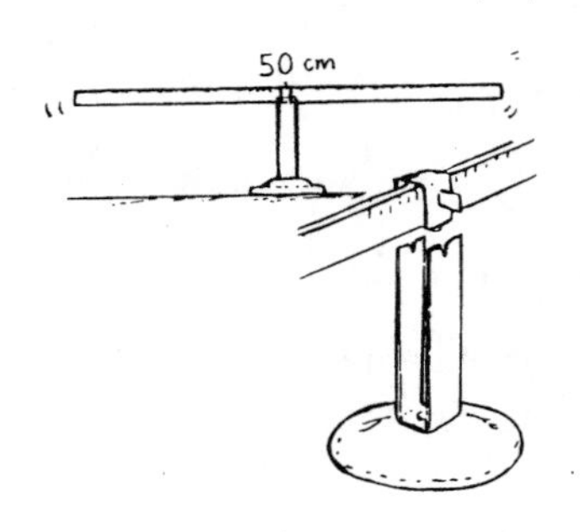

Step 1: Carefully balance a meterstick horizontally on a wedge or knife-edge. Suspend a 200-g mass 10 cm from the fulcrum. Suspend a 100-g mass on the opposite side of the fulcrum at the point that balances the meterstick. Record the masses and distances from the fulcrum in Data Table A.

1. Can a heavier mass be balanced by a lighter one? Explain how.

Yes. It can be balanced if it is closer to the fulcrum than the lighter mass.

Step 2: Make more trials to fill in Data Table A. You can do this by using the masses of Step 1 and changing their positions. For instance, you can move the heavier mass to a new location 5 cm farther away, and then rebalance the meterstick with the lighter mass.

You can also change the magnitudes of the masses. Replace the heavier mass with another mass and rebalance the lever by moving the lighter mass. Record the masses and distances from the fulcrum in Data Table A. Be sure to take into account the mass of any hanger or clamp.

Remind your students that torque is the product of force and distance (see answer to Step 4).

Data Table A

TRIAL	SMALL MASS (g)	DISTANCE FROM FULCRUM (cm)	LARGE MASS (g)	DISTANCE FROM FULCRUM (cm)

Analyze data for pattern.

Step 3: Use any method you can devise to discover a pattern in the data of Data Table A. You can try graphing the large mass vs. its distance from the fulcrum, the small mass vs. its distance from the fulcrum, or another pair of variables. You can also try forming ratios or products.

Step 4: After you have convinced yourself and your laboratory partners that you have discovered a pattern, convert this pattern into a word statement.

Express pattern as statement.

The product of one mass and its distance from the fulcrum equals the product of the other mass and its distance from the fulcrum. (Students may state this in terms of the equivalent ratios: the first mass divided by the second mass equals the second distance divided by the first distance.) Since torque is the product of *force* and distance, and the force on each mass (its weight) is proportional to its mass, this word statement is equivalent to saying that the torques on the two masses are equal.

Step 5: Now convert this word statement into a mathematical equation. Be sure to explain what each symbol stands for.

Convert statement into equation.

One equation is $m_1d_1 = m_2d_2$, where m_1 is one mass, d_1 is its distance from the fulcrum, m_2 is the other mass, and d_2 is its distance from the fulcrum. Equivalently, $m_1/m_2 = d_2/d_1$.

Step 6: With the help of your partners or your teacher, use your equation to find the distance a 50-kg person should stand from the fulcrum in order to balance the 600-kg elephant. Show your work (neglect the mass of the supporting board).

Solve for unknown distance.

If *d* is the person's distance from the fulcrum, then

(2310 kg) × (2 m) = (50 kg) × (*d*)

Solving for *d* gives a value of 92 m. You would need a very long plank!

$d =$ ________________ m

Analysis

2. Why must the mass of the hangers and clamps be taken into account in this experiment?

These items exert forces and torques on the meterstick just as the masses do.

3. If you are playing seesaw with your younger sister (who weighs much less than you), what can you do to balance the seesaw? Mention at least two things.

   You can move closer to the fulcrum and/or she can move farther away from the fulcrum. If the fulcrum is not fixed, you can *move* the fulcrum closer to you.

4. Taking account of the fact that the board holding up the elephant and the trainer (see sketch in the Discussion section) has weight, would the actual position of the trainer be farther from or closer to the fulcrum than calculated in Step 6?

   Closer, since some of the torque needed to balance the elephant is provided by the board.

Name Period Date

34 Keeping in Balance

Purpose

To use the principle of balanced torques to find the value of an unknown mass.

Setup: <1

Lab Time: 1

Learning Cycle: concept development

Conceptual Level: moderate

Mathematical Level: moderate

Required Equipment/Supplies

meterstick
standard mass with hook
rock or unknown mass
triple-beam balance
fulcrum
fulcrum holder
string
masking tape

Discussion

Gravity pulls on every part of an object. It pulls more strongly on the more massive parts of objects and more weakly on the less massive parts. The sum of all these pulls is the weight of the object. The average position of the weight of an object is its center of gravity, or CG.

The whole weight of the object is effectively concentrated at its center of gravity. The CG of a uniform meterstick is at the 50-cm mark. In this experiment, you will balance a meterstick with a known and an unknown mass, and compute the mass of the unknown. Then you will simulate a "solitary seesaw."

Procedure

Step 1: Balance the meterstick horizontally with nothing hanging from it. Record the position of the CG of the meterstick.

Balance the meterstick.

position of meterstick CG = ________________

Using a string, attach an object of unknown mass, such as a rock, at the 90-cm mark of the meterstick, as shown in Figure A. Place a known mass on the other side to balance the meterstick. Record the mass used and its position.

mass = ______________ position = ______________

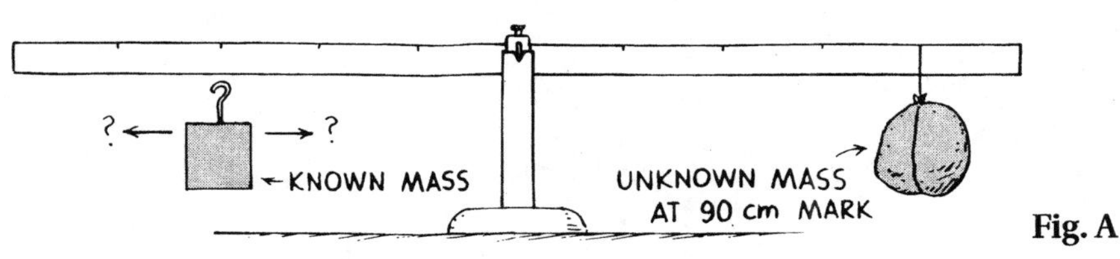

Fig. A

Compute unknown mass.

Step 2: Compute the distances from the fulcrum to each object.

distance from fulcrum to unknown mass = ___________

distance from fulcrum to known mass = ___________

These two distances are known as lever arms. The lever arm is the distance from the fulcrum to the place where the force acts.

In the following space, write an equation for balanced torques, first in words and then with the known values. Compute the unknown mass.

(unknown mass) × (its distance to fulcrum) = (known mass) × (its distance to fulcrum)

$\text{mass}_{\text{computed}}$ = ___________

Measure mass of unknown.

Step 3: Measure the unknown mass, using a triple-beam balance.

$\text{mass}_{\text{measured}}$ = ___________

Calculate percent error.

Step 4: Compare the measured mass to the value you computed in Step 2, and calculate the percentage difference.

$$\% \text{ error} = \frac{|\text{mass}_{\text{computed}} - \text{mass}_{\text{measured}}|}{\text{mass}_{\text{measured}}} \times 100\%$$

= ___________

Set up "solitary seesaw."

Step 5: Place the fulcrum exactly on the 85-cm mark. Balance the meterstick using a single mass hung between the 85-cm and 100-cm marks, as in Figure B. Record the mass used and its position.

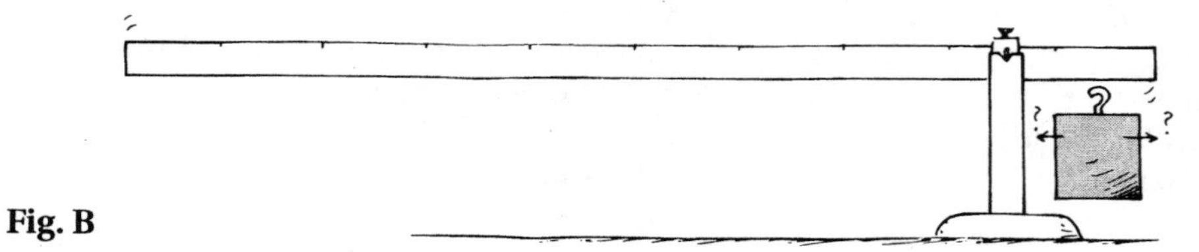

Fig. B

mass = ___________ position = ___________

Name Period Date

Step 6: Draw a lever diagram of your meterstick system in the following space. Be sure to label the fulcrum, the masses whose weights give rise to torques on each side of the fulcrum, and the lever arm for each mass.

Draw lever diagram.

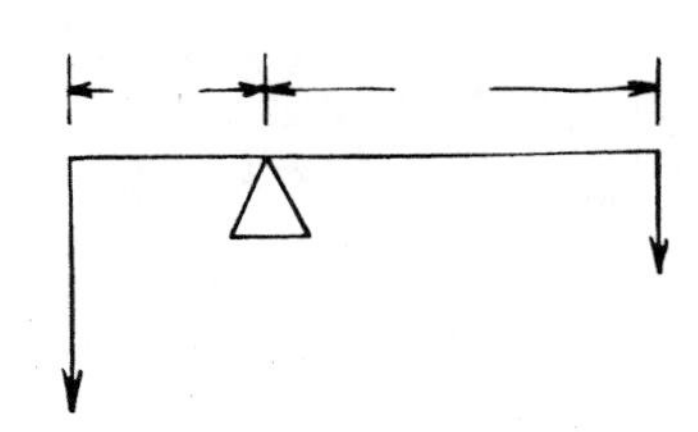

1. Where is the mass of the meterstick effectively located?

The mass of the meterstick is effectively at its CG (near the 50-cm mark).

Compute the mass of the meterstick. Show your work in the following space.

(meterstick mass) × (distance from CG to fulcrum) = (hung mass) × (its distance)

$\text{mass}_{\text{computed}}$ = ____________

Step 7: Remove the meterstick and measure its mass on a triple-beam balance.

Measure mass of meterstick.

$\text{mass}_{\text{measured}}$ = ____________

Step 8: Find the percent error for the computed value of the mass of the meterstick.

Calculate percent error.

$$\%\ \text{error} = \frac{|\text{mass}_{\text{computed}} - \text{mass}_{\text{measured}}|}{\text{mass}_{\text{measured}}} \times 100\%$$

= ____________________

Going Further

For a uniform, symmetrical object the CG is located at its geometrical center. The CG of a uniform meterstick is at the 50-cm mark. But for an asymmetrical object such as a baseball bat, the CG is nearer the heavier end. In this part of the experiment, you will learn how to find the location of the center of gravity for an asymmetrical rigid object.

Balance meterstick.

Step 9: If a rock is attached to your meterstick away from the midpoint, the new CG of the *combined* meterstick plus rock is *not* in the center. Use masking tape to attach the rock to the meterstick between the 0-cm and 50-cm mark. Move the fulcrum to the 60-cm mark, as shown in Figure C. Hang a known mass between the 60-cm and 100-cm mark to balance the meterstick. Record the mass used and its position.

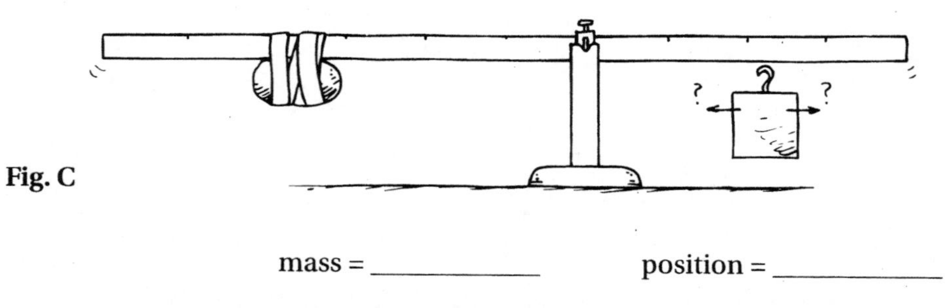

Fig. C

mass = ____________ position = ____________

Caution your students not to remove the rock from the meterstick when using the triple-beam balance. Tell them to think of the rock as a "growth" on the meterstick.

Step 10: Find the mass of the meterstick/rock combination with a triple-beam balance.

mass of meterstick/rock combination = ____________

Find the CG of the meterstick/rock combination using an equation for balanced torques. In the following space, write down the equation first in words and then with known values.

(mass of combination) × (distance from CG of combination to fulcrum) = (known mass) × (its distance from fulcrum)

distance from CG to fulcrum = ____________

computed position of CG = ____________

Step 11: Verify the location of the CG by removing the added known mass and placing the fulcrum at the predicted CG point.

2. Does the meterstick balance?

Answers will vary.

3. How far was your predicted location of the CG from its actual location?

They should agree within a few percent.

35 Rotational Derby

Purpose

To observe how round objects of various shapes and masses roll down an incline and how their rotational inertias affect their rate of rotation.

Setup: <1

Lab Time: >1

Learning Cycle: concept development

Conceptual Level: moderate

Mathematical Level: moderate

Thanks to John Davis for ideas and suggestions for this lab.

Required Equipment/Supplies

smooth, flat board, 1 m in length
ring stand or metal support with clamp and rod
balance
meterstick
3 solid steel balls of different diameter (3/4" minimum)
3 empty cans of different diameter, with both the top and the bottom removed
3 unopened cans of different diameter, filled with nonsloshing contents (such as chili or ravioli)
2 unopened soup cans filled with different kinds of soup, one liquid (sloshing) and one solid (nonsloshing)

Activity

Discussion

Why is most of the mass of a flywheel, a gyroscope, or a Frisbee concentrated at its outer edge? Does this mass configuration give these objects a greater tendency to resist changes in rotation? How does their "rotational inertia" differ from the inertia you studied when you investigated linear motion? Keep these questions in mind as you do this activity.

The *rotational speed* of a rotating object is a measure of how fast the rotation is taking place. It is the angle turned per unit of time and may be measured in degrees per second or in radians per second. (A radian is a unit similar to a degree, only bigger; it is slightly larger than 57 degrees.) The *rotational acceleration,* on the other hand, is a measure of how quickly the rotational speed changes. (It is measured in degrees per second squared or in radians per second squared.) The rotational speed and the rotational acceleration are to rotational motion what speed and acceleration are to linear motion.

Procedure

Step 1: Make a ramp with the board and support. Place the board at an angle of about 10°.

Set up ramp.

Roll balls down ramp.

Step 2: Select two balls. Predict which ball will reach the bottom of the ramp in the shorter time.

predicted winner: ____________

Place a meterstick across the ramp near the top, and rest the balls on the stick. Quickly remove the stick to allow the balls to roll down the ramp. Record your results.

actual winner: ______________

Repeat the race for the other pair of balls.

1. What do you conclude about the time it takes for two solid steel balls of different diameter to roll down the same incline?

Solid steel balls all take about the same time to roll down the same incline, regardless of diameter (and mass).

Roll hollow cylinders.

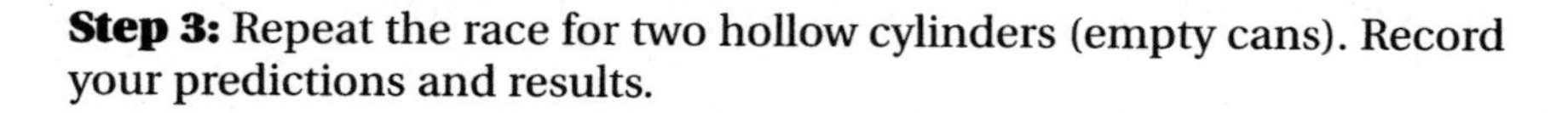

Step 3: Repeat the race for two hollow cylinders (empty cans). Record your predictions and results.

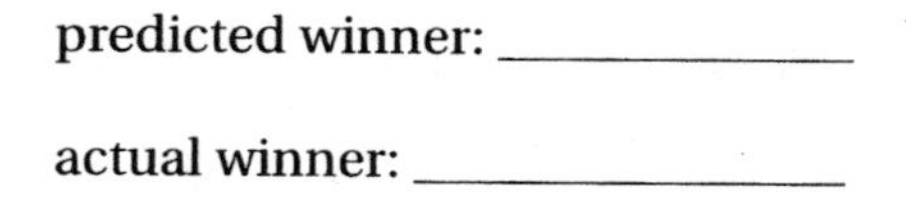

predicted winner: ____________

actual winner: ______________

Repeat the race for the other pairs of hollow cylinders.

2. What do you conclude about the time it takes for two hollow cylinders of different diameter to roll down the same incline?

Hollow cylinders all take about the same time to roll down the same incline, regardless of diameter (and mass).

Roll solid cylinders.

Step 4: Repeat the race for two solid cylinders (filled cans with non-sloshing contents). Record your predictions and results.

predicted winner: ____________

actual winner: ______________

Repeat the race for the other pairs of solid cylinders.

3. What do you conclude about the time it takes for two solid cylinders of different diameter to roll down the same incline?

Solid cylinders all take about the same time to roll down the same incline, regardless of diameter (and mass).

Name Period Date

Step 5: Repeat the race for a hollow cylinder and a solid one. Before trying it, predict which cylinder will reach the bottom of the ramp first.

Roll hollow and solid cylinders.

predicted winner: ______________

Now try it, and record your results.

actual winner: ______________

Repeat for other pairs of hollow vs. solid cylinders.

4. What can you conclude about the time it takes for a hollow and a solid cylinder to roll down the same incline?

A solid cylinder beats a hollow cylinder down the same incline, regardless of diameter (and mass).

5. How do you explain the results you observed for the hollow and solid cylinders?

Although neither mass nor diameter affects the rolling time, the distribution of mass does. In the hollow cylinder, the mass is all concentrated at the greatest distance from the turning axis. In the solid cylinder, much of the mass is close to the turning axis.

Step 6: Repeat the race for a solid ball and a solid cylinder. Record your prediction and result.

Roll balls and solid cylinders.

predicted winner: ______________

actual winner: ______________

6. What do you conclude about the time it takes for a solid ball and a solid cylinder to roll down the same incline?

A solid steel ball and a solid cylinder may be too close to call. Theoretically, the ball will beat the cylinder.

Step 7: Repeat the race for a solid ball and a hollow cylinder. Record your prediction and result.

Roll balls and hollow cylinders.

predicted winner: ______________

actual winner: ______________

7. What can you conclude about the time it takes for a solid ball and a hollow cylinder to roll down the same incline?

Students should be able to predict correctly that the steel ball will beat the hollow cylinder, regardless of diameter (and mass).

Roll cans of different kinds of soup.

Step 8: Repeat the race for two soup cans, one with liquid (sloshing) contents and the other with solid (nonsloshing) contents. Record your prediction and results.

predicted winner: ________________

actual winner: __________________

8. How can you explain the results you observed for the sloshing vs. nonsloshing kinds of soup?

Soup cans do not take the same time to reach the bottom because the thicker, more viscous (nonsloshing) kinds of soup roll with the can, whereas the liquidy (sloshing) soups allow the can to slide around them. Thus, the distribution of the *rolling* mass can be markedly different. A can of Bean with Bacon beats a can of Beef Bouillon.

Analysis

9. Of all the objects you tested, which took the least time to roll down the incline?

The balls reached the bottom in the shortest time.

10. Gravity caused the objects to turn faster and faster—that is, they had rotational acceleration. Of the objects you tested, what shape of objects had the greatest rotational inertia—that is, the greatest *resistance* to rotational acceleration?

The hollow cylinders had the greatest rotational inertia (for their mass and diameter).

36 Acceleration of Free Fall

Experiment

Purpose

To measure the acceleration of an object during free fall with the help of a pendulum.

Setup: >1

Lab Time: 1

Learning Cycle: concept development

Conceptual Level: moderate

Mathematical Level: difficult

Required Equipment/Supplies

Free Fall apparatus from Arbor Scientific Co. or
meterstick with 2 eye hooks in one end
3-inch-long, 3/4-inch-diameter wooden dowel with 2 nails in it
ring stand
condenser clamp
C clamp
stopwatch
drilled steel or brass ball
string
masking tape
carbon paper
white paper

Thanks to Herb Gottlieb for ideas and suggestions for this lab.

Discussion

Measuring the acceleration g of free fall is not a simple thing to do. Galileo had great difficulty in his attempts to measure g because he lacked good timing devices and the motion was too fast. His measurements were done on inclined planes, to slow the motion down.

Galileo would have appreciated the technique you will use in this lab. He was the first to note that the time it takes a pendulum to swing to and fro depends only on the length of the pendulum (provided the angle of swinging is not too great). The time it takes for a pendulum to make a complete to-and-fro swing is called its *period.* How long does a pendulum take to swing from its lowest position to an angle of maximum deflection? The answer is one fourth of its period. If the period is known, it can be used as a unit of time. If a freely falling object and a pendulum are released from elevated positions at the same time, their times can be compared.

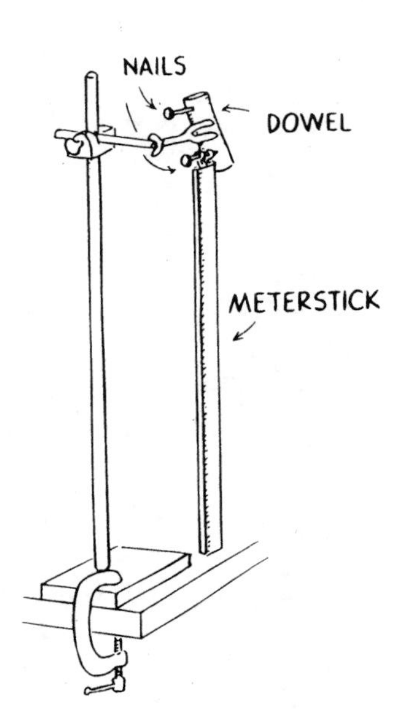

Fig. A

Procedure 

Step 1: Assemble the meterstick pendulum shown in Figure A, attaching the dowel to the ring stand with a condenser clamp. Suspend the meterstick from the lower of the two nails in the wooden dowel.

Assemble pendulum.

Measure period.

Step 2: Displace the bottom of the meterstick 20 cm from the vertical. With a stopwatch, measure and record the time it takes for the meterstick to swing back and forth 10 times.

time for 10 complete swings = __________

Compute the period of the meterstick. Show your computation and record the period.

period = ____________

Attach ball.

Step 3: Using a string 210 cm long, connect one end of the string to the bottom of the meterstick and attach the other end to a drilled steel or brass ball. Loop the string over the upper nail in the dowel, as in Figure B.

Check alignment.

Step 4: Pull the ball up so that it barely touches the 0-cm mark on the meterstick (see Figure C). Check the alignment of the apparatus by slowly lowering the ball along the meterstick. It should just graze the stick all the way down.

Make trial run.

Step 5: Hold the string with the center of the ball at the 0-cm mark of the meterstick and the bottom of the meterstick displaced 20 cm from the vertical (see Figure C). With the apparatus in this position, you will release the string to simultaneously release the ball and start the meterstick swinging. The instant that the meterstick reaches its vertical position, the ball and the meterstick will collide (Figure D). The interval between the time of release and the time of collision is exactly one-fourth the period of the meterstick. The distance that the ball falls is measured from the top of the meterstick to the point of collision

Make a trial run to determine the approximate point of the collision between the ball and the meterstick. At this location, tape a strip of white paper to the meterstick and cover it with a strip of carbon paper, taping the carbon side *against the white paper.* When the experiment is repeated, the ball will leave a mark on the white paper where it collides with the meterstick. This provides a more exact measurement.

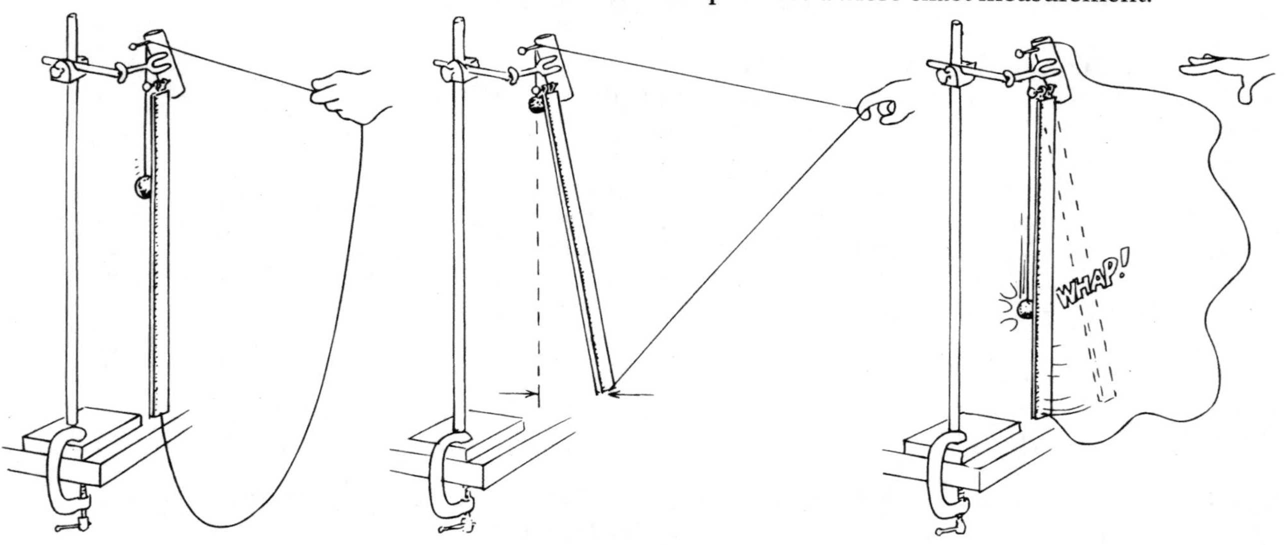

Fig. B Fig. C Fig. D

Name Period Date

Step 6: With the carbon paper in position on the meterstick, make three trials. Remove the carbon paper and measure the distances from the 0-cm mark on the meterstick to the marks on the white paper. Record the distances in Data Table A.

Measure distance.

Data Table A

TRIAL	DISTANCE		ACCELERATION OF GRAVITY
	(cm)	(m)	(m/s^2)
1			
2			
3			
			AVERAGE:

1. When the ball and the meterstick collide, the mark left on the white paper is a vertical line approximately 0.5 cm long, rather than a dot. In measuring the distance that the ball fell, should you measure to the top, the middle, or the bottom of this line?

Measure to the top because that is where it first hits.

Step 7: The equation for the distance *d* traveled by an object that starts from rest and undergoes the acceleration *g* of gravity for time *t* is

Compute g.

$$d = \frac{1}{2} gt^2$$

When this equation is solved for *g*, it becomes

$$g = \frac{2d}{t^2}$$

Compute the acceleration of gravity for three trials and record in Data Table A. Record your average value for *g*.

Analysis

2. What effect does the friction of the cord against the upper support have on your value of *g*?

It provides a force that acts opposite to gravity and thus decreases your value of *g*. The direct effect is to decrease the distance the ball falls.

3. Why doesn't the air resistance on the meterstick affect your value of g?

The air resistance was already taken into account when the period was measured. Therefore, your value of g is *unaffected* by air resistance.

Name Period Date

37 Computerized Gravity

Experiment

Purpose

To measure the acceleration due to gravity using the computer.

Setup: >1
Lab Time: >1
Learning Cycle: concept development
Conceptual Level: moderate
Mathematical Level: difficult

Required Equipment/Supplies

computer
light probe with interface
light source
2 ring stands
2 clamps
cardboard or metal letter "g"
Plexiglas® acrylic plastic strip with black electrical tape on it
printer

Thanks to Dewey Dykstra and Marvin De Jong for their ideas and suggestions for this lab.

Part A: The Letter "g" Method

Discussion

If you know the initial and final speeds of a falling object, and the time interval between them, you can compute the object's acceleration. In this experiment, you will exploit the computer's ability to time events accurately using its game port. The falling object is a square letter "g" cut out of stiff cardboard or sheet metal (see Figure A). The computer has been programmed to clock the time t_1 it takes for the figure to fall the first distance, d_1, and the time t_2 it takes to fall the second distance, d_2.

The measurement of how long it takes to fall each distance past a light probe enables you to compute the average speed for each interval. The average acceleration is then easily computed. From the definition of acceleration as the *change* in velocity per second, the acceleration g of a freely falling object is

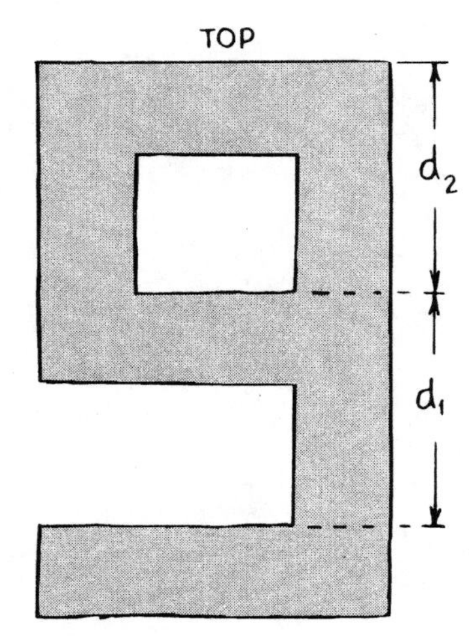

Fig. A

$$g = a = \frac{(v_{2av} - v_{1av})}{t_{av}}$$

where $v_{1av} = \dfrac{d_1}{t_1}$

$$v_{2av} = \frac{d_2}{t_2}$$

$$t_{av} = \frac{(t_1 + t_2)}{2}$$

You might wonder if these equations are exact, since they use finite distances, average speeds, and average times. In fact, they are exact so long as the distances d_1 and d_2 are adjacent.

Procedure

Measure d_1 and d_2.

Step 1: Measure the distances d_1 and d_2 of the letter "g" with a meter-stick to the nearest thousandth of a meter (millimeter).

Set up light probe.

Step 2: Set up the computer with a light probe. Drop the letter "g" through the light beam, as in Figure B. If your computed values for g are not within 2% of the accepted value, repeat the experiment.

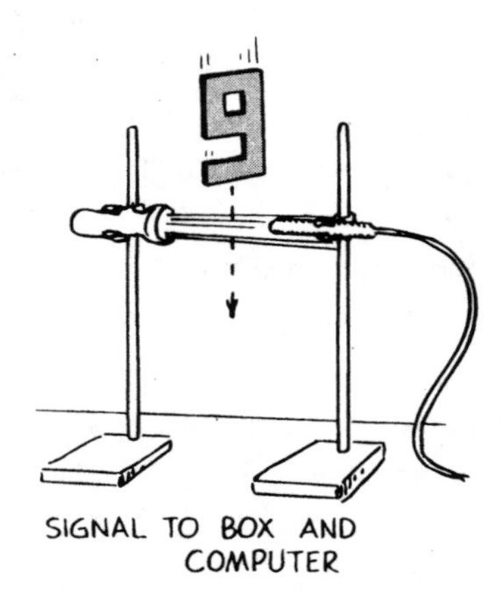

Fig. B

Analysis

1. List the possible sources of error in this experiment. Assume that the computer has been programmed with 100% accuracy!

   Sources of error are (1) uncertainties in the measurements of d_1 and d_2 and (2) inaccuracy in timing due to spreading of light around the edges of the letter "g."

2. Why would a letter "g" made of sheet metal probably be better than one made of cardboard?

   The effect due to air resistance would be less for the sheet metal.

3. What are the advantages of using a square letter "g" instead of a round one?

   For a square letter "g" the distances d_1 and d_2 are the same regardless of the position in the light beam (provided the letter is not tilted).

4. If the letter "g" were held at one corner, so that the edges were no longer vertical and horizontal, and then dropped through the light beam, how would the calculation of the acceleration due to gravity be affected?

   The times would be longer, and the computed acceleration due to gravity would be less than the actual value. (If the angle of the letter is known, one could eliminate the error by using larger values for d_1 and d_2.)

Name Period Date

Part B: The Picket Fence Method

Discussion

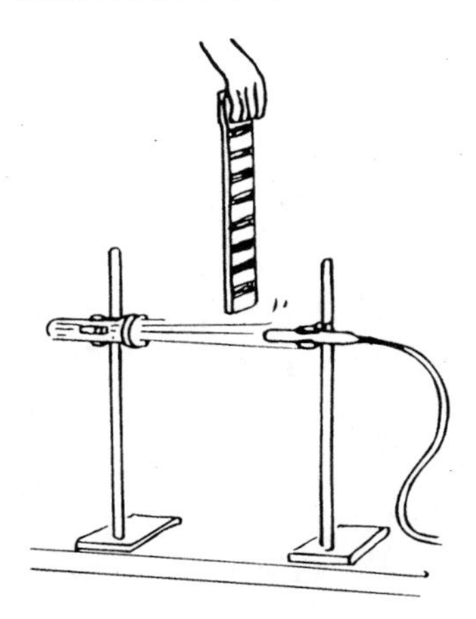

Fig. C

This method of measuring g is an extension of the method used in Part A. A long strip of Plexiglas acrylic plastic is dropped past the light probe instead of a letter "g". The Plexiglas strip has eight dark strips spaced 5 cm from one edge to the next, as shown in Figure C. The strips of tape create alternating opaque and transparent regions ("picket fence") as the Plexiglas strip falls past the light probe. The same light probe arrangement as in Part A is used here.

The computer measures the time it takes for the Plexiglas strip ("fence") to fall from the top of one dark strip ("picket") to the top of the next. If the spaces between the dark strips of tape are all the same (equidistant), this experiment becomes the "undiluted" incline of Experiment 4, "Merrily We Roll Along." In that experiment, you rolled a steel ball down an incline. In this one, you simply drop the picket fence through the light probe! (Wouldn't Galileo have loved to see you do this!)

Procedure

Step 3: Measure the distance from picket to picket. Enter this distance in meters into the program. Position something squishy such as a piece of foam under the light probe to cushion the stop of the dropped fence.

Step 4: Drop the picket fence through the light probe. Be sure to pinch the top of the fence and release it so that it falls as nearly vertically as possible. For best results, hold the bottom of the fence as close as possible to the light probe without triggering it.

Step 5: After you successfully acquire data, save it and then plot distance (vertical axis) vs. time (horizontal axis) using data plotting software.

Use data plotting software to plot d vs. t^2.

5. Describe your graph.

The distance vs. time graph is a parabola.

Save your data and make a printout of your graph.

Step 6: Plot velocity (vertical axis) vs. time (horizontal axis).

6. Describe your graph.

The velocity vs. time graph is a straight line.

Save your data and make a printout of your graph; include it with your report.

Analysis

7. Are either of your graphs straight lines? If so, ask your teacher how to measure the slope of the graph. What is the significance of the slope of your graph?

The slope of the straight line in the velocity vs. time graph is equal to the acceleration of gravity.

Name Period Date

38 Apparent Weightlessness

Purpose

To observe the effects of gravity on objects in free fall.

Setup: 1
Lab Time: 1
Learning Cycle: exploratory
Conceptual Level: easy
Mathematical Level: none

Required Equipment/Supplies

2 plastic foam or paper cups
2 long rubber bands
2 washers or other small masses
masking tape
large paper clip
water

Thanks to Tim Cooney for his ideas and suggestions for this lab.

Discussion

Some people believe that *because* astronauts aboard an orbiting space vehicle appear weightless, the pull of gravity upon them is zero. This condition is commonly referred to as "zero-*g*." While it is true that they *feel* weightless, gravity *is* acting upon them. It acts with almost the same magnitude as on the earth's surface.

The key to understanding this condition is realizing that both the astronauts and the space vehicle are in free fall. It is very similar to how you would feel inside an elevator with a snapped cable! The primary difference between the runaway elevator and the space vehicle is that the runaway elevator has no *horizontal velocity* (relative to the earth's surface) as it falls toward the earth, so it eventually hits the earth. The horizontal velocity of the space vehicle ensures that as it falls *toward* the earth, it also moves *around* the earth. It falls without getting closer to the earth's surface. Both cases involve free fall.

Activity

Fig. A

Procedure 

Attach washers.

Step 1: Knot together two rubber bands to make one long rubber band. Knot each end around a washer, and tape the washers to the ends. Bore a small hole about the diameter of a pencil through the bottom of a Styrofoam or paper cup. Fit the rubber bands through the hole from the inside. Use a paper clip to hold the rubber bands in place under the bottom of the cup (see Figure A). Hang the washers over the lip of the cup. The rubber bands should be taut.

Drop cup.

Step 2: Drop the cup from a height of about 2 m.

1. What happens to the washers?

The washers move into the cup.

Fill cup with water.

Step 3: Remove the rubber bands from the cup and fill the cup half-full with water, using your finger as a stopper over the hole. Hold the cup directly over a sink. Drop the cup into the sink.

2. What happens to the water as the cup falls?

Hardly any water comes out of the hole while the cup is falling.

Fig. B

Step 4: Repeat Step 3 for a second cup half-filled with water with two holes poked through its *sides* (Figure B). You may wish to place a piece of masking tape over the hole on the bottom of the cup.

3. What happens to the water as the cup falls?

Hardly any water comes out of the hole while the cup is falling.

Analysis

4. Explain why the washers acted as they did in Step 2.

In free fall the tension in the rubber bands pulls the "weightless" washers into the cup as the rubber bands return to their unstretched length.

5. Explain why the draining water acted as it did in Steps 3 and 4.

There is (nearly) no draining because the cup and water fall together.

6. Suppose you were standing on a bathroom scale inside an elevator. Based on your observations in this activity, predict what would happen to your weight reading when the elevator:

a. accelerated upward.

Your weight reading would increase.

b. accelerated downward at an acceleration less than g.

Your weight reading would decrease.

c. moved upward at a constant speed.

Your weight reading would be the same as normal.

d. moved downward at a constant speed.

Your weight reading would be the same as normal.

e. accelerated downward at an acceleration greater than g.

The reading of the scale would be the force needed to keep the scale accelerating downward faster than free fall.

Name Period Date

39 Getting Eccentric

Purpose

To get a feeling for the shapes of ellipses and the locations of their foci by drawing a few.

Setup: <1
Lab Time: 1
Learning Cycle: exploratory
Conceptual Level: easy
Mathematical Level: none

To save time, you might have students do this lab at home.

Required Equipment/Supplies

20 cm of string
2 thumbtacks or small pieces of masking tape
pencil
graph paper

Discussion

The path of a planet around the sun is an ellipse, the shadow of a sphere cast on a flat table is an ellipse, and the trajectory of Halley's comet around the sun is an ellipse.

An ellipse is an oval-shaped curve for which the distances from any point on the curve to two fixed points (the *foci*) in the interior have a constant sum. One way to draw an ellipse is to place a loop of string around two thumbtacks and pull the string taut with a pencil. Then slide the pencil along the string, keeping it taut.

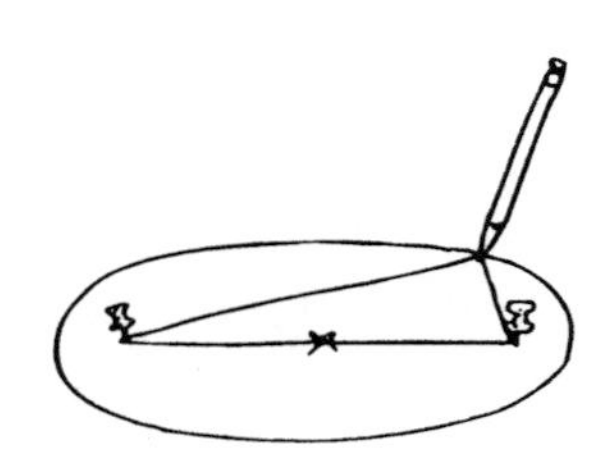

Activity

Procedure 

Step 1: Using graph paper, a loop of string, and two thumbtacks (you may wish to use a small piece of masking tape instead), draw an ellipse. Label the two foci of your ellipse.

Step 2: Repeat twice, using different separation distances for the foci.

Step 3: A circle is a special case of an ellipse. Determine the positions of the foci that would give you a circle, and construct a circle.

Step 4: A straight line is a special case of an ellipse. Determine the positions of the foci that would give you a straight line, and construct a straight line.

Analysis

1. Which of your drawings is closest to the earth's nearly circular orbit around the sun?

The earth's orbit is nearly circular. The two foci are very close together compared with the distance across the ellipse.

2. Which of your drawings is closest to the orbit of Halley's comet around the sun?

The orbit of Halley's comet is very eccentric (elongated) because the foci are very far apart.

40 Trial and Error

Experiment

Purpose

To discover Kepler's third law of planetary motion through a procedure of trial and error using the computer.

Setup: <1

Lab Time: 1

Learning Cycle: concept development

Conceptual Level: difficult

Mathematical Level: difficult

Required Equipment/Supplies

computer
data plotting software

Thanks to Jay Obernolte for developing the software *Data Plotter.*

Discussion

Pretend you are a budding astronomer. In order to earn your Ph.D. degree, you are doing research on planetary motion. You are looking for a relationship between the time it takes a planet to orbit the sun (the *period*) and the average radius of the planet's orbit around the sun. Using a telescope, you have accumulated the planetary data shown in Table A.

You have access to a program that allows you to plot data easily on a computer. It can also plot many different relations between the variables. You can plot the period T vs. the radius R. You can also plot T vs. R^2, T vs. R^3, and so on. You can even plot R vs. T, R vs. T^2, and R vs. T^3, and so on, just as easily.

PLANET	PERIOD (YEARS)	AVERAGE RADIUS (AU)
MERCURY	0.241	0.39
VENUS	0.615	0.72
EARTH	1.00	1.00
MARS	1.88	1.52
JUPITER	11.8	5.20
SATURN	29.5	9.54
URANUS	84.0	19.18
NEPTUNE	165.	30.06
PLUTO	248.	39.44

Table A

An astronomical unit (AU) is the average distance between the sun and the earth.

Procedure

Step 1: Using data plotting software, try to make a graph involving T and R that is a straight line. A linear (straight-line) relationship shows that the two plotted variables are directly proportional to each other. Try plotting T vs. R, T vs. R^2, T vs. R^3, R vs. T, R vs. T^2, and R vs. T^3 to see whether any of these relationships form a straight line when graphed.

Step 2: If the relationship between T and R cannot be discovered by modifying the power of only one variable at a time, try modifying the power of both variables at the same time.

If you find the exact relationship between T and R during this lab period, feel *good.* It took Johannes Kepler (1571–1630) ten years of painstaking effort to discover the relationship. Computers were not around in the seventeenth century!

Stress the importance of discovering functional relationships rather than Kepler's third law.

This is such a powerful data reduction technique that you may wish to kick off your course with this activity.

1. Did you discover a linear relationship between some power of T and some power of R using the program? What powers of T and R graph as a straight line?

Using data plotting software, your students can discover in minutes that T^2 is proportional to R^3—an amazing feat considering that it took Kepler years.

2. Make a printout of the graph closest to a straight line and include it with your report.

41 Flat as a Pancake

Purpose

To estimate the diameter of a BB.

Setup: ≤1
Lab Time: 1
Learning Cycle: exploratory
Conceptual Level: easy
Mathematical Level: moderate

Required Equipment/Supplies

75 mL of BB shot
100-mL graduated cylinder
tray
rulers
micrometer

The computer simulation by the same name on *Laboratory Simulations-II* complements this lab nicely.

Activity

Discussion

Consider 512 cubes, each with sides one centimeter long. If all the cubes are packed closely with no spaces between them to make a large cube with sides 8 cubes long, the volume of the large cube is (8 cm) × (8 cm) × (8 cm), or 512 cm^3. If the cubes are stacked to make a block 4 cm by 16 cm by 8 cm, the volume remains the same but the outside surface area becomes greater. In a block 2 cm by 16 cm by 16 cm, the area of the surface is greater still. If the cubes are spread out so that the stack is only one cube high, the area of the surface is the greatest it can be.

The different configurations have different surface areas, but the volume remains constant. The volume of pancake batter is also the same whether it is in the mixing bowl or spread out on a surface (except that on a hot griddle the volume increases because of the expanding bubbles that form as the batter cooks). The volume of a pancake equals the surface area of one flat side multiplied by the thickness. If both the volume and the surface area are known, then the thickness can be calculated from the following equations.

$$\text{volume} = \text{area} \times \text{thickness} \qquad \text{thickness} = \frac{\text{volume}}{\text{area}}$$

Instead of cubical blocks or pancake batter, consider a shoe box full of marbles. The total volume of the marbles equals the volume of the box (length times width times height). Suppose you computed the volume and then poured the marbles onto a large tray. Can you think of a way to *estimate* the diameter (or thickness) of a single marble without measuring the marble itself? Would the same procedure work for smaller size balls such as BB's? Try it in this activity and see. It will be simply another step smaller to consider the size of molecules.

"d"

Procedure 

Step 1: Use a graduated cylinder to measure the volume of the BB's. (Note that 1 mL = 1 cm^3.)

volume = ____________

Step 2: Spread the BB's to make a compact layer one pellet thick on the tray. If trays are not available, try taping three rulers together in a U-shape on your lab table. Determine the area covered by the BB's. Describe your procedure and show your computations.

area = ____________ cm^2

Step 3: Using the area and volume of the BB's, estimate the diameter of a BB. Show your computations.

estimated diameter = __________ cm

An effective way to teach your students how to use a micrometer is to make several student "experts" and have them in turn teach the other students.

Step 4: Check your estimate by using a micrometer to measure the diameter of a BB.

measured diameter = __________ cm

Analysis

1. What assumptions did you make when estimating the diameter of the BB?

It is assumed that the total volume occupied by the BB's did not change. Since the BB's are spheres, they pack together more compactly than cubes of the same width. This compaction applies only when they are at least several layers deep, so spreading them into a single layer increases the volume they occupy.

2. How do the measured and estimated diameters of the BB compare?

Answers will vary.

3. Oleic acid is an organic substance that is insoluble in water. When a drop of oleic acid is placed on water, it usually spreads out over the water surface, creating a layer one molecule thick. Describe a method to estimate the size of an oleic acid molecule.

Drop a known volume of oleic acid on water and measure the area it covers. Divide the volume by the area to estimate the thickness. See the next experiment.

42 Extra Small

Experiment

Purpose

To estimate the size of a molecule of oleic acid.

Setup: <1
Lab Time: 1
Learning Cycle: application
Conceptual Level: moderate
Mathematical Level: difficult

Required Equipment/Supplies

chalk dust or lycopodium powder
oleic acid solution
10-mL graduated cylinder
tray
water
eyedropper

Thanks to Dave Olsen for his ideas and suggestions for this lab.

Discussion

An oleic acid molecule is not spherical but elongated like a hot dog. One end is attracted to water, and the other end points away from the water surface.

During this investigation, you will estimate the length of a single molecule of oleic acid, to determine for yourself the extreme smallness of a molecule. The length can be calculated by dividing the volume of oleic acid used by the area of the *monolayer*, or layer one molecule thick. The length of the molecule is the depth of the monolayer.

$$\text{volume} = \text{area} \times \text{depth}$$

$$\text{depth} = \frac{\text{volume}}{\text{area}}$$

Procedure    

Set up tray.

Step 1: Pour water into the tray to a depth of about 1 cm. Spread chalk dust or lycopodium powder very lightly over the surface of the water; too much will hem in the oleic acid.

Compute area of film.

Step 2: Using the eyedropper, gently add a single drop of the oleic acid solution to the surface of the water. When the drop touches the water, the alcohol in it will dissolve in the water, but the oleic acid will not. The oleic acid spreads out to make a circle on the water. Measure the diameter of the oleic acid circle in three places, and compute the average diameter of the circle. Also, compute the area of the circle.

average diameter = ____________ cm

area of circle = ______________ cm^2

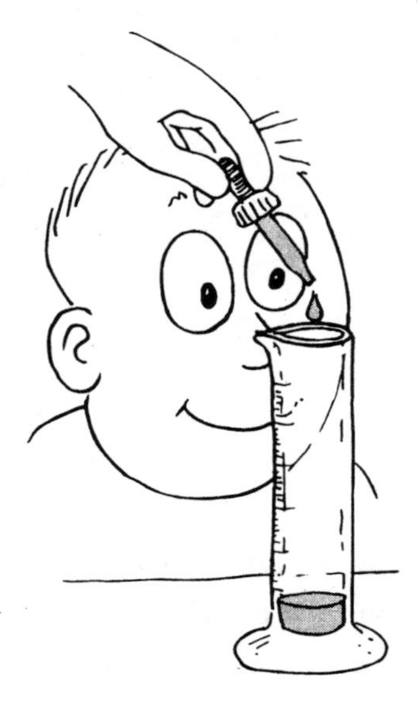

Step 3: Count the number of drops of solution needed to occupy 3 mL (or 3 cm^3) in the graduated cylinder. Do this three times, and find the average number of drops in 3 cm^3 of solution.

number of drops in 3 cm^3 = ________

Divide 3 cm^3 by the number of drops in 3 cm^3 to determine the volume of a single drop.

volume of drop = __________ cm^3

Step 4: The volume of the oleic acid alone in the circular film is much less than the volume of a single drop of the solution. The concentration of oleic acid in the solution is 5 cm^3 per 1000 cm^3 of solution. Every cubic centimeter of the solution thus contains only 5/1000 cm^3 of oleic acid. The ratio of oleic acid to total solution is 0.005 for any volume. Multiply the volume of a drop by 0.005 to find the volume of oleic acid in the drop. This is the volume of the layer of oleic acid in the tray.

volume of oleic acid = __________ cm^3

Compute length of molecule.

Step 5: Estimate the length of an oleic acid molecule by dividing the volume of oleic acid by the area of the circle.

length of molecule = __________

Analysis

The computer simulation "Extra Small" will help students with the computations. If students get bogged down in the calculations, remind them that "Flat as a Pancake" and "Extra Small" are conceptually identical.

1. What is meant by a monolayer?

A monolayer is a layer one molecule thick.

2. Why is it necessary to dilute the oleic acid?

At full strength, a single drop would cover a huge area (such as a swimming pool).

3. Which substance forms the monolayer film—the oleic acid or the alcohol?

The oleic acid molecules form the monolayer.

4. The shape of oleic acid molecules is more like that of a rectangular hot dog than a cube or marble. Furthermore, one end is attracted to water so that the molecule actually "floats" vertically like a log with a heavy lead weight at one end. If each of these rectangular molecules is 10 times as long as it is wide, how would you compute the volume of one oleic acid molecule?

The volume of one rectangular molecule would be the length times one-tenth the length times one-tenth the length.

43 Stretch

Experiment

Purpose

To verify Hooke's law and determine the spring constants for a spring and a rubber band.

Setup: <1

Lab Time: 1

Learning Cycle: concept development

Conceptual Level: moderate

Mathematical Level: moderate

Required Equipment/Supplies

ring stand or other support with rod and clamp
3 springs
paper clip
masking tape
meterstick
set of slotted masses
large rubber band
graph paper
computer, printer, and data plotting software (optional)

Discussion

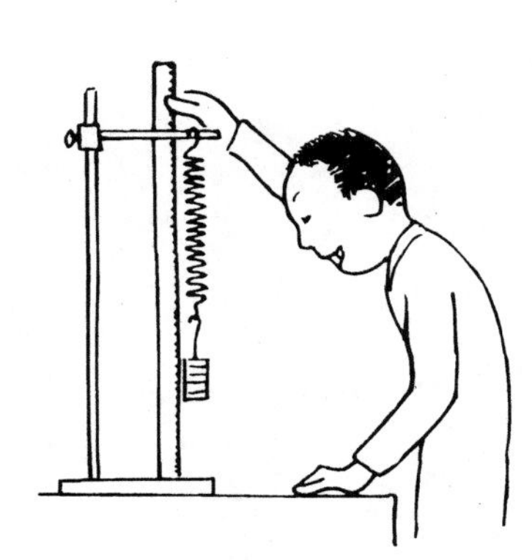

When a force is applied, an object may be stretched, compressed, bent, or twisted. The internal forces between atoms in the object resist these changes. These forces become greater as the atoms are moved farther from their original positions. When the outside force is removed, these forces return the object to its original shape. Too large a force may overcome these resisting forces and cause the object to deform permanently. The minimum amount of stretch, compression, or torsion needed to do this is called the *elastic limit.*

Hooke's law applies to changes below the elastic limit. It states that the amount of stretch or compression is directly proportional to the applied force. The proportionality constant is called the *spring constant, k.* Hooke's law is written $F = kx$, where x is the displacement (stretch or compression). A stiff spring has a high spring constant and a weak spring has a small spring constant.

Procedure 

Set up apparatus.

Step 1: Hang a spring from a support. Attach a paper clip to the free end of the spring with masking tape. Clamp a meterstick in a vertical position next to the spring. Note the position of the bottom of the paper clip relative to the meterstick. Place a piece of masking tape on the meterstick with its lower edge at this position.

Attach masses to spring.

Step 2: Attach different masses to the end of the spring. With your eye level with the bottom of the paper clip, note its position each time. The stretch in each case is the difference between the positions of the paper clip when a load is on the spring and when no load is on the spring. Be careful not to exceed the elastic limit of the spring. Record the mass and stretch of each trial in the first section of Data Table A.

Data Table A

	MASS	FORCE	STRETCH	SPRING CONSTANT
SPRING 1				
SPRING 2				
SPRING 3				
RUBBER BAND				
SPRINGS IN SERIES				

Repeat using different springs.

Step 3: Repeat Steps 1 and 2 for two more springs. Record the masses and strains in the next two sections of Data Table A.

Name Period Date

Step 4: Repeat Steps 1 and 2 using a large rubber band. Record the masses and the corresponding stretches in the fourth section of Data Table A.

Repeat using rubber band.

Step 5: Calculate the forces, and record them in Data Table A. On graph paper, make a graph of force (vertical axis) vs. stretch (horizontal axis) for each spring and the rubber band. If available, use data plotting software to plot your data and print out your graph.

Graph data.

Step 6: For each graph that is an upward sloping straight line, draw a horizontal line through one of the lowest points on the graph. Then draw a vertical line through one of the highest points on the graph. Now you have a triangle. The slope of the graph equals the vertical side of the triangle divided by the horizontal side. The slope of a force vs. stretch graph is equal to the spring constant. By finding the slope of each of your graphs, determine the spring constant k for each spring, and record your values in Data Table A.

Measure slope of graphs.

1. How is the value of k related to the stiffness of the springs?

The spring constant *k* should be larger for stiffer springs.

2. Are all your graphs straight lines? If they are not, can you think of a reason why?

The points for the springs should lie on nearly straight lines. Rubber bands deviate from a straight-line graph principally because of their molecular structure, in contrast to the relatively simple lattice structure of metals.

Going Further

Step 7: Repeat Steps 1 and 2 for two springs connected in series (end to end). Record the masses and stretches in the last section of Data Table A.

Connect several springs in series.

Step 8: Repeat Steps 5 and 6 to determine the spring constant for the combination.

Graphically determine the spring constant.

3. How does the spring constant of two springs connected in a series compare with that of a single spring?

The spring constant for two springs in series is less than that for either spring by itself. (For two identical springs, it is half the value for one spring by itself.)

Name Period Date

44 Geometric Physics

Purpose

To investigate ratios of surface area to volume.

Setup: 1
Lab Time: >1
Learning Cycle: exploratory
Conceptual Level: difficult
Mathematical Level: easy

Required Equipment/Supplies

Styrofoam (plastic foam) balls of diameter 1", 2", 4", 8"
4 sheets of paper
stopwatch or clock with second hand
50-mL beaker
hot plate
2 200-mL beakers
400-mL beaker
800-mL beaker
water
10 g granulated salt
10 g rock salt
2 stirrers
thermometer or
computer
2 temperature probes with interface
printer

Activity

Discussion

Why do elephants have big ears? Why does a chunk of coal burn, while coal dust explodes? Why are ants not the size of horses? This activity will give you insights into some secrets of nature that may at first appear to have no direct connection to physics.

Part A

Procedure 

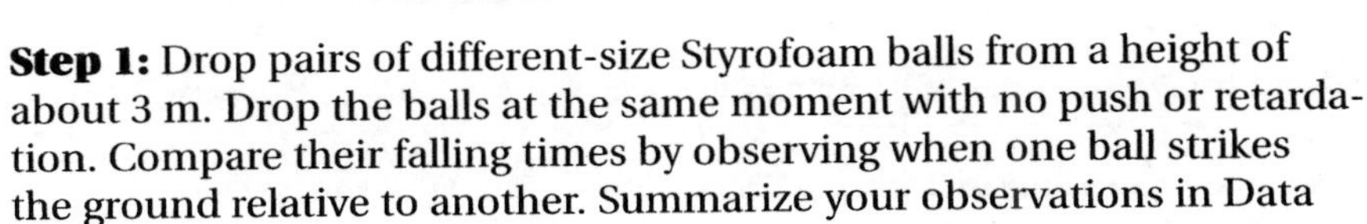
Step 1: Drop pairs of different-size Styrofoam balls from a height of about 3 m. Drop the balls at the same moment with no push or retardation. Compare their falling times by observing when one ball strikes the ground relative to another. Summarize your observations in Data Table A.

Drop Styrofoam balls.

FALLING TIME	BALL DIAMETER
HITS THE GROUND 1ST	
HITS THE GROUND 2ND	
HITS THE GROUND 3RD	
HITS THE GROUND LAST	

Data Table A

1. Describe any regularity you observe.

Larger Styrofoam balls reach the ground sooner.

Analysis

2. Which size Styrofoam ball had the greatest average speed when dropped?

The largest Styrofoam ball had the greatest average speed.

3. Which size Styrofoam ball had the least average speed when dropped?

The smallest Styrofoam ball had the least average speed.

4. Which size Styrofoam ball has the greatest surface area?

The largest Styrofoam ball has the greatest surface area.

5. Which size Styrofoam ball has the greatest volume?

The largest Styrofoam ball has the greatest volume.

6. Which size Styrofoam ball has the greatest ratio of surface area to volume?

The smallest Styrofoam ball has the greatest ratio of surface area to volume. (Since surface area depends on the square (r^2) of the radius r, and the volume depends on the cube (r^3) of the radius, the ratio of surface area to volume is proportional to (r^2/r^3) = 1/r, or inversely proportional to, the radius.)

7. Which size Styrofoam ball has the smallest ratio of surface area to volume?

The largest Styrofoam ball has the smallest ratio of surface area to volume (since the ratio is inversely proportional to the radius).

Name Period Date

8. A Styrofoam ball falling through the air has two forces acting on it. One is the downward force due to gravity—its weight. The weight of the ball is proportional to its volume. The other force is the upward force of air resistance—drag—which opposes the fall. Drag is proportional to the surface area of the ball.

a. To what is the upward force due to drag proportional?

The upward force due to drag is proportional to the surface area.

b. Which size Styrofoam ball should have experienced the greatest upward force due to drag?

The largest Styrofoam ball should have experienced the greatest upward force due to drag.

c. Which size Styrofoam ball should have experienced the greatest *net* downward force per unit mass (due to gravity *and* drag)—that is, which should show the greatest acceleration?

The largest Styrofoam ball should have experienced the greatest *net* downward force per unit mass. The gravitational force per unit mass is constant. The drag force per unit mass is proportional to $r^2/r^3 = 1/r$, which decreases as r gets larger.

d. Does your answer to (c) agree with your observations?

Yes; the ball with the greatest average speed has the greatest acceleration. (Since acceleration is proportional to net force per unit mass, the ball with the greatest net force per unit mass has the greatest acceleration.)

9. Predict which will fall to the ground faster—a heavier raindrop or a lighter one. Why?

The heavier (and larger) raindrop will fall faster because it has the smaller ratio of area to weight and thus less upward drag compared to its weight.

Part B

Procedure

Dissolve salt in water.

Step 2: Record the number of seconds it takes to dissolve 10 g of granulated salt in 100 mL of water at room temperature that is being stirred vigorously. Repeat, substituting 10 g of rock salt for the granulated salt.

time for granulated salt = ________

time for rock salt = ________

Analysis

10. Which salt dissolved more quickly—the granulated salt or the rock salt? Why?

The granulated salt dissolves faster than the rock salt because it has more surface area per gram.

11. Suggest a relationship for predicting dissolving times for salt granules of different size, such as rock salt and table salt.

The dissolving time is inversely proportional to the size of the granules—a smaller granule of salt has a larger ratio of surface area to volume.

Part C

Procedure

Step 3: Make similar paper airplanes from whole, half, and quarter sheets of paper. Record which one flies the farthest.

airplane that flies farthest: ____________

Analysis

12. Which paper airplane generally traveled farthest? Why?

Larger paper airplanes will generally travel farther. Their surface-to-volume (or surface-to-mass) ratio is less than for smaller planes, so, just as for the larger Styrofoam balls, they experience less retarding force per unit mass. (During most of a typical flight, the paper airplane is being decelerated. The magnitude of its acceleration is approximately proportional to the inverse of its linear dimension.)

Name Period Date

Part D

Procedure

Step 4: Heat 500 mL of water to 40°C. Then pour 50 mL into a 50-mL beaker and 400 mL into a 400-mL beaker. Allow the beakers to cool down by themselves on the lab table. Measure the temperature of both every 30 seconds, using either a thermometer or temperature probes connected to a computer. (If you are using the computer, use two temperature probes to monitor the temperatures of the two beakers of water. Save your data, and print a copy of your graph; include it with your lab report.) Record your findings in Data Table B.

Monitor cooling water.

Data Table B

TIME (s)	TEMPERATURE (°C)	
	50 mL	400 mL
0		
30		
60		
90		
120		
150		
180		
210		
240		
270		
300		

The temperature drop in the 50-mL beaker is quite dramatic compared with the 400-mL beaker; it is best to add the water to the 400-mL beaker first.

Analysis

13. Which beaker of warm water cooled faster?

The smaller beaker should cool faster.

14. Which beaker has more surface area?

The larger beaker has more surface area.

15. Which beaker has the greater volume?

The larger beaker has the greater volume.

16. Which beaker has the larger ratio of surface area to volume?

The smaller beaker has the larger ratio of surface area to volume.

17. Which beaker has the smaller ratio of surface area to volume?

The larger beaker has the smaller ratio of surface area to volume.

18. The cooling of a beaker of warm water takes place at the surface. The total amount of heat that must leave the water for it to cool to room temperature depends on the volume. Therefore, on what ratio does the rate of cooling (the temperature drop with time) depend?

The rate of cooling depends on the ratio of surface area to volume.

19. Suppose that a small wading pool is next to a swimming pool. Predict which pool will heat up faster during the day. Why?

The smaller pool will heat up faster because it has a larger ratio of surface area (heating surface) to volume (water to be heated).

Name Period Date

45 Eureka!

Purpose

To explore the displacement method of finding volumes of irregularly shaped objects and to relate their masses to their volumes.

Setup: <1
Lab Time: >1
Learning Cycle: exploratory
Conceptual Level: moderate
Mathematical Level: moderate

Required Equipment/Supplies

two 35-mm film canisters (prepared by the teacher): one of which is filled with lead shot, and the other with a bolt just large enough to cause the canister to sink when placed in water
triple-beam balance
string
graduated cylinder
water
masking tape
5 steel bolts of different size
irregularly shaped piece of scrap iron
1000-mL beaker

Thanks to Rog Lucido for his ideas and suggestions for this lab.

Activity

Discussion

The volume of a block is easy to compute. Simply measure its length, width, and height. Then, multiply length times width times height. But how would you go about computing the volume of an irregularly shaped object such as a bolt or rock? One way is by the displacement method. Submerge the object in water, and measure the volume of water displaced, or moved elsewhere. This volume is also the volume of the object. Go two steps further and (1) measure the mass of the object; (2) divide the mass by the volume. The result is an important property of the object—its *density*.

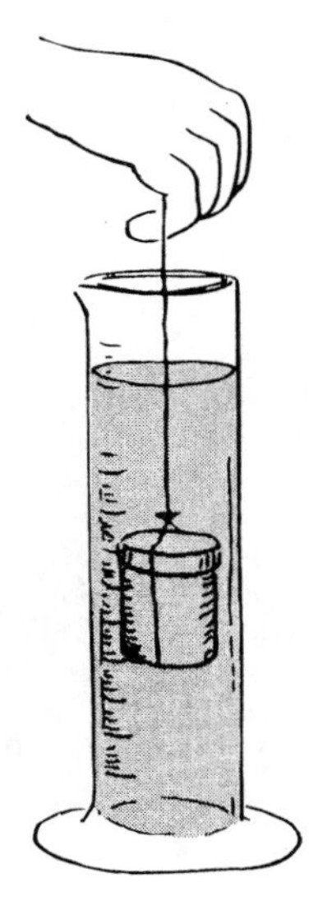

Fig. A

Part A

Procedure   

Find the mass of each canister.

Step 1: Your teacher has prepared two film canisters for you. Each canister should have a piece of string attached to it. Use a triple-beam balance to find the mass of each one.

mass of lighter canister = ____________

mass of heavier canister = ____________

Observe rise in water level.

Step 2: Place a strip of masking tape vertically at the water level of a graduated cylinder about 2/3 full of water. Mark the water level on the tape. Submerge the lighter canister in the water, as shown in Figure A. Observe the rise in water level. Mark the new water level on the tape. Remove the canister.

Predict rise in water level.

Step 3: Predict what you think the rise in the water level for the heavier canister will be when it is submerged. Do you think it will be less than, the same as, or greater than the rise in water level for the lighter canister?

prediction: ____________________

Observe rise in water level.

Step 4: Test your prediction by submerging the heavier canister. Write down your findings.

findings: ____________________

Analysis

1. How do the amounts of water displaced by each canister compare?

   Since the rise in water level is the same for each canister, they must each displace the same amount of water.

2. Does the amount of water displaced by a submerged canister depend on the mass of the canister? On the volume of the canister?

   The amount of water displaced by a submerged canister depends on its volume but not its mass.

Part B

Procedure

Measure mass and volume of bolts.

Step 5: Measure the masses of five different-sized bolts and then measure their volumes using the displacement method. Record your data in Data Table A.

Data Table A

BOLT NUMBER	MASS (g)	VOLUME (mL)	_______ (g/mL OR g/cm^3)
1			
2			
3			
4			
5			

Name Period Date

Step 6: For each bolt, divide its mass by its volume and enter the results in the last column of Data Table A.

Analysis

3. How do the ratios of mass to volume compare for each of the bolts?

   All the bolts should have about the same ratio of mass to volume.

4. Would you expect the ratio of mass to volume to be the same for bolts of different metals?

   Bolts of different metals would have different ratios of mass to volume.

5. What name is given to the ratio of mass to volume? (Fill this name in at the top of the last column in Data Table A.)

   The ratio of mass to volume is the density.

Part C

Going Further

Pretend that you are living in ancient Greece during the time of Archimedes. The king has commissioned a new crown to be made of pure gold. The king is worried that the goldsmith may have cheated him by replacing some of the gold with less precious metals. You are asked to devise a way to tell, without damaging the crown, whether the king was cheated or not.

Unfortunately, your school cannot supply you with an actual gold crown for this activity. A less valuable piece of scrap iron will simulate the crown.

Write down your procedure and any measurements you make.

Measure the mass of the crown and divide by the volume (via displacement) to get the density. If the crown has the same mass *and* density as the original gold, the goldsmith is vindicated.

46 Sink or Swim

Purpose

To introduce Archimedes' principle and the principle of flotation.

Setup: <1
Lab Time: >1
Learning Cycle: concept development
Conceptual Level: difficult
Mathematical Level: moderate

Thanks to Art Farmer for his ideas and suggestions for this lab.

Required Equipment/Supplies

spring scale
triple-beam balance
string
rock or hook mass
600-mL beaker
500-mL graduated cylinder
clear container
water
masking tape
chunk of wood
modeling clay
toy boat
50-g mass
2 100-g masses

Discussion

An object submerged in water takes up space and pushes water out of the way. The water is said to be displaced. Interestingly enough, the water that is pushed out of the way, in effect, pushes back on the submerged object. For example, if the object pushes a volume of water with a weight of 10 N out of its way, then the water reacts by pushing back on the object with a force of 10 N. We say that the object is buoyed upward with a force of 10 N. In this experiment, you will investigate what determines whether an object sinks or floats in water.

Procedure 

Weigh an object in air and submerged in water.

Step 1: Use a spring scale to determine the weight of an object (rock or hook mass) first in air and then underwater. The difference in weights is the buoyant force. Record the weights and the buoyant force.

weight of object in air = ____________

apparent weight of object in water = ________

buoyant force on object = ____________

Measure the water displaced.

Step 2: Devise an experimental setup to find the volume of water displaced by the object. Record the volume of water displaced. Compute the mass and weight of this water. (Remember, 1 mL of water has a mass of 1 g and a weight of 0.01 N.)

volume of water displaced = __________

mass of water displaced = ____________

weight of water displaced = ___________

1. How does the buoyant force on the submerged object compare with the weight of the water displaced?

The buoyant force should equal the weight of the water displaced (or nearly so).

Float a piece of wood.

Step 3: Find the mass of a piece of wood with a triple-beam balance, and record the mass in Data Table A.

NOTE: To keep the calculations simple, from here on in this experiment, you will measure and determine masses, without finding the equivalent weights. Keep in mind, however, that an object floats because of a buoyant force. This force is due to the *weight* of the water displaced.

Measure the volume of water displaced when the wood floats. Record the volume and mass of water displaced in Data Table A.

Data Table A

OBJECT	MASS (g)	VOLUME OF WATER DISPLACED (mL)	MASS OF WATER DISPLACED (g)
WOOD			
WOOD AND 50-g MASS			
CLAY BALL			
FLOATING CLAY			

2. What is the relation between the buoyant force on any floating object and the weight of the object?

The buoyant force on any floating object equals the weight of the object. (If it were less, the object would sink; if it were more, the object would rise higher.)

3. How does the mass of the floating wood compare to the mass of water displaced?

The mass of the floating object should equal the mass of water displaced.

4. How does the buoyant force on the wood compare with the weight of water displaced?

The buoyant force on the wood, equal to the weight of the wood, also equals the weight of water displaced, since the masses are equal.

Step 4: Add a 50-g mass to the wood so that the wood displaces more water *but still floats.* The 50-g mass needs to float on top of the wood. Measure the volume of water displaced and calculate its mass, recording them in Data Table A.

Float wood plus 50-g mass.

5. How does the combined buoyant force on the wood and the 50-g mass compare with the weight of water displaced?

The combined buoyant force, equal to the combined weight, also equals the weight of water displaced, since the masses are equal.

Step 5: Find the mass of a ball of modeling clay. Measure the volume of water it displaces when it sinks to the bottom. Calculate the mass of water displaced, and record all volumes and masses in Data Table A.

Measure dipslacement of clay.

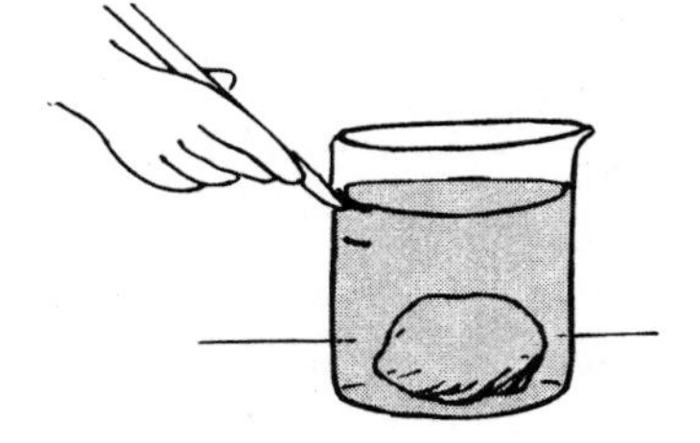

6. How does the mass of water displaced by the clay compare to the mass of the clay?

The mass of water displaced is less than the mass of the clay.

7. Is the buoyant force on the submerged clay greater than, equal to, or less than its weight in air? Explain.

The buoyant force on the submerged clay is less than its weight in air. Otherwise, the clay could not remain submerged.

Step 6: Retrieve the clay from the bottom, and mold it into a shape that allows it to float. Sketch or describe this shape.

Mold the clay so that it floats.

Measure the volume of water displaced by the floating clay. Calculate the mass of the water, and record in Data Table A.

8. Does the clay displace more, less, or the same amount of water when it floats as it did when it sank?

The clay displaces more water when it floats.

9. Is the buoyant force on the floating clay greater than, equal to, or less than its weight in air?

The buoyant force on the floating clay equals its weight in air.

10. What can you conclude about the weight of water displaced by a floating object compared with the weight in air of the object?

The weight of water displaced by any floating object equals the weight in air of the object.

Predict new water level in canal lock.

Step 7: Suppose you are on a ship in a canal lock. If you throw a ton of bricks overboard into the canal lock, will the water level in the canal lock go up, down, or stay the same? Write down your answer *before* you proceed to Step 8.

prediction for water level in canal lock: ___________

Step 8: Float a toy boat loaded with "cargo" (such as one or two 100-g masses) in a container filled with water deeper than the height of the masses. Mark and label the water levels on masking tape placed on the container and on the sides of the boat. Remove the masses from the boat and put them in the water. Mark and label the new water levels.

For best results, use dense masses such as lead weights as cargo.

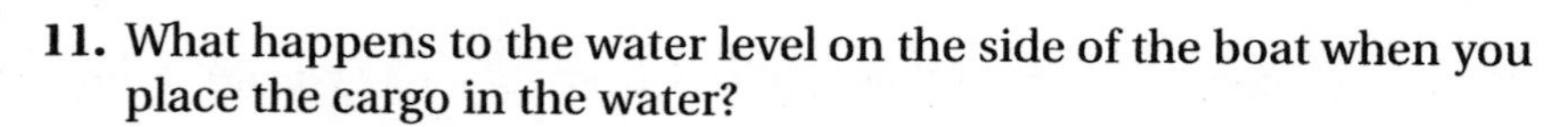

11. What happens to the water level on the side of the boat when you place the cargo in the water?

The water level on the side of the boat goes down; the boat rides higher in the water.

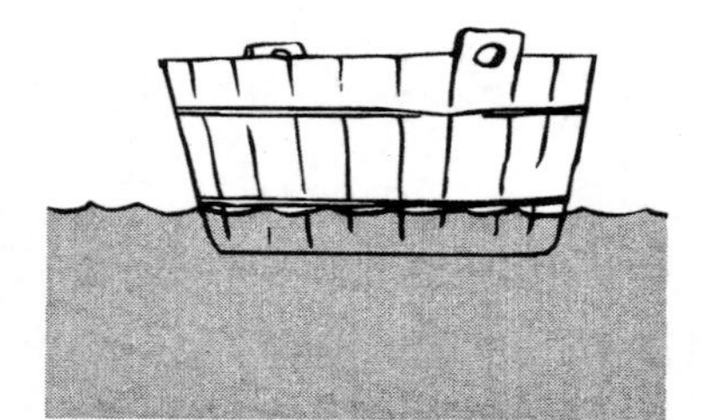

12. If a large freighter is riding high in the water, is it carrying a relatively light or heavy load of cargo?

A freighter riding high is carrying a relatively light cargo.

13. What happens to the water level in the container when you place the cargo in the water? Explain why it happens.

It drops, since the cargo displaces less water when it sinks than when it is floating.

14. What will happen to the water level in the canal lock when the bricks are thrown overboard?

The water level in the canal lock will drop.

Name Period Date

47 Weighty Stuff

Purpose

To recognize that air has weight.

Setup: <1
Lab Time: 1
Learning Cycle: exploratory
Conceptual Level: moderate
Mathematical Level: moderate

Required Equipment/Supplies

centigram balance
sheet of paper
basketball
hand pump
needle valve
tire pressure gauge (preferably scaled in 1-lb increments)

Discussion

If you hold up an empty bottle and ask your friends what's in it, they will probably say that nothing is in the bottle. But strictly speaking, there is something in the "empty" bottle: air. Perhaps a fish would say similarly that there is nothing in an "empty" bottle at the bottom of the ocean. We don't ordinarily notice the presence of air unless it is moving. And we don't notice that air has mass and weight, like all matter on earth. In this activity, you will focus attention on the mass and weight of air.

Procedure

Find mass of deflated basketball.

Step 1: Flatten out a basketball so that there is no air in it. Place a sheet of paper on a centigram balance. (In Step 3, you will crumple the paper and use it to keep the basketball from rolling off the balance.) *Carefully* measure the mass of the flattened basketball and paper.

mass of flattened basketball and paper = ___________

Inflate basketball.

Step 2: Remove the basketball from the balance. Inflate it with air to a pressure of 7.5 lb/in.^{2}, as measured by the pressure gauge.

Find mass of inflated basketball.

Step 3: Crumple the piece of paper so that it will keep the inflated basketball from rolling off the balance. Measure the mass of the inflated basketball and paper.

mass of inflated basketball and paper = ___________

Compute mass of air in basketball.

Step 4: Compute the mass of the air pumped into the basketball.

mass of air = ___________

Measure pressure of deflated basketball.

The gauge pressure should read zero.

Step 5: Insert a needle valve in the inflated basketball, and allow the air to escape without flattening the ball. Wait until you no longer hear air escaping. Then, measure the final pressure inside the basketball.

pressure after air escapes = ___________ lb/in.^{2}

1. A pressure gauge measures the *difference* between pressures at the two ends of the gauge. Is there still air in the basketball?

Yes, there is still air in the basketball. It is at normal air pressure.

Compute volume of extra air.

Step 6: You really inflated the flattened basketball in two stages. First you inflated it out to its full spherical shape. When it reached a spherical shape, the basketball displaced an amount of air *equal* to the air added. Therefore, *the buoyant force on the ball equaled the weight of the air added.* The reading on the balance would have been the same as for the flattened basketball. In the second stage, you increased the pressure above normal air pressure (14.7 lb/in.^{2}) by forcing *extra* air into the basketball without appreciably changing its volume. The basketball did not displace any additional air, so the buoyant force did not increase any further. The reading on the balance increased by the mass of the air added during the second stage.

If you had, hypothetically, raised the pressure—the amount by which the pressure exceeds atmospheric pressure—to 14.7 lb/in.^{2}, then the *extra* air pumped into the basketball during the second stage would have been just enough to inflate a second flattened basketball to full volume at zero final gauge pressure. However, you actually inflated the basketball to a gauge pressure of about 7.5 lb/in.^{2}. Compute what size flattened ball would just be inflated to full size at zero gauge pressure by the *extra* air in the basketball.

CAUTION: *Do NOT attempt to inflate a basketball to a gauge pressure of 14.7 lb/in.^{2}! It may burst violently.*

The volume should be roughly half the volume of the basketball.

volume of extra air pumped into
basketball at zero gauge pressure = ___________

2. When air is added to a flattened basketball, why does a balance measure only the mass of the *extra* air that increases the gauge pressure above zero?

The air needed to fill out a flattened basketball to its full spherical shape weighs the same as the air it displaces, so the buoyant force equals the additional weight. The basketball presses down no harder on the balance than when it is flattened. The balance will detect only the mass of the *extra* air that increases the pressure *above* surrounding atmospheric pressure.

3. Does the air in your classroom have weight as well as mass?

Yes; the earth's gravity pulls on all air, just as it pulls on anything else with mass.

Name Period Date

48 Inflation

Purpose

To distinguish between pressure and force, and to compare the pressure that a tire exerts on the road with the air pressure in the tire.

Setup: <1
Lab Time: 1
Learning Cycle: application
Conceptual Level: moderate
Mathematical Level: moderate

Thanks to Evan Jones and Ken Ford for their ideas and suggestions for this lab.

Required Equipment/Supplies

automobile
owner's manual for vehicle (optional)
graph paper
tire pressure gauge (preferably scaled in 1-lb increments)

Discussion

Activity

People commonly confuse the two words *force* and *pressure*. Tire manufacturers add to this confusion by saying "inflate to 45 pounds" when they really mean "inflate to 45 pounds *per square inch*." The amount of pressure depends on how a force is distributed over a certain area. It decreases as the area increases.

$$\text{pressure} = \frac{\text{force}}{\text{area}}$$

In this activity, you will determine the pressure that an automobile tire exerts on the road. Then you'll find out how that compares with the pressure of the air in the tire. If it weren't for the support given by the sidewalls of the tire, the two pressures would be the same. You can directly measure the air pressure in the tire with a pressure gauge. You can measure and calculate the pressure of the tire against the road by dividing the weight of the car by the area of contact for the four tires (the tires' "footprints"). Although there are gaps between the treads that don't support the car, their role is small compared with that of the sidewalls, so will be neglected.

Procedure 

Step 1: Position a piece of graph paper directly in front of the right front tire. Roll the automobile onto the paper.

Step 2: Trace the outline of the tire where it makes contact with the graph paper. Roll the vehicle off the paper. Compute the area inside the trace in square inches. Record your data in Data Table A.

Data Table A

TIRE	AREA OF CONTACT (in^2)	WEIGHT/4	PRESSURE CALCULATED (lb/in^2)	PRESSURE MEASURED (lb/in^2)
LEFT FRONT				
RIGHT FRONT				
LEFT REAR				
RIGHT REAR				

Repeat for each tire.

Step 3: Repeat Steps 1 and 2 for the other tires. Record your data in Data Table A.

Determine the force.

Step 4: Ascertain the weight of the vehicle from the owner's manual or the dealer. Often times the weight of the vehicle is stamped onto the front door jam on the driver's side. If information on the weight distribution between front and rear tires is available, use it to find the weight supported by each tire. Otherwise, assume that each tire supports one fourth of the weight. Enter the force exerted by each tire in Data Table A.

Compute the pressures.

Step 5: For each tire, divide the force exerted by the tire by the "footprint" area of the tire to get the actual pressure exerted on the ground by the tire. Record in Data Table A.

Measure tire pressure.

Step 6: Use a tire gauge to measure the pressure in each tire in pounds per square inch. Record the pressures of the tires in Data Table A.

Compare pressures.

Step 7: Compare the pressure exerted by each tire with the air pressure in the tire. What percentage of the pressure exerted by the tire is accounted for by the air pressure? Air pressure as a percentage of tire pressure on ground for the four tires:

______________, ______________, ______________, ______________

Analysis

1. The pressure exerted by the tire on the road is greater than the air pressure in the tire. Do you think this would be the case if the tire were a membrane with no strength of its own?

The structure of the tire, especially the sidewalls, supports some of the weight. (If the tire were a membrane with no strength of its own, the tire pressure on the road would be the same as the air pressure. If the tire were a strong, rigid structure that required little or no air, its pressure on the road would not be related to any internal air pressure. An actual automobile tire falls between these two extremes.)

2. Consider the other extreme. Suppose the tire were a strong rigid structure that required little or no air. Would air pressure in the tire correspond to the pressure the tire exerts against the road? Explain.

Air pressure in a steel tire would have no bearing on the tire's pressure against the road. Pressure is only the weight of the car (and that "extra air" that may be in the steel tire) divided by the tires' footprints.

3. How might the weight given in the owner's manual be different from the actual weight of the car?

A load in the trunk, a full tank of gas, and extra installed equipment may make the actual weight greater.

4. We ignored the tire tread in computing pressures. One should not ignore the role of tire treads in practice—particularly in rainy weather. Why?

Good tread provides better contact with the road because the sharp tread (smaller area of contact than when the tire is bald) exerts a larger pressure on the road surface. As a result, the tire squishes the water into the tread channels where it can escape.

5. A tire gauge measures the *difference* between the pressure in the tire and atmospheric pressure (14.7 $lb/in.^2$) outside the tire (*for example, a perfectly "flat tire" reads zero on the gauge, but atmospheric pressure exists in it*). What is the total pressure *inside* the front tire?

The total pressure equals the measured pressure (gauge pressure) plus the atmospheric pressure.

6. Why was the atmospheric pressure (14.7 $lb/in.^2$) not added to the pressure of the tire gauge when the tire pressure and air pressures were compared?

There is as much atmospheric pressure inside as outside. The result is not net force. Atmospheric pressure does not contribute to supporting the weight of the car. When you have a flat tire, there is still 14.7 $lb/in.^2$ of pressure *inside* the tire (as well as *outside* the tire).

Going Further

Repeat the above with one of the tires deflated, say to half its normal air pressure. You'll note its footprint is larger. Find out if the lower air pressure accounts for a greater or lesser percentage of the actual tire pressure on the ground.

49 Heat Mixes: Part I

Purpose

To predict the final temperature of a mixture of cups of water at different temperatures.

Setup: <1
Lab Time: 1
Learning Cycle: exploratory
Conceptual Level: easy
Mathematical Level: easy

Required Equipment/Supplies

3 Styrofoam (plastic foam) cups
liter container
thermometer (Celsius)
pail of cold water
pail of hot water

Thanks to Verne Rockcastle for his ideas and suggestions for this lab.

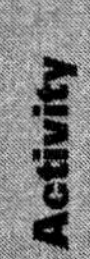

Discussion

If you mix a pail of cold water with a pail of hot water, the temperature of the mixture will be between the two initial temperatures. What information would you need to predict the final temperature? This lab investigates what factors are involved in changes of temperature.

Your goal is to find out what happens when you mix equal masses of water at different temperatures. Before actually doing this, imagine a cup of hot water at 60°C and a pail of room-temperature water at 20°C.

(circle one)

1. Which do you think is hotter—the cup or the pail? (cup) pail
2. Which do you think has more thermal energy? cup (pail)
3. Which would take longer to change its temperature by 10°C if the cup and pail were set outside on a winter day? cup (pail)
4. If you put the same amount of red-hot iron into the cup and the pail, which one would change temperature more? (cup) pail

Procedure  

Step 1: Two pails filled with water are in your room, one with cold water and one with hot water. Fill one cup 3/4 full with water from the cold-water pail. Mark the water level along the inside of the cup. Pour the cup's water into a second cup. Mark it as you did the first one. Pour the cup's water into a third cup, and mark it as before. Now all three cups have marks that show nearly equal measures.

Mark cups.

Marking the inside of the cup eliminates guessing where the actual water level is.

5. Why don't the marks show exactly equal measures?

Some water might be lost or spilled.

Measure temperature of two cups.

Step 2: Fill the first cup to the mark with water from the hot-water pail. Measure and record the temperature of both cups of water.

temperature of cold water = ________°C

temperature of hot water = ________°C

Predict temperature of mixture.

The correct prediction is the average of the two temperatures.

Step 3: What will be the temperature if you mix the two cups of water in the liter container? Record your prediction.

predicted temperature of mixture = ________°C

Measure temperature of mixture.

Step 4: Pour the two cups of water into the liter container, stir the mixture, and measure and record its temperature.

actual temperature of mixture = ______°C

6. If there was a difference between your prediction and your observation, what might have caused it?

Students may have waited too long after thermal equilibrium was reached or may not have measured the temperatures carefully.

Pour the mixture into the sink or waste pail. Do not pour it back into the pail of either cold or hot water!

Measure temperature of three cups.

Step 5: Fill one cup to its mark with water from the cold-water pail. Fill the other two cups to their marks with hot water from the hot-water pail. Measure and record their temperatures.

temperature of cold water = ________°C

temperature of hot water (cup 1) = _____°C

temperature of hot water (cup 2) = _____°C

Predict temperature of mixture.

The correct prediction is an average of the three temperatures.

Step 6: What will be the temperature when all three cups of water are mixed in one container? Record your prediction.

predicted temperature of mixture = ________°C

Measure temperature of mixture.

Step 7: Pour the three cups of water into the liter container, stir the mixture, and measure and record its temperature.

actual temperature of mixture = ______°C

7. How did your observation compare with your prediction?

Answers will vary.

8. Which of the water samples (cold or hot) changed more when it became part of the mixture? Why do you think this happened?

The cold water changes more because there was less of it, and it absorbs heat from both cups of hot water. The temperature of the mixture was closer to that of the hot water.

Pour the mixture into the sink or waste pail. Do *not* pour it back into the pail of either cold or hot water!

Step 8: Fill one cup to its mark with *hot* water. Fill the other two cups to their marks with *cold* water. Measure and record their temperatures.

Reverse Step 5.

temperature of hot water = ___________°C

temperature of cold water (cup 1) = ______°C

temperature of cold water (cup 2) = ______°C

Step 9: Record your prediction of the temperature when these three cups of water are mixed in one container.

Predict temperature of mixture.

The correct prediction is the sum of the three temperatures divided by 3. In general, the students will probably recognize that the final temperature is an average of the individual temperatures of the cups. This is true *only* if the cups have *equal amounts of water.* Differing amounts of hot and cold water will have an average nearer the temperature of the greater initial amount of water.

predicted temperature of mixture = ______°C

Step 10: Pour the three cups of water into the liter container, stir the mixture, and measure and record its temperature.

actual temperature of mixture = ________°C

9. How did this observation compare with your prediction?

Answers will vary.

10. Which of the water samples (cold or hot) changed more when it became part of the mixture? Why do you think this happened?

The hot water changes more because there was less of it, and it gives off heat to both cups of cold water. The temperature of the mixture was closer to that of the cold water.

Name Period Date

50 Heat Mixes: Part II

Purpose

To predict the final temperature of water and nails when mixed.

Setup: <1
Lab Time: 1
Learning Cycle: exploratory
Conceptual Level: moderate
Mathematical Level: moderate

Required Equipment/Supplies

Harvard trip balance
2 large insulated cups
bundle of short, stubby nails tied with string
thermometer (Celsius)
hot and cold water
paper towels

Thanks to Verne Rockcastle for his ideas and suggestions for this lab.

Discussion

Activity

If you throw a hot rock into a pail of cool water, you know that the temperature of the rock will go down. You also know that the temperature of the water will go up—but will its rise in temperature be more than, less than, or the same as the temperature drop of the rock? Will the temperature of the water go up as much as the temperature of the rock goes down? Will the changes of temperature instead depend on how much rock and how much water are present and how much energy is needed to change the temperature of water and rock by the same amount?

You are going to study what happens to the temperature of water when hot nails are added to it. Before doing this activity, think about the following questions.

1. Suppose that equal masses of water and iron are at the same temperature. Suppose you then add the same amount of heat to each of them. Would one change temperature more than the other?

(circle one) yes no

If you circled "yes," which one would warm more?

(circle one) water iron

2. Again, suppose that equal masses of water and iron are at the same temperature. Suppose you then take the same amount of heat away from each of them. Would one cool more than the other?

(circle one)  yes no

If you circled "yes," which one would cool more?

(circle one) water 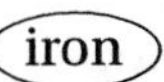 iron

Procedure   

Balance nails with cold water.

Step 1: Place a large cup on each pan of the balance. Drop the bundle of nails into one of the cups. Add cold water to the other cup until it balances the cup of nails. When the two cups are balanced, the same mass is in each cup—a mass of nails in one, and an equal mass of water in the other.

Step 2: Set the cup of cold water on your work table. Lift the bundle of nails out of its cup and place it beside the cup of cold water.

Warm nails in hot water.

Step 3: Fill the empty cup 3/4 full with hot water. Lower the bundle of nails into the hot water and leave it there for two minutes to allow the nails to come to the temperature of the hot water.

Measure temperature of cold and hot water.

Step 4: Measure and record the temperature of the cold water and the temperature of the hot water around the nails.

temperature of cold water = ________°C

temperature of hot water = ________°C

3. Is the temperature of the hot water equal to the temperature of the nails? Why do you think it is or is not? Can you think of a better way to heat the nails to a known temperature?

Yes, the temperature of the hot water should equal the temperature of the nails. (They should be in thermal equilibrium.) This is the best way to heat the nails to a known temperature.

Predict temperature of mixture.

Step 5: Predict what the temperature of the mixture will be when the hot nails are added to the cold water.

Although some may recognize that the final temperature will be closer to the initial temperature of the cold water than to the initial temperature of the nails, most students will probably give the average temperature.

predicted temperature of mixture = ________°C

Measure temperature of mixture.

Step 6: Lift the nails from the hot water and put them quickly into the cold water. When the temperature of the mixture stops rising, record it.

actual temperature of mixture = ______°C

4. How close is your prediction to the observed value?

The observed value should be closer to the initial temperature of the water than to the initial temperature of the nails.

Balance nails with hot water.

Step 7: Now you will repeat Steps 1 through 6 for hot water and cold nails. First, dry the bundle of nails with a paper towel. Then, balance a cup with the dry bundle of nails with a cup of *hot* water.

Step 8: Remove the nails and fill the cup 3/4 full with *cold* water. Record the temperature of the hot water in the first cup.

Record temperature.

temperature of hot water = ________°C

Step 9: Lower the bundle of nails into the cup of cold water, wait one minute (why?), and then record the temperature of the water around the nails.

Cool nails in cold water.

temperature of cold water = ________°C

Step 10: Predict what the temperature of the mixture will be when the cold nails are added to the hot water.

Predict temperature of mixture.

predicted temperature of mixture = ________°C

Step 11: Lift the nails from the cold water and put them quickly into the hot water. When the temperature of the mixture stops changing, record it.

Measure temperature of mixture.

actual temperature of mixture = ________°C

5. How close is your prediction to the observed value?

Again, the observed value should be closer to the initial temperature of the water than to the initial temperature of the nails.

Analysis

6. Discuss your observations with the rest of your team, and write an explanation for what happened.

In each case, the amount of heat given off by the warmer material equals the amount of heat absorbed by the cooler material. However, equal amounts of heat do not cause the same temperature change in equal masses of different materials. The nails change temperature more easily than the water. The water has a greater capacity for absorbing (or releasing) heat without changing its temperature as greatly.

7. Suppose you have equal masses of water and nails at the same temperature. Suppose you then light similar candles and place a candle under each of the masses, letting the candles burn for equal times. Would one of the materials change temperature more than the other?

(circle one) **(yes)** no

If your answer to the question is "yes," which one would reach a higher temperature?

(circle one) water **(nails)**

8. Suppose you have cold feet when you go to bed, and you want something to warm your feet. Would you prefer to have a hot-water bottle filled with hot water, or one filled with an equal mass of nails at the same temperature as the water? Explain.

It would be better to have the bottle filled with water because it would take longer to cool down due to its higher specific heat capacity. That is, water gives off more heat than an equal mass of nails for the same change in temperature.

9. Why does the climate of a mid-ocean island stay nearly constant, getting neither very hot nor very cold?

Water has a much larger specific heat capacity than most other substances, including soil and air. This means that the oceans warm the soil in the winter and cool it during summer. For a mid-ocean island, this same effect will occur during day and night as well as from one season to another.

Name Period Date

51 Antifreeze in the Summer?

Experiment

Purpose

To investigate what effect antifreeze (ethylene glycol) has on the cooling of a car radiator during the summer.

Setup: <1
Lab Time: >1
Learning Cycle: application
Conceptual Level: difficult
Mathematical Level: difficult

Adapted from PRISMS.

Required Equipment/Supplies

400 mL (400 g) water
timer
380 mL (400 g) 50% mixture of antifreeze and water
500-mL graduated cylinder
2 600-mL beakers
single-element electric immersion heater
thermometer (Celsius) or
  computer
  temperature probe with interface

Part A

Discussion

Put an iron frying pan on a hot stove, and very quickly it will be too hot to touch. But put a pan of water on the same hot stove, and the time for a comparable rise in temperature is considerably longer, even if the pan contains a relatively small amount of water. Water has a very high capacity for storing internal energy. We say that water has a high *specific heat capacity*. The specific heat capacity c of a substance is the quantity of heat required to increase the temperature of one gram of the substance by one degree Celsius. This means that the amount of heat Q needed to increase the temperature of a mass m of material by a temperature difference ΔT is

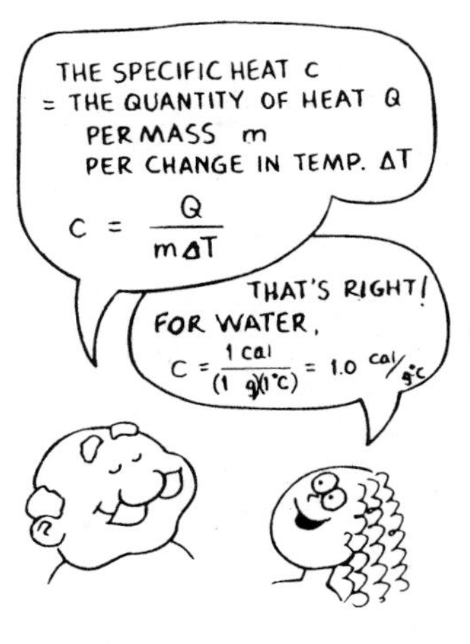
THE SPECIFIC HEAT c
= THE QUANTITY OF HEAT Q
PER MASS m
PER CHANGE IN TEMP. ΔT
c = Q / mΔT
THAT'S RIGHT!
FOR WATER,
c = 1 cal / (1 g)(1°C) = 1.0 cal/g°C

$$Q = mc\Delta T$$

Turning this formula around, the specific heat capacity is related to the quantity of heat absorbed, the mass of the material, and the temperature change by

$$c = \frac{Q}{m\Delta T}$$

Each substance has its own specific heat capacity, which is characteristic of that substance. The specific heat capacity of pure water is 1.0 cal/g°C. So, if 1 calorie (cal) of heat is absorbed by 1 g of water, its temperature will rise by 1°C; or if 1 calorie of heat is extracted from 1 gram of water, its temperature will fall by 1°C.

Water has a higher specific heat capacity than most other materials. The high specific heat capacity of water makes it an excellent coolant. That's why it is used in an automobile to keep the engine from overheating. But water has a striking disadvantage in winter. It freezes at 0°C, and, what's even worse, it expands when it freezes. This expansion can crack the automobile engine block or fracture the radiator. To prevent such a mishap, antifreeze (ethylene glycol) is mixed with the water. The mixture has a much lower freezing point than water. The mixture has its lowest freezing point when the proportions are about 50% water and 50% antifreeze. But what effect does adding the antifreeze have on the mixture's specific heat capacity? In Part A of this experiment, you will determine the specific heat capacity of a 50% mixture of antifreeze and water.

To find the specific heat capacity, you will simply measure the quantity of heat that is absorbed and the corresponding temperature change. To supply heat, you will use an electric immersion heater.

An immersion heater is a small coil of nichrome wire commonly used to heat small amounts of liquids. It heats up much like the wire in a toaster. This is a very efficient device, for unlike a hot plate or flame, it transfers all of its energy to the material being heated, in this case a liquid. The energy dissipated by a heater is the same regardless of whether the liquid is pure water or a mixture of water and antifreeze.

You can use a computer with a temperature probe instead of a thermometer. Hook up a temperature probe to an interface box. Be sure to calibrate the temperature probe.

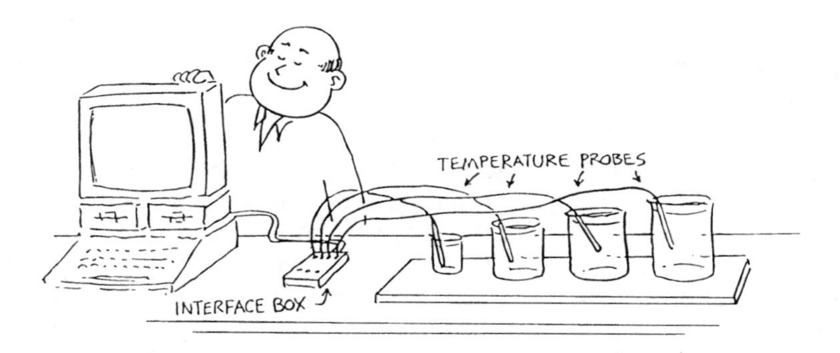

Procedure 

Heat water.

Step 1: Heat 400 mL (400 g) of water with the heater for 200 seconds (a little over 3 minutes).

Point out the cautionary note regarding the immersion heaters *before* students use them.

CAUTION: *Plug and unplug the immersion heater only when the heater element is submerged in liquid. Operating it in air destroys it immediately.*

Record the initial and final temperatures, and compute the change in temperature.

initial temperature = __________°C

final temperature = __________°C

change in temperature (ΔT) = ________°C

Step 2: Find the quantity of heat transferred to the water from the equation

Compare energy dissipated.

$$Q = mc\Delta T$$

where m = mass of water

c = specific heat capacity of water = 1 cal/g°C

ΔT = change in temperature

Q = __________ calories

Step 3: Pour 380 mL of a mixture of 50% antifreeze and 50% water into a beaker. Since antifreeze is slightly denser than water, 380 mL of 50% antifreeze mixture has a mass of 400 g. Heat the 400 g of antifreeze mixture for 200 seconds. Record the initial and final temperatures and the change in temperature.

Heat antifreeze mixture.

initial temperature = __________°C

final temperature = __________°C

change in temperature (ΔT) = ________°C

Step 4: With these data, compute the specific heat capacity of 50% antifreeze mixture from the equation

Compare specific heat capacities of antifreeze mixture.

$$c = \frac{Q}{m\Delta T}$$

specific heat c of antifreeze mixture = ________cal/g°C

A good value is 0.8 cal/g°C.

Analysis

1. Which liquid has the lower specific heat capacity—pure water or a 50% mixture of antifreeze?

   The 50% antifreeze mixture has lower specific heat capacity than pure water.

2. Which would warm from 25°C to 40°C faster with the same rate of energy input—pure water or a 50% mixture of antifreeze?

   The 50% antifreeze mixture heats up more quickly.

Part B

Discussion

By now you have found that the specific heat capacity of a mixture of antifreeze and water is lower than that of pure water. This suggests it is a poorer coolant than pure water. Yet it is advisable to continue using antifreeze in summer months. Is there an advantage, other than convenience, to leaving antifreeze in the radiator during summer, or should it be replaced with pure water?

To answer this question, you need to understand the role of the coolant in an automobile. Coolant draws heat from the engine block and then circulates it to the radiator, where it is dissipated to the atmosphere. The coolant is then recycled back to the engine.

The temperature of the coolant increases as it absorbs heat from the engine. But if it reaches its boiling point, the system boils over. So a coolant is effective only at temperatures below boiling. If the coolant is pure water at atmospheric pressure, this temperature is 100°C. (With a pressure cap on the radiator, both the pressure and the boiling point are higher.) Could an antifreeze mixture have a higher boiling point than pure water? If so, this would lessen the likelihood of boil-overs when the engine is overworked. You will experiment to find out.

Heat antifreeze mixture to its boiling point.

Step 5: Heat a sample of 50% antifreeze mixture with the immersion heater, and measure its boiling point. Record this temperature.

boiling point of antifreeze mixture = ________ °C

Analysis

3. What effect does the boiling point of the antifreeze mixture have on the mixture's ability to act as a coolant?

The higher boiling point allows the liquid to reach a higher temperature before it boils in a car radiator.

4. Would it be appropriate to call ethylene glycol "antiboil" instead of "antifreeze" in climates where the temperature never goes below freezing?

Yes, a mixture of ethylene glycol and water has a higher boiling point than pure water, so the mixture will reach a higher temperature before it boils.

Name Period Date

52 Gulf Stream in a Flask

Purpose

To observe liquid movement due to temperature differences.

Setup: 1
Lab Time: 1
Learning Cycle: exploratory
Conceptual Level: easy
Mathematical Level: none

Adapted from PRISMS.

Required Equipment/Supplies

safety goggles
paper towels
250-mL flask
food coloring
stirring rod
hot plate
thermometer (Celsius)
2-hole rubber stopper (with glass tubing as shown in Figure A)
battery jar or a large pickle jar
water

Activity

Discussion

The air near the floor of a room is cooler than the air next to the ceiling. The water at the bottom of a swimming pool is cooler than the water at the surface. You can feel these different temperatures that result from fluid motion, but you cannot see the fluid motion in these cases. In this activity, you will be able to see the movements of two liquids with different temperatures.

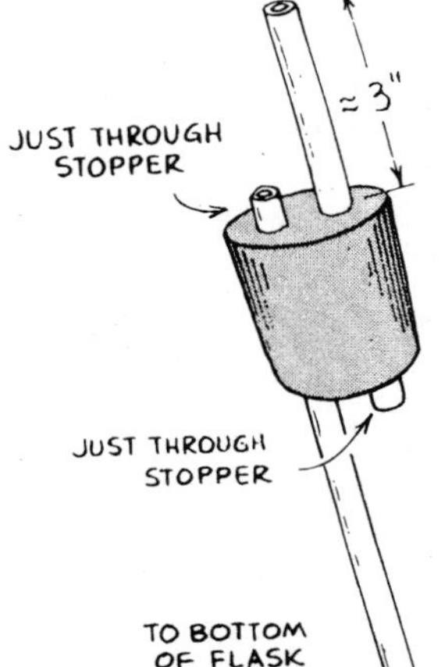

Fig. A

Procedure     

Step 1: Read Steps 2 through 5 completely before conducting the activity, and predict what you expect to observe.

____

____

____

____

Step 2: Put on safety goggles. Mix some food coloring into a flask filled completely with water. Stir well. Heat the colored water to a temperature of 70°C.

Heat colored water.

Insert stopper.

Step 3: Make sure that the glass tubing in the stopper arrangement is as shown in Figure A. With a damp paper towel, firmly grasp the neck of the flask and insert the stopper into the flask.

CAUTION: *The flask will be very hot. Be careful not to burn yourself.*

Immerse flask.

Step 4: As shown in Figure B, immerse the hot flask carefully into a large jar filled with plain water at room temperature. The water level should be above both pieces of tubing.

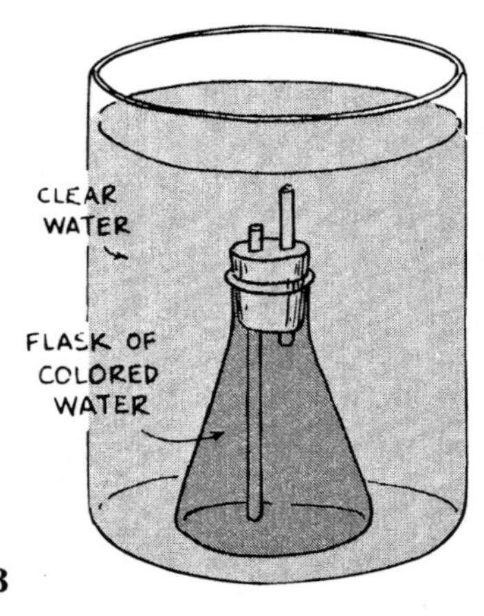

Fig. B

Record observations.

Step 5: Record your observations over the next 10 minutes.

Hot, colored water rises out of the higher, shorter tube in the stopper, spreading into the surrounding water. Clear, cooler water enters the flask through the lower, longer tube.

1. What caused the movements of liquid you observed?

Denser, cooler water pushed warmer, less dense water toward the top of the container.

53 The Bridge Connection

Purpose

To calculate the minimum length of the expansion joints for the Golden Gate Bridge.

Setup: 1
Lab Time: >1
Learning Cycle: application
Conceptual Level: difficult
Mathematical Level: difficult

Required Equipment/Supplies

safety goggles
hollow steel rod
thermal expansion apparatus (roller form)
micrometer
steam generator
bunsen burner
rubber tubing

Discussion

Gases of the same pressure and volume expand equally when heated, and contract equally when cooled—regardless of the kinds of molecules that compose the gas. That's because the molecules of a gas are so far apart that their size and nature have practically no effect on the amount of expansion or contraction. However, liquids and solids are individualists! Molecules and atoms in the solid state are close together in a variety of definite crystalline structures. Different liquids and solids expand or contract at different rates when their temperature is changed.

Expansions or contractions of metal can be critical in the construction of bridges and buildings. Temperatures at the Golden Gate Bridge in San Francisco can vary from –5°C in winter to 40°C in the summer. On this very long bridge, the change in length from winter to summer can be almost 2 meters! Clearly, engineers must keep thermal expansion in mind when designing bridges and other large structures.

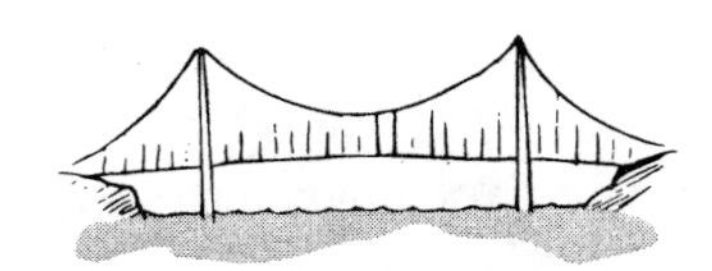

Suppose you owned an engineering firm *before* the Golden Gate Bridge was built and you were asked for advice on the minimum length to make expansion joints for the proposed bridge. (See a typical expansion joint in Figure 21.9 in your textbook.) Consider two expansion joints that connect the bridge to land at each end. The steel to be used is the same as that in the steel rods you will use in this experiment. You are to measure how much the steel expands per meter for each degree increase in temperature. This ratio is called the *coefficient of linear expansion.* Then, compute the amount that the 2740-m structure will expand as its temperature increases. Now you can advise a minimum size of the expansion joints needed, to assure the success of the bridge.

You will determine the coefficient of linear expansion for steel by measuring the expansion of a steel rod when it is heated with steam at 100°C. The amount of expansion ΔL depends on the original length L, the change in temperature ΔT, and the coefficient of linear expansion α (which is characteristic of the type of material).

$$\Delta L = \alpha L \Delta T$$

The temperature change ΔT is simply 100°C minus room temperature. By measuring L, the original length of the rod before heating, and ΔL, the change of the length after heating, you can compute α, the coefficient of linear expansion, by rearranging the equation.

$$\alpha = \frac{\Delta L}{L \Delta T}$$

It is difficult to measure how much the rod expands because the expansion is small. One way to measure it is shown in Figure A. The end of the steel rod is arranged to be perpendicular to the axle of the pointer. When the steel rod resting on the axle expands, it rotates the axle and the pointer through an angle.

The pointer turns 360° (one full rotation) when the axle has rotated a distance equal to its circumference. The circumference C of the axle equals its diameter d times the number *pi.*

$$C = \pi d$$

If the expansion rotates the pointer through 180° (a half rotation), then the increase ΔL in length of the rod is equal to half the circumference of the axle. Ratio and proportion give you ΔL for other expansions. The ratio of the actual distance ΔL rotated to the distance πd rotated in one complete rotation equals the ratio of the angle θ through which the pointer rotates through to 360°.

$$\frac{\Delta L}{\pi d} = \frac{\theta}{360°}$$

Solve this equation for ΔL. Substituting this value into the second equation enables you to compute the coefficient of linear expansion. Once you have computed the value of α, you can compute the minimum length of the bridge and the estimated temperature difference for summer and winter.

Procedure    

Measure axle diameter.

Some students may need instruction on how to use a micrometer.

Step 1: Using a micrometer, measure the diameter of the axle of the pointer.

$d =$ ______ mm

Assemble apparatus.

Step 2: Assemble your apparatus as in Figure A, with the steam generator ready to go. Make sure the rod has a place to drain. Measure the distance in millimeters from the groove in the steel rod to the axle of the pointer (since you are sampling expansion between these two points). Do *not* measure the length of the entire rod.

$L =$ ______ mm

Name Period Date

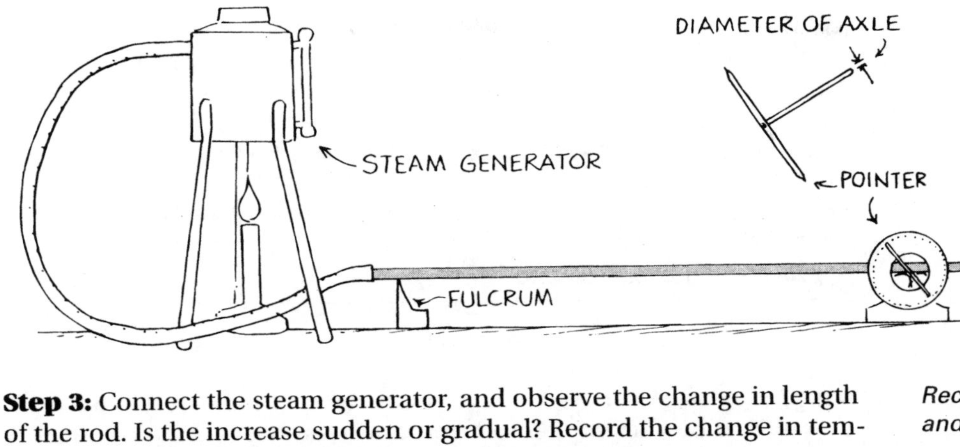

Fig. A

Step 3: Connect the steam generator, and observe the change in length of the rod. Is the increase sudden or gradual? Record the change in temperature of your rod, assuming it started at room temperature.

Record temperature change and angle pointer turns.

$\Delta T =$ ________ °C

Record the angle θ through which the pointer turned due to the expansion of the rod.

angle $\theta =$ ______ °

Step 4: Compute the increase ΔL in length of the rod by substituting your measured values of θ and d into the following equation.

Compute increase in length of rod.

$$\Delta L = \frac{\pi \theta d}{360°} = \text{\_\_\_\_\_\_\_\_ mm}$$

Step 5: Compute the coefficient of linear expansion using your measured values of L, ΔT, and ΔL.

Compute coefficient of linear expansion.

$$\alpha = \frac{\Delta L}{L \Delta T} = \text{\_\_\_\_\_\_\_\_}$$

Step 6: Express the coefficient of linear expansion in scientific notation (that is, as a number times a power of ten).

$\alpha =$ ______

Step 7: Compute the minimum length of the expansion joint(s). Show your work.

Compute length of expansion joint.

$\Delta L =$ ______

Analysis

1. Why do you measure the diameter of the axle of the pointer in Step 1 instead of the diameter of the steel rod?

   It is the axle that rotates to move the pointer though an angle of θ.

2. Why do you measure the distance L from the groove in the steel rod to the point of contact with the axle instead of the entire length of the rod?

The rod is held fixed at the groove and expands on both sides of it. We are measuring only how much the portion of the rod between the groove and the axle expands.

3. Suppose you placed the pointer device in the middle of the rod. Would the coefficient of linear expansion for the steel rod be larger, less, or the same? Why?

The value would be the same—for instance, half the change in *L* for half the *L*.

4. What are the units of α?

The units of α are inverse temperature, or "per °C."

5. What are the sources of error in your experiment? List them along with an estimate of the percent contribution of each.

The sources of error include inaccurate measurements of

(a) *L* (error of 1 mm in 600 mm, or under 1%)

(b) the change in *L* (error of 1 degree in 20 degrees, or 5%)

(c) ΔT (error of 1°C in 80°C, or about 1%)

(d) the diameter *d* of the axle (error of about 2%)

6. In your consultation, suppose you experimentally determined that the expansion joints had to be a minimum of 2 m long. How large would you recommend them to be, taking your margin of error into account for safety?

Usually a safety factor of 2 or more is advised, but cutting it close would be plus or minus 10% so the expansion joints should be at least 2.2 m long.

7. What do you think would be a good design for an expansion joint?

Answers will vary. (The design in the text is two wide-teethed combs that interlock.)

54 Cooling Off

Purpose

To compare the rates of cooling of objects of different colors and surface reflectances.

Setup: 1
Lab Time: >1
Learning Cycle: application
Conceptual Level: difficult
Mathematical Level: difficult

Thanks to John Layman for his ideas and suggestions for this lab.

Required Equipment/Supplies

4 empty soup cans of the same size (one covered with aluminum foil, one painted black, one white, and one any other color)
100-watt lightbulb and receptacle
4 thermometers (Celsius) or
  computer
  temperature probes with interface
  printer
large test tube
one-hole rubber stopper, with thermometer or temperature probe mounted in it, to fit test tube
600-mL beaker
graduated cylinder
Styrofoam (plastic foam) cups with covers
hot water
crushed ice and water mixture
variety of different size, color, and shape containers made of various materials
meterstick
graph paper
computer and data plotting software (optional)

Activity

Discussion

When you are confronted with a plate of food too hot to eat, you may have noticed that some things cool faster than others. Mashed potatoes can be comfortably eaten when boiled onions are still too hot. And blueberries in a blueberry muffin are still hot when the rest of the muffin has cooled enough to eat. Different materials cool at different rates. Explore and see.

Procedure   

Step 1: Place the four cans 20 cm from an unshielded lightbulb. Place a thermometer in each can, and turn on the light. Record the temperature of each can after 1 minute, 2 minutes, and 3 minutes in Data Table A.

CAN	TEMPERATURE (°C)		
	1 MIN	2 MIN	3 MIN
1			
2			
3			
4			

Data Table A

As an alternative, if a computer is available, use a temperature probe to read your data. Calibrate your probes, and monitor the temperature rise of each can. Save your data and make a printout.

1. For which can was the temperature rise the fastest?

   The black can should heat up the fastest.

2. For which can was the temperature rise the slowest?

   The slowest can to heat up will probably be the foil-covered can.

3. After several minutes the temperatures of all the cans level off and remain constant. Does this mean that the cans stop absorbing radiant energy from the lightbulb?

   When the temperature levels off, each can is absorbing and emitting energy, with no *net* absorption of energy.

4. For a can at constant temperature, what is the relationship between the amount of radiant energy being absorbed by the can and the amount of radiant energy being radiated and convected away from the can?

   The amount of radiant energy absorbed equals the amount of energy radiated and convected away.

Monitor cooling of water in test tube.

Step 2: Place 25 mL of hot water from the tap in a test tube. Insert the thermometer mounted in the rubber stopper. Place the test tube in a beaker filled with crushed ice and water. Record the temperatures over the next 5 minutes every 30 seconds in Data Table B. (Once again, if you have a computer, use a temperature probe instead of a thermometer.) Draw a graph of temperature (vertical axis) vs. time (horizontal axis).

TIME (MIN)	0	0.5	1.0	1.5	2.0	2.5	3.0	3.5	4.0	4.5	5.0
TEMP. (°C)											

Data Table B

5. Describe your graph.

   Temperature drops exponentially with time. The curve drops most quickly at first and then levels off.

Monitor cooling of water in Styrofoam cup.

Step 3: Fill a Styrofoam (plastic foam) cup with hot tap water. Monitor the temperature drop with either a thermometer or temperature probe. Make a graph of temperature (vertical axis) vs. time (horizontal axis).

Step 4: Repeat, but cover the cup with a plastic lid. Poke a hole (if it doesn't already have one), and insert the probe in the water to monitor the temperature.

6. How do the two cooling curves compare?

The cup with the lid cools at a much slower rate.

7. What cooling process is primarily responsible for the difference?

Convection is primarily responsible for the difference. It is impeded by the lid.

Going Further

Step 5: Place 200 mL of hot tap water in each of a variety of containers. You may want to wrap them with various materials. List the containers used.

Predict cooling rates of a variety of containers

Predict which container will cool fastest and which container will cool most slowly.

Predicted fastest: ___________

Predicted slowest: ___________

The container that cools the fastest is likely to have the greatest surface area of water exposed to the air. Convection usually has the greatest effect, while color and type of insulating material have lesser effect. A container with a lid is likely to cool most slowly.

Step 6: Monitor the cooling with either temperature probes or thermometers. Record which cooled the fastest and which the most slowly.

Observed fastest: ___________

Observed slowest: ___________

8. What factors determine the rate of cooling, based on your data from Step 6?

Factors that determine the rate of cooling include: (1) exposed surface area of water; (2) dark vs. light color; (3) shiny vs. dull surface; (4) container covered or not; (5) type of insulation material; (6) wall thickness of container.

Extra for Experts

Plot temperature vs. time.

Step 7: Using data plotting software, plot data from Step 6 for each of your samples. Let time be the independent variable (horizontal axis), and, for convenience, set $t = 0$ at the origin. Let $T - T_2$ be the dependent variable (vertical axis), where T is the temperature of the sample and T_2 is room temperature. Can you think of a way to make your graph a straight line? *Hint:* Try using some function of the vertical variable (temperature difference), such as its square or its square root or its logarithm.

9. What does your graph look like? What mathematical relationship exists between the cooling rate and time? (The answer to this question requires some heavy-duty math—ask your teacher for help on this one!)

The graph of log $(T - T_2)$ vs. time is a straight line sloping downward. The temperature of a sample cooling from an initial temperature T_1 to the final temperature T_2 of its surroundings is given by $T = T_2 + (T_1 - T_2)e^{-kt}$, where k is a constant. In other words, the temperature difference $T - T_2$ declines exponentially from its initial value of $T_1 - T_2$ to its final value of 0. The logarithm of $T - T_2$, when plotted against time, is then a straight line with negative slope. (This exponential cooling follows mathematically from Newton's law of cooling, which states that the rate of temperature change dT/dt is proportional to the temperature difference $T - T_2$ between the sample and its surroundings.)

Name Period Date

55 Solar Equality

Experiment

Purpose

To measure the sun's power output and compare it with the power output of a 100-watt lightbulb.

Setup: 1
Lab Time: >1
Learning Cycle: application
Conceptual Level: difficult
Mathematical Level: difficult

Required Equipment/Supplies

2-cm by 6-cm piece of aluminum foil with one side thinly coated with flat black paint
clear tape
meterstick
two wood blocks
glass jar with a hole in the metal lid
one-hole stopper to fit the hole in the jar lid
thermometer (Celsius, range –10°C to 110°C)
glycerin
100-watt lightbulb with receptacle

Discussion

If you told some friends that you had measured the power output of a lightbulb, they would not be too excited. However, if you told them that you had computed the power output of the whole sun using only household equipment, they would probably be quite impressed (or not believe you). In this experiment, you will estimate the power output of the sun. You will need a sunny day to do this experiment.

Wind or convective cooling can affect your results significantly.

Procedure  

Step 1: With the blackened side facing out, fold the middle of the foil strip around the thermometer bulb, as shown in Figure A. The ends of the metal strip should line up evenly.

Assemble apparatus.

METAL ENDS EVEN

Fig. A

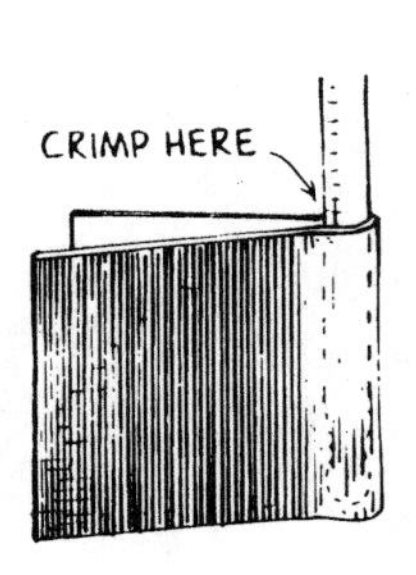

Fig. B

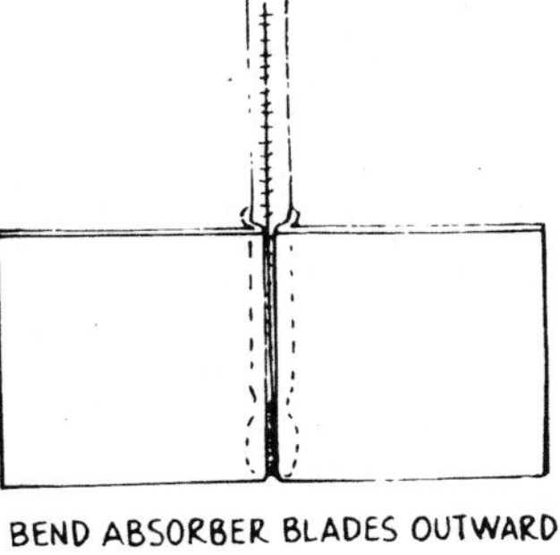

Fig. C

Step 2: Crimp the foil so that it completely surrounds the bulb, as in Figure B. Bend each end of the foil strip outward (Figure C). Use a meterstick to make a flat, even surface. Use a piece of clear tape on the unpainted side to hold the foil on the thermometer.

Step 3: Use glycerin to insert the free end of the thermometer into the one-hole stopper. Remove the lid from the jar, and place the stopper in the lid from the bottom side. Slide the thermometer until the foil strip is located in the middle of the jar. Place the lid on the jar.

Place apparatus outdoors in sun.

Step 4: Take your apparatus outdoors. Prop it at an angle so that the blackened side of the foil strip is exactly perpendicular to the rays of the sun.

Record maximum temperature reached.

Step 5: Leave the apparatus in this position in the sunlight until the maximum temperature is reached. Record this temperature.

maximum temperature = __________

Allow apparatus to cool to room temperature.

Step 6: Return the apparatus to the classroom, and allow the thermometer to cool to room temperature.

Set up apparatus using lightbulb.

Step 7: Set the meterstick on the table. Place the lightbulb with its filament located at the 0-cm mark of the meterstick (Figure D). Center the jar apparatus at the 95-cm mark with the blackened side of the foil strip exactly perpendicular to the light rays from the bulb. You may need to put some books under the jar apparatus.

Match outdoor temperature with lightbulb.

Step 8: Turn the lightbulb on. Slowly move the apparatus toward the lightbulb, 5 cm at a time, allowing the thermometer temperature to stabilize each time. As the temperature approaches that reached in Step 5, move the apparatus only 1 cm at a time. Adjust the distance of the apparatus 0.1 cm at a time until the temperature obtained in Step 5 is maintained for two minutes. Turn the lightbulb off.

Fig. D

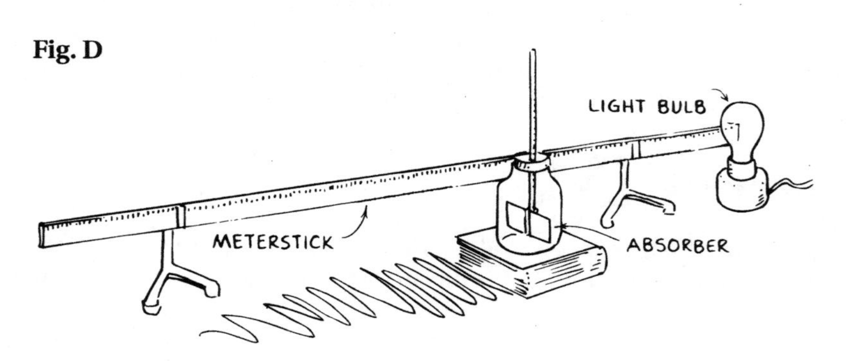

Measure distance to lightbulb.

Step 9: Measure as exactly as possible the distance in meters between the absorber strip of the apparatus and the filament of the lightbulb. Record this value.

distance from light filament to foil strip = ________ m

Name Period Date

Step 10: For the distance obtained, use the following equation to compute the wattage of the sun. The sun's distance in meters is 1.5×10^{11} m. Show your work.

Compute wattage of the sun.

$$\text{sun's wattage} = (\text{bulb's wattage}) \times \frac{(\text{sun's distance})^2}{(\text{bulb's distance})^2}$$

sun's wattage = ____________ W

Step 11: Use the value of the sun's wattage to compute the number of 100-watt lightbulbs needed to equal the sun's power. Show your work.

Compute number of lightbulbs.

The number of 100-watt bulbs equals the sun's wattage divided by 100 W.

number of 100-watt lightbulbs = __________

Analysis

1. Re-express the equation in Step 10 as a relationship between two ratios. Express the new equation as a sentence that begins, "The ratio of . . ."

$$\frac{\textbf{sun's wattage}}{(\textbf{sun's distance})^2} = \frac{\textbf{bulb's wattage}}{(\textbf{bulb's distance})^2}$$

The ratio of the sun's wattage to the square of the sun's distance equals the ratio of the bulb's wattage to the square of the bulb's distance. (Accept equivalent equations, especially one stating that the ratio of the wattages is the ratio of the squares of the distances.)

2. Would it be possible to turn on the number of 100-watt lightbulbs you computed in Step 11 at once? Explain.

No—even if all the operating electrical generators in the world were diverted toward just lighting bulbs, they would still produce only a tiny fraction of the energy radiated by the sun.

3. The accepted value for the sun's wattage is 3.83×10^{26} W. List at least three factors that might account for the difference between your experimental value and the accepted value.

Possible errors include: (a) Bulb's wattage is not exactly 100 watts. (b) Distance to bulb is not measured accurately. (c) Not all of the sun's energy gets through—atmospheric absorption. (d) Distance to the sun is not always 1.5×10^{11} m; the sun is closer in January than in July. (e) Did not get radiation perpendicular to black vanes in either measurement. (f) Energy radiating from the lightbulb is not the same in all directions; it depends on the orientation of the filament.

56 Solar Energy

Purpose

To find the daily amount of solar energy reaching the earth's surface and relate it to the daily amount of solar energy falling on an average house.

Setup: 1
Lab Time: >1
Learning Cycle: application
Conceptual Level: moderate
Mathematical Level: moderate

Adapted from PRISMS.

Required Equipment/Supplies

2 Styrofoam (plastic foam) cups
graduated cylinder
water
blue and green food coloring
plastic wrap
rubber band
thermometer (Celsius)
meterstick

Discussion

Activity

How do we know how much total energy the sun emits? First, we assume that it emits energy equally in all directions. Imagine a heat detector so big that it completely surrounds the sun—like an enormous basketball with the sun at its center. Then, the amount of heat reaching the detector would be the same as the total solar output. Or if our detector were half a basketball and caught half the sun's energy, then we would multiply the detector reading by 2 to compute the total solar output. If our detector encompassed a quarter of the sun and caught one-fourth its energy, then the sun's total output would be four times the detector reading.

Now that you have the concept, suppose that our detector is the water surface area of a full Styrofoam cup here on earth facing the sun. Then it comprises only a tiny fraction of the area that surrounds the sun at this distance. If you figure what that fraction is and also measure the amount of energy captured by your cup, you can tell how much total energy the sun emits. That's how it's done! In this activity, however, you will measure the amount of solar energy that reaches a Styrofoam cup and relate it to the amount of solar energy that falls on a housetop. You will need a sunny day to do this activity.

Procedure  

Step 1: Fill a Styrofoam cup, adding small equal amounts of blue and green food coloring to the water until it is dark (and a better absorber of solar energy). Then measure and record the amount of "water" in your cup.

Add food coloring.

volume of water = ________

mass of water = __________

Nest the cup in a second Styrofoam cup (for better insulation).

Measure temperature.

Step 2: Measure the water temperature and record it. Cover the cup with plastic wrap sealed with a rubber band.

initial water temperature = ________

Step 3: Put the cup in the sunlight for 10 minutes.

Step 4: Remove the plastic wrap. Stir the water in the cup gently with the thermometer, and record the final water temperature.

final water temperature = ________

Step 5: Find the difference in the temperature of the water before and after it was set in the sun.

temperature difference = ________

Measure surface area of cup.

Step 6: Measure and record the diameter in centimeters of the top of the cup. Compute the surface area of the top of the cup in square centimeters.

diameter = ___________ cm

surface area of water = ______ cm^2

Step 7: Compute the energy in calories that was collected in the cup. You may assume that the specific heat of the mixture is the same as the specific heat of water. Show your work.

THE QUANTITY Q OF HEAT ENERGY COLLECTED BY THE WATER = MASS OF THE WATER × ITS SPECIFIC HEAT (c = 1 cal/g·°C) × ΔT, ITS CHANGE IN TEMPERATURE

energy = ___________ cal

Compute solar energy flux.

Step 8: Compute the solar energy flux, the energy collected per square centimeter per minute. Show your work.

A typical value obtained for the solar energy flux is 1.1 cal/cm^2•min.

solar energy flux = _________ cal/cm^2•min

Name Period Date

Step 9: Compute how much solar energy reaches each square meter of the earth per minute at your present time and location. Show your work. (*Hint:* There are 10 000 cm^2 in 1 m^2.)

Since there are 10 000 cm^2 in 1 m^2, the solar energy flux obtained in Step 8 should be multiplied by 10 000 cm^2/m^2. A typical value would be 11 000 $cal/m^2 \cdot min$.

solar energy flux = __________ $cal/cm^2 \cdot min$

Step 10: Use your data to compute the rate at which solar energy falls on a flat 6-m by 12-m roof located in your area at the time when you made your measurement. Obtain the answer first in calories per second, then in watts. Show your work.

Compute solar energy received by roof.

The area of a 6-m by 12-m roof is 72 m^2. This is then multiplied by the number of calories per square meter per minute obtained in Step 9. That quantity is then divided by 60 s/min. A typical value is 1.1×10^4 cal/s. Finally, this is multiplied by 4.18 J/cal to give the power in J/s, or watts (W). A typical value is 45 000 W, or 45 kW, which is larger than the power likely to be consumed within the house.

solar power received by roof = __________ cal/s

solar power received by roof = __________ W

1. How does this solar power compare with typical power consumption within the house?

____

____

Analysis

Scientists have measured the amount of solar energy flux outside our atmosphere to be 2 calories per square centimeter per minute on an area perpendicular to the direction of the sun's rays. This energy flux is called the *solar constant.* Only 1.5 calories per square centimeter per minute reaches the earth's surface after passing through the atmosphere. What factors could affect the amount of sunlight reaching the earth's surface and decrease the flux of solar energy that you measure?

The factors that might affect the amount of solar energy reaching a location on the earth's surface are: (a) cloud cover; (b) humidity; (c) air pollution; (d) time of day; (e) season of year; (f) latitude; (g) nearby obstructions. If the surface area being illuminated is not perpendicular to the direction of the radiation, the measured energy per unit area will also be reduced by that factor.

57 Boiling Is a Cooling Process

Purpose

To observe water changing its phase as it boils and then cools.

Setup: <1
Lab Time: 1
Learning Cycle: exploratory
Conceptual Level: easy
Mathematical Level: none

Required Equipment/Supplies

safety goggles
paper towels
ring stand
wire gauze
2 plastic-coated test-tube clamps
1000-mL Florence flask (flat-bottomed) or Franklin's flask (crater-bottomed)
one-hole rubber stopper to fit flask, fitted with short (15-cm) thermometer or temperature probe
water
crushed ice or beaker of cool tap water
Bunsen burner
thermometer (Celsius), regular length (30 cm), or
  computer
  temperature probe with interface
  printer
pan or tub
computer with data plotting software (optional)
graph paper (if computer is not used)

Activity

Discussion

When water evaporates, the more energetic molecules leave the liquid. This results in a lowering of the average energy of the molecules left behind. The liquid is cooled by the process of evaporation. Is this also true of boiling? Try it and see.

Procedure    

Step 1: Attach an empty flask to the ring stand with a test-tube clamp. Insert the teacher-prepared stopper with the thermometer or temperature probe into the neck of the flask. Loosen the wing nut on the clamp so that when the flask is rotated upside down, the end of the thermometer clears the table top by about 3 cm.

Adjust height of flask.

Step 2: Before filling the flask with water, you need to practice the procedure you will be performing later. Remove the stopper from the flask. Attach a ring and gauze below the flask, and place an unlit Bunsen

Prepare setup for dry run.

burner below the ring. Attach a second thermometer or probe to the ring stand, and suspend it inside the flask.

Practice inverting flask.

Step 3: Imagine that you have been boiling water in the flask, and have just turned off the burner. Move the burner aside. Remove the second thermometer or probe and its clamp from the stand. Loosen the screw on the ring, and lower it and the gauze to the tabletop. Firmly grasp the clamp and neck of the "hot" flask with a damp paper towel. Insert the stopper into the neck. Loosen the wing nut on the clamp. Holding the flask firmly with a damp paper towel, rotate it in the clamp until it is upside down. Tighten the wing nut to keep it in this position.

Practice this procedure a few times so that you will be comfortable manipulating the apparatus in this way when the flask is filled with boiling-hot water.

Place the stopper aside for use in Step 5.

Monitor water temperature.

Step 4: Put on safety goggles. Fill the flask half full of water, and attach it to the ring stand. The ring and gauze should be below it for safety.

If you are using thermometers, attach a thermometer to the ring stand so that its bulb is in the water but not touching the flask. Make a data table so that you can record the temperature every 30 seconds until 3 minutes after the water begins boiling vigorously. Heat the water, and record your data.

Alternatively, if you are using a computer, attach a temperature probe to the ring stand in place of the thermometer. Calibrate your probe before heating the water. If available, use data plotting software to make a graph of temperature vs. time. Print out your graph.

Insert rubber stopper.

Step 5: When you stop heating the flask, remove the second thermometer. Drop the ring. Firmly grasp the clamp and neck with a damp paper towel. Insert the stopper with the thermometer or probe into the flask. Atmospheric pressure will make it fit snugly. Loosen the wing nut on the clamp. *Carefully* hold the flask, using a damp paper towel, and invert it by rotating it in the clamp. Tighten the wing nut on the clamp to keep the flask in the inverted position, as shown in Figure A.

Fig. A

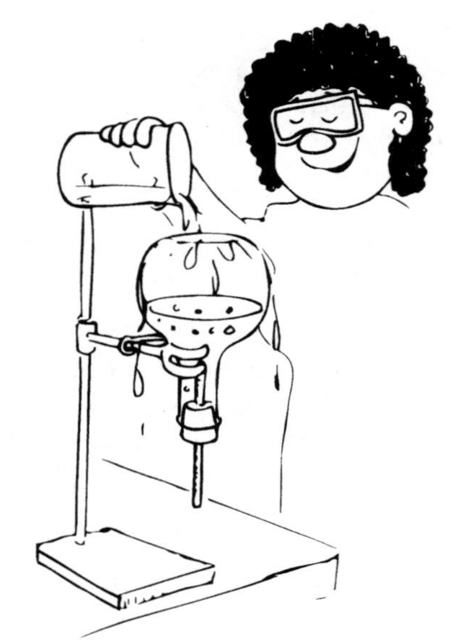

Step 6: Place a pan or tub under the flask. Pour cool tap water over the top half of the flask, or place crushed ice on it. Repeat the pouring several times. If you are using a computer, monitor the temperature of the water in the flask.

Cool flask.

1. What do you observe happening inside the flask?

   The water in the inverted flask boils when the flask is cooled with water or ice.

2. What happens to the temperature of the water in the flask?

   The temperature of the water drops dramatically when it boils.

Step 7: Prepare a graph of temperature (vertical axis) vs. time (horizontal axis), either plotting data from your table or making a printout from the computer.

Make graph.

Analysis

3. Does the temperature of boiling water increase when heat continues to be applied?

   The water remains at constant temperature (100°C at standard atmospheric pressure).

4. Explain your observations of the temperature of the water as it continues to boil.

   The water temperature does not increase because the heat energy absorbed from the flame is exactly equal to the energy that leaves the water with the newly formed vapor (steam).

5. How does the amount of heat energy absorbed by a pot of water boiling on a stove compare with the amount of energy removed from the water by boiling?

   They are the same.

6. Explain your observations of the inside of the inverted flask after cool water or ice was put on it.

The water boils because the air pressure in the flask is reduced when the air is suddenly cooled. Water boils at a temperature lower than 100°C when the air pressure is reduced (this can be observed at high altitude). The boiling process removes energy from the remaining water, so its temperature drops. (Unlike the first part of this activity, no heat is being supplied by an outside source as the water boils, so there *is* a temperature drop during boiling.) Boiling is a cooling process!

Name Period Date

58 Melting Away

Purpose

To measure the heat of fusion of water.

Setup: 1
Lab Time: >1
Learning Cycle: concept development
Conceptual Level: difficult
Mathematical Level: moderate

Thanks to Lisa Kappler for her ideas and suggestions for this lab.

Required Equipment/Supplies

250-mL graduated cylinder
8-oz. Styrofoam (plastic foam) cup
water
ice cube (about 25 g)
paper towel
thermometer (Celsius) or
  computer
  temperature probe with interface
  computer and data plotting software (optional)
  printer
graph paper (if computer is not used)

Discussion

If you put heat into an object, will its temperature increase? Don't automatically say yes, for there are exceptions. If you put heat into water at 100°C, its temperature will not increase until all the water has become steam. The energy per gram that goes into changing the phase from liquid to gas is called the *heat of vaporization.* And when you put heat into melting ice, its temperature will not increase until all the ice has melted. The energy per gram that goes into changing the state from solid to liquid is called the *heat of fusion.* That's what this experiment is about.

Procedure 

Step 1: Use a graduated cylinder to measure 200 mL of water, and pour it into a Styrofoam cup. The water should be about 5°C warmer than room temperature. If you are using a thermometer, make a data table for temperature and time. With the thermometer or temperature probe, measure and record the temperature of the water every 10 seconds for 3 minutes.

Monitor temperature of water.

Step 2: Dry an ice cube by patting it with a paper towel. Add it to the water. Continue monitoring the temperature of the water every 10 seconds while gently stirring it until 3 minutes *after* the ice cube has melted. Record the data in a table or with the computer. Note the time at which the ice cube has *just* melted.

Determine the final volume of water.

Step 3: Determine the final volume of the water.

final volume = __________

1. What was the mass of the water originally? What was the mass of the ice cube originally? Explain how you determined these masses.

The mass of the water was 200 g, since each milliliter of water has a mass of one gram. To find the mass of the ice cube, start with the mass of the final volume of water and melted ice and subtract 200 g.

Make graph.

Step 4: Plot your data to make a temperature (vertical axis) vs. time (horizontal axis) graph. If you are using a computer with data plotting software, print a copy for your analysis and report.

Study your graph. Draw vertical dashed lines to break your graph into three distinct regions. Region I covers the time before the ice cube was placed in the water. Region II covers the time while the cube was melting. Region III covers the time after the cube has melted.

2. What was the total temperature change of the water while the cube was melting (Region II)?

The total temperature change while the ice was melting equals the vertical drop on the graph from the beginning to the end of Region II.

3. How did placing the ice cube in the water affect the rate at which the water was cooling?

In Region II the temperature should drop at a faster rate than in Region I, as evidenced by the steeper downward slope of the graph.

Compute the energy lost by water.

Step 5: Compute the total amount of heat energy Q lost by the water as the ice cube was melting. Use the relation $Q = mc\Delta T$, where m is the initial mass of the water, c is the specific heat capacity of water (1.00 cal/g•°C), and ΔT is the magnitude of the water's temperature change from the beginning of Region II to the end of Region III where the temperature is stabilized.

heat energy lost by original water = ______ cal

Compute heat energy absorbed during warming.

Step 6: Compute the amount of heat energy absorbed as the water from the melted ice warmed from 0°C to its final temperature.

heat energy absorbed by melted ice during warming = ____ cal

Name Period Date

Step 7: From the difference between the values found in Steps 5 and 6, compute the amount of heat energy that was absorbed by the ice as it melted.

Compute heat energy absorbed during melting.

heat energy absorbed
by ice during melting = ______ cal

Step 8: Compute the heat by fusion by dividing the value found in Step 7 by the mass of ice that melted in the cup.

Compute heat of fusion.

heat of fusion = __________ cal/g

4. Compare this value to the standard value, 80 cal/g, and calculate the percentage difference.

The computed value should be within a few percent of the standard value. One source of error is using ice colder than 0°C. Then part of the energy goes into warming the ice up to melting temperature. To minimize or prevent this, use ice that is quite wet on the outside—that is in the process of melting when your experiment begins. Ice taken directly from a freezer is most often significantly cooler than 0°C and will affect the accuracy of this experiment. Let ice taken from a freezer sit at room temperature for several minutes before use.

Analysis

5. In order for the ice cube to melt, it has to extract heat energy from the warmer water, first a small amount of heat to warm up to 0°C (which we neglect in this experiment), then a larger amount to melt. The melting ice absorbs heat energy, and thus cools the water. The heat energy absorbed per gram when a substance changes from a solid to a liquid is called the heat of fusion. How is the amount of heat energy absorbed by the solid related to the heat of fusion and the mass of the solid?

The amount of heat energy absorbed in melting equals the mass of the solid times the heat of fusion.

6. The total amount of heat energy lost by the water is equal to the amounts of heat energy it took to do what things?

The total amount of heat energy lost by the water equals the amounts of heat energy to (1) melt the ice, and (2) warm the melted ice water from 0°C to the final temperature of the water.

7. What are some sources of error in this experiment?

Sources of error include: (a) film of water on ice cube; (b) temperature of ice cube below 0°C if taken directly from freezer; (c) inaccurate thermometer; (d) inaccurate measurement of initial volume of water and final volume of water plus melted ice cube; (e) heat lost to surroundings due to evaporation and radiation.

Name Period Date

59 Getting Steamed Up

Purpose

To determine the heat of vaporization of water.

Setup: 1
Lab Time: >1
Learning Cycle: concept development
Conceptual Level: moderate
Mathematical Level: moderate

Required Equipment/Supplies

safety goggles
Bunsen burner
steam generator with rubber tubing and steam trap
3 large Styrofoam (plastic foam) cups
water
balance
thermometer (Celsius)

Discussion

The condensation of steam liberates energy that warms many buildings on cold winter days. In this lab, you will make steam by boiling water and investigate the amount of heat energy given off when steam condenses into liquid water. You will then determine the heat energy released per gram, or *heat of vaporization*, for the condensation of steam.

This quantity is relatively small but should not be neglected.

Procedure

Step 1: Make a calorimeter by nesting 3 large Styrofoam cups. Measure the mass of the empty cups and record it in Data Table A.

Measure mass of cups.

MASS OF EMPTY CUPS	
MASS OF CUPS AND COOL WATER	
MASS OF COOL WATER	
TEMPERATURE OF COOL WATER	
TEMPERATURE OF WARM WATER	
MASS OF WARM WATER	
SPECIFIC HEAT OF WATER	$1.00\ \frac{\text{cal}}{\text{g}\cdot{}^\circ\text{C}}$

Data Table A

Measure initial temperature and mass.

Step 2: Fill the inner cup half-full with cool water. Measure the mass of the triple cup and water to the nearest 0.1 g and record it in Data Table A. Also record the original temperature of the water to the nearest 0.5°C.

Assemble apparatus.

Step 3: Fill the steam generator half-full with water, and put its cap on. Place the end of the rubber tubing in the water in the calorimeter (Figure A).

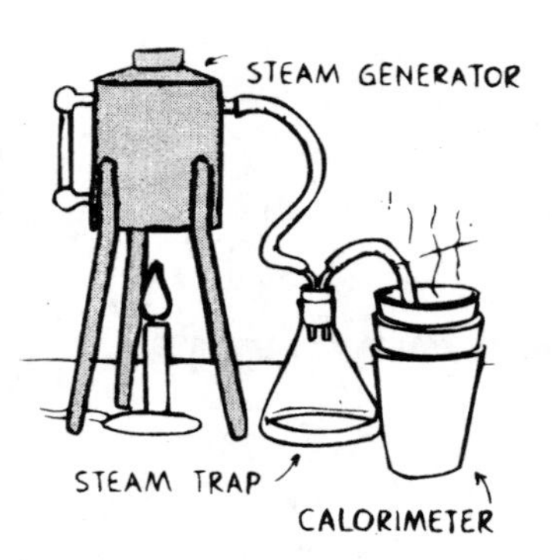

Fig. A

Ignite Bunsen burner.

Step 4: Put on safety goggles. Place the burner under the steam generator and ignite it to heat the water inside.

Adjust steam generator.

Step 5: Once the water in the steam generator starts boiling, adjust the rate at which steam comes out of the tubing so that it does not gurgle water out of the calorimeter. The steam should make the water bubble vigorously, but not violently. Gently stir the water with the thermometer. Bubble steam into the calorimeter until the water reaches a temperature no higher than 50°C.

Measure final temperature and mass.

Step 6: Shut off the burner, and remove the tubing from the water. Immediately determine the final mass and temperature of the warm water, and record them in Data Table A.

Compute mass of steam.

Step 7: Compute the mass of steam that condensed in your calorimeter from the change in mass of the water in the calorimeter.

mass of steam = __________ g

Compute heat energy gained.

Step 8: Compute the total amount of heat energy Q gained by the cool water using the relation $Q = mc\Delta T$, where m is the initial mass of the water, c is the specific heat of water (1.00 cal/g•°C), and ΔT is the change in the temperature of the water. This heat comes in two steps: from the condensation of steam, and then from the condensed steam (now water) as it cools from 100°C to the final temperature. Show your work.

heat energy gained = ________ cal

Compute the energy lost during cooling.

Step 9: Compute the amount of heat energy lost by the water from the condensed steam as it cooled from 100°C down to its final temperature. Show your work.

heat energy lost during cooling
of water from the condensed steam = _____ cal

Name Period Date

Step 10: From the difference between the values found in Steps 8 and 9, compute the amount of heat energy that was released by the steam as it condensed.

heat energy released during condensation = ______ cal

Step 11: Compute the heat of vaporization by dividing the value found in Step 10 by the mass of steam that condensed in the cup.

Compute heat of vaporization.

heat of vaporization = ______ cal/g

1. Compare this value to the standard value, 540 cal/g, and calculate the percentage difference.

If students are careful and use steam traps, their values for the heat of fusion should come out within 10% of 540 cal/g.

Analysis

2. There are two major energy changes involved in this experiment. One happens in the generator, and the other in the calorimeter. Where does the energy come from or go to during these changes?

In the generator, the water absorbs energy from the heat source and changes into steam. In the calorimeter, the steam condenses into water and gives off energy that warms the cool water.

3. Does the amount of steam that escapes into the air make any difference in this experiment?

Only the steam that actually condenses and becomes part of the water warms up the water. The steam that bubbles up and out of the water does not significantly warm it, nor does it condense and become part of it.

4. Why does keeping the cap on when heating the water in the steam generator cause the water to boil more quickly?

Keeping the cap on reduces convection, which would warm the surrounding air instead of the water. It also reduces evaporation.

60 Changing Phase

Purpose

To recognize, from a graph of the temperature changes of two systems, that energy is transferred in changing phase even though the temperature remains constant.

Setup: 1
Lab Time: >1
Learning Cycle: concept development
Conceptual Level: moderate
Mathematical Level: easy

Adapted from PRISMS.

Required Equipment/Supplies

hot plate
2 15-mm × 125-mm test tubes
heat-resistant beaker
paraffin wax
2 ring stands
2 test-tube clamps
hot tap water
2 thermometers (Celsius) in #3 stoppers to fit test tubes or
  computer
  2 temperature probes in slotted corks to fit test tubes
  printer
graph paper (if computer is not used)

Activity

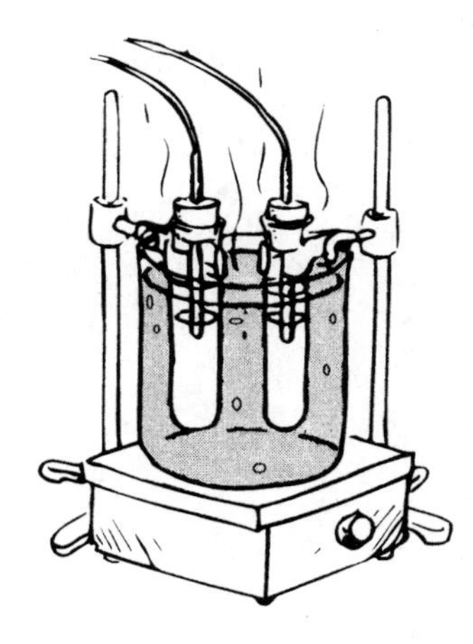

Discussion

When water is at its freezing point, cooling the water no longer causes its temperature to drop. The temperature remains at 0°C until all the water has frozen. After the water has frozen, continued cooling lowers the temperature of the ice. Is this behavior also true of other substances with different freezing temperatures?

Procedure    

Step 1: Fill one test tube 3/4 full with wax (moth flakes). Fill another test tube with water to the same level. Fasten the test tubes in clamps fastened to a ring stand, and place them in hot tap water in a heat-resistant beaker. The beaker should be heated with a hot plate. Place a stopper with a thermometer or a cork with a temperature probe in each of the test tubes. Calibrate the temperature probes before using them. The thermometer or probe in wax should reach only just below the surface.

Set up apparatus.

Step 2: Heat the water until all the wax has melted. Both thermometers or temperature probes should show nearly the same temperature. Turn the heat off, and record the temperatures.

Melt wax.

Monitor temperature of wax and water.

Step 3: Record the times and temperatures of the wax and water every 30 seconds until 5 minutes after the wax has entirely solidified. Note the temperature at which the wax begins to solidify (freeze).

Plot your data.

Step 4: On a single piece of graph paper, plot the temperature (vertical axis) vs. time (horizontal axis) of (1) the water and (2) the wax. Use different colors for the curve of each material. Alternatively, use the computer to plot the data. Print out the data and graph, and include them with your lab report.

Analysis

1. How is the temperature curve of the wax similar to that of the water?

   Both temperature curves are falling.

2. How do the temperature curves of the wax and the water differ?

   The water continuously cools while the wax curve has a flat, nearly horizontal plateau.

3. What physical change occurred while the wax temperature remained constant?

   The wax froze, or solidified.

4. When the temperature of the wax was constant, it was higher than that of the surrounding water bath. What was the heat source responsible for this?

   The heat source is the portion of the wax that is changing state from liquid to solid. Freezing is a warming process!

61 Work for Your Ice Cream

Purpose

To measure the energy transfers occurring during the making and freezing of homemade ice cream.

Setup: 1

Lab Time: >1

Learning Cycle: concept development

Conceptual Level: moderate

Mathematical Level: easy

Required Equipment/Supplies

homemade ice cream machine
ice
rock salt
thermometer
insulated cups
ice cream mix
triple-beam balance
spring scale with maximum capacity of 50 N or 10 lb

Thanks to David Vernier for his ideas and suggestions for this lab.

Discussion

Energy must be supplied to a home to heat it on a cold winter day. Energy must also be supplied to an air conditioner that is cooling a home on a hot summer day. The air conditioner is transferring heat from the cooler indoors to the warmer outdoors. Energy input is always required to move heat from a region of lower temperature to a region of higher temperature.

Energy must be taken away from a liquid to make it a solid. If the liquid is cooler than the surrounding air, additional energy must be added to move heat from the liquid to the warmer air, just as extra energy is needed to move heat from a room to the warmer outdoors in summer. When the liquid is sweet cream, the solid that results is ice cream. (Making ice cream also involves swirling in air; ice cream is actually a mixture of a solid and a gas.)

The freezing of homemade ice cream involves several energy transfers.

1. Energy (work) is expended in turning the crank to overcome inertia and friction.
2. The slush of ice and rock salt takes up energy as it melts.
3. The ice cream mix cools from its original to its final temperature.
4. The metal container cools from its original to its final temperature.
5. The ice cream mix gives up energy as it freezes.

Procedure 

Each group in the class should determine how it will make the measurements to determine the amount of energy transferred in one of the five processes above. If some constants for ice cream are needed, a group also should do a separate experiment to determine those constants. The constants might be the specific heat of the ice cream mix and the heat of fusion of ice cream. Each group should also carry out its method to determine the energy transferred.

You may have to sacrifice some of the ice cream you make to determine these constants and the energy transferred. Don't forget that many of the advances of science came only after great sacrifices! Organize your data to share with the class to make a profile of all the energy transfers.

Describe your procedure.

Questions

1. Which process of the five listed is the greatest absorber of energy?

Melting the slush of ice and rock salt is the biggest absorber of energy.

2. How did the energy absorbed by the melting slush of ice and rock salt compare with the energy released by the other processes?

The energy absorbed should equal the energy released, but quite a bit of energy is involved in chemical changes here, and this energy is not being tracked.

3. Salt is put on icy roads to promote melting. When you make homemade ice cream, salt is used to help promote freezing. Is this practice paradoxical, or does it make good physics sense?

Two effects are at work: (1) salt dissolves in the *solid* (as well as liquid) water (ice!) thereby extracting energy from its environment; (2) the melting point is lowered by pressure.

62 The Drinking Bird

Purpose

To investigate the operation of a toy drinking bird.

Setup: <1
Lab Time: 1
Learning Cycle: application
Conceptual Level: moderate
Mathematical Level: none

Thanks to Willa Ramsey for her ideas and suggestions for this lab.

Required Equipment/Supplies

toy drinking bird
cup of water

Discussion

Toys often illustrate fundamental physical principles. The toy drinking bird is an example.

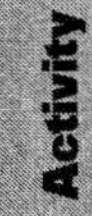

Procedure

Step 1: Set up the toy drinking bird with a cup of water as in Figure A. Position the cup to douse the bird's beak with each dip. The fulcrum may require some adjustment to let it cycle smoothly.

Fig. A

Step 2: After the bird starts "drinking," study its operation carefully. Your goal is to explain how it works. Answering the questions in the "Analysis" section will help you understand how it works.

Analysis

1. What causes the fluid to rise up the toy bird's neck?

   Cooling of the head by evaporation from the beak lowers the vapor pressure in the head. The higher vapor pressure in the lower reservoir forces the liquid up the neck.

2. What causes the bird to dip?

   The head gains fluid and becomes heavier.

3. What causes the bird to make itself erect?

   When the bird dips, the bottom of the tube pokes above the surface of the liquid in the lower reservoir, allowing fluid to flow back into the reservoir.

4. What is the purpose of the fuzzy head?

   The fuzzy head enhances evaporation.

5. Will the bird continue to drink if the cup is removed?

   At first it will continue its motion, but as the beak dries out, the up and down motion slows and then stops.

6. Will the relative humidity of the surrounding air affect the rate of dipping?

   Yes, as the water will not evaporate quickly from the beak if the surrounding air is humid.

7. Will the bird dip faster indoors or outdoors? Why?

   It will probably dip faster outdoors; a breeze increases evaporation.

8. Under what conditions will the bird fail to operate?

   Some conditions are: saturated air; fulcrum not adjusted properly; water level in cup too low.

Name Period Date

9. Brainstorm with your lab partners about practical applications of the drinking bird.

Student answers will vary. Studies have been made to see whether modified drinking birds could be used in Saudi Arabia as pumps.

10. Explain how the bird works.

The toy drinking bird operates by the evaporation of ether inside its body and by the evaporation of water from the outer surface of its head. The lower body contains liquid ether, which evaporates rapidly at room temperature. As the vapor accumulates above the liquid, it creates pressure on the surface of the liquid, which pushes the liquid up the tube. Ether in the upper body does not vaporize because the head is cooled by the evaporation of water from the outer felt-covered beak and head. When the weight of ether in the head is sufficient, the bird pivots forward, and the ether spills back into the lower reservoir. Each pivot wets the felt surface of the beak and head, and the cycle is repeated.

63 The Uncommon Cold

Experiment

Purpose

To use linear extrapolation to estimate the Celsius value of the temperature of absolute zero.

Setup: 1
Lab Time: 1
Learning Cycle: concept development
Conceptual Level: moderate
Mathematical Level: moderate

Adapted from PRISMS.

Required Equipment/Supplies

safety goggles
paper towel
Bunsen burner
wire gauze
ring stand with ring
250-mL Florence flask
flask clamp
short, solid glass rod
one-hole rubber stopper to fit flask
500-mL beaker
water
thermometer (Celsius)
ice bucket or container (large enough to submerge a 250-mL flask)
ice
large graduated cylinder
graph paper

Optional Equipment/Supplies

computer
data plotting software

Discussion

Under most conditions of constant pressure, the volume of a gas is proportional to the absolute temperature of the gas. As the temperature of a gas under constant pressure increases or decreases, the volume increases or decreases, in direct proportion to the change of absolute temperature. If this relationship remained valid all the way to absolute zero, the volume of the gas would shrink to zero there. This doesn't happen in practice because all gases liquefy when they get cold enough. Also, the finite size of molecules prevents liquids or solids from contracting to zero volume (which would imply infinite density).

In this experiment, you will discover that you can find out how cold absolute zero is even though you can't get close to that temperature. You will cool a volume of air and make a graph of its temperature-volume

The idea that an extrapolated graph can give an accurate answer even when the graph itself is not valid in the region where it is extrapolated is a difficult idea for the student to grasp. Point out that the important part of the student's graph is its *slope in the region where the measurements are made,* and that it is this slope that determines where absolute zero is.

relation. Then you will extrapolate the graph to zero volume to predict the temperature, in degrees Celsius, of absolute zero.

Procedure   

Set up ice bath.

Step 1: Make an ice bath of ice and water, using a bucket or container large enough to submerge a 250-mL flask. (Do not put the flask in it yet.)

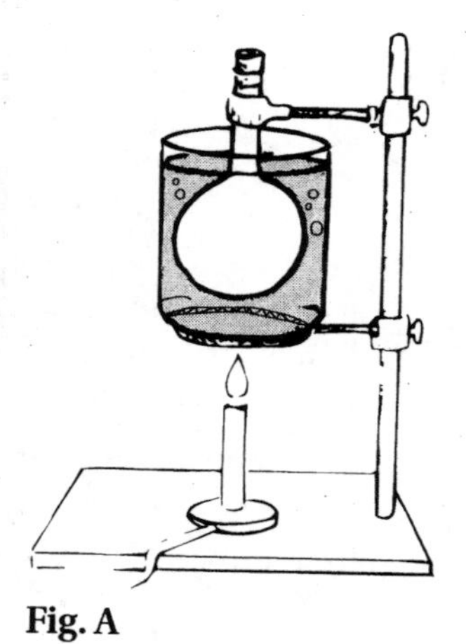

Fig. A

Step 2: Put on safety goggles. Select a *dry* 250-mL Florence flask, and fit a *dry* single-hole stopper into the flask. Half-fill a 500-mL beaker with water, set it on the ring stand, and clamp the flask to the ring stand in the water with the water level approximately 4 cm below the top of the beaker (see Figure A). Boil the water for at least 3 minutes to make sure that the air in the flask is heated to the water temperature. Measure and record the water temperature.

temperature of boiling water = ______________

Step 3: Turn off the burner. Use a damp paper towel to grasp the clamp, and quickly place a solid glass rod into the hole of the rubber stopper to trap all the air molecules. Do this quickly while the air in the flask has the same temperature as the boiling water. Loosen the clamp from the ring stand, and lift the flask, stopper, and clamp assembly. Allow the flask to cool for a minute or so until it can be handled comfortably. Remove the clamp from the flask, and lower the flask upside down into the ice bath. Hold the flask and stopper below the water surface of the ice bath, and remove the glass rod.

Record bath temperature.

Step 4: Hold the flask upside down under the water surface for at least 3 minutes. Measure and record the temperature of the ice bath.

temperature of ice bath = ______________

Remove flask from bath.

Step 5: Some water has entered the flask to take the place of the contracted air. The air in the flask now has the same temperature as the ice bath. With the flask totally submerged and the neck of the flask just under the water surface, place the glass rod or your finger over the hole of the stopper, and remove the flask from the ice bath.

Measure volumes of air.

Step 6: Devise a method to determine the volume of the trapped air at both temperatures. Record the volumes.

volume of air at temperature of boiling water = ______

volume of air at temperature of ice bath = ______

Graph and extrapolate.

Step 7: Plot the volume of air in the flask (vertical axis) vs. the temperature (horizontal axis) for the two conditions. Use a scale of –400 degrees Celsius to +100 degrees Celsius on the horizontal axis and a scale of 0 mL to 250 mL on the vertical axis. Draw a straight line through the two points and to the x-axis (where the extrapolated value for the volume of the gas is zero).

Name Period Date

Analysis

1. Why did water flow into the flask in the ice bath?

   As the temperature of the air in the flask drops, so does the pressure and volume of the air in the flask. The pressure of the surrounding water is greater than that of the air in the flask, and water gets pushed in until the pressure inside the flask equals that outside the flask.

2. What is your predicted temperature for absolute zero in degrees Celsius? Explain how the graph is used for the prediction.

   Answers should be –273°C ± 30°C. The line in the graph is extended until it crosses the zero-volume line. The temperature at this intersection point is absolute zero because volume is directly proportional to absolute temperature at the temperatures where the measurements were made. Using great care will yield more accurate results.

3. In what way does this graph suggest that temperature cannot drop below absolute zero?

   If the straight-line volume-temperature relationship were valid all the way to zero volume, that would set a lower limit on temperature, for compression beyond infinite density is certainly not possible. A material can't take up less space than no space at all!

4. What can be done to improve the accuracy of this experiment?

   Possibly the greatest source of error in this experiment is that a graph defined by only two points is extrapolated a long way. If either point is slightly off, then the value of absolute zero obtained from the extrapolation will be off, too. The single most important improvement that could be made is to take more data points to improve the accuracy of the graph and its slope. For example, the student could find the volume of the gas at other temperatures.

5. What are some assumptions that you made in conducting the experiment and analyzing the data?

Various answers include: (a) The pressure remained constant. (b) The volume of the glass flask remained constant. (c) Volume varies in direct proportion to absolute temperature.

6. What happens to air when it gets extremely cold?

The various gases that make up air condense to their liquid state (nitrogen at −196°C and oxygen at −183°C).

Going Further

Use data plotting software to plot your data. Use the computer-drawn graph to check your prediction of the temperature of absolute zero.

7. How does the computer prediction compare with your earlier prediction?

Although answers will vary, it's conceivable that the computer-drawn graph might give a more accurate value for absolute zero because of errors students make drawing graphs.

Name Period Date

64 Tick-Tock

Purpose

To construct a pendulum with a period of one second.

Setup: <1
Lab Time: <1
Learning Cycle: exploratory
Conceptual Level: moderate
Mathematical Level: easy

Required Equipment/Supplies

ring stand
pendulum clamp
pendulum bob
string
stopwatch or wristwatch
balance
meterstick

You will capitalize on student interest by allowing them about 15-20 minutes to complete this lab.

Activity

Discussion

A simple pendulum consists of a small heavy ball (the bob) suspended by a lightweight string from a rigid support. The bob is free to oscillate (swing back and forth) in any direction. A pendulum completes one cycle or oscillation when it swings forth from a position of maximum deflection and then back to that position. The time it takes to complete one cycle is called its *period*.

If, during a 10-second interval, a pendulum completes 5 cycles, its period T is 10 seconds divided by 5 cycles, or 2 seconds. The period is then 2 seconds for this pendulum.

What determines the period of a pendulum? You will find out the factors by trying to make your pendulum have a period of exactly one second.

Use of a computer for this lab is not recommended.

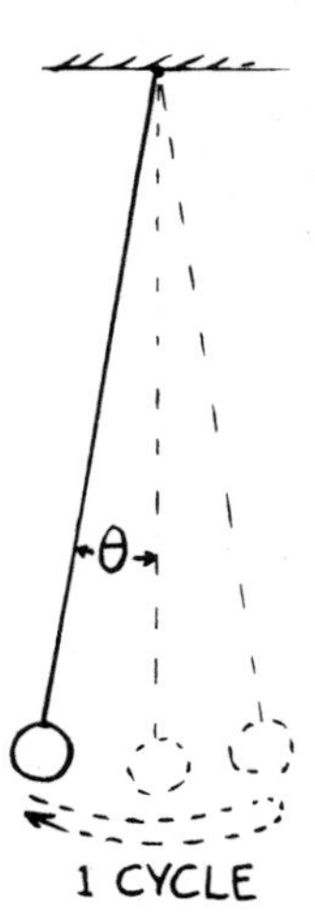

Procedure

Construct a pendulum with a period of *exactly* one second. To do this, change one variable at a time and keep track of which ones affect the period and which ones do not.

Questions

1. Briefly describe the method you used to construct your pendulum.

Answers will vary.

2. What was the mass of your pendulum?

Answers will vary.

3. What effect, if any, does mass have on the period of a pendulum?

The period is independent of the mass.

4. What effect, if any, does amplitude (size of swing) have on the period of a pendulum?

For a given pendulum length, large amplitudes have *slightly* longer periods than smaller ones. But for small angles, the period is essentially independent of amplitude.

5. What was the length of your pendulum?

The length should be close to 25 cm. (Theoretically, the period T and length L are related by the equation $T = 2\pi\sqrt{L/g}$. Using $g = 9.80\ m/s^2$ and solving for L when T is 1.00 s gives $L = 0.248$ m, or 24.8 cm.)

6. What effect, if any, does length have on the period of a pendulum?

The longer the length, the longer the period, but the relation is not a simple direct proportion. (Students will explore the relation more precisely in Experiment 65, "Grandfather's Clock.")

7. If you set up your pendulum atop Mt. Everest, would the period be less than, the same as, or greater than it would be in your lab? Why?

Weaker gravity atop Mt. Everest would cause the period to be longer than at sea level.

8. If you set up your pendulum aboard an orbiting space vehicle, would the period be less than, the same as, or greater than it would be in your lab?

In the free fall environment of an orbiting space vehicle, the pendulum would not oscillate; it would have no period.

Name Period Date

65 Grandfather's Clock

Experiment

Purpose

To investigate how the period of a pendulum varies with its length.

Setup: <1
Lab Time: >1
Learning Cycle: exploratory
Conceptual Level: moderate
Mathematical Level: moderate

Required Equipment/Supplies

ring stand
pendulum clamp
pendulum bob
string
graph paper or data plotting software and printer
stopwatch or
  computer
  light probe with interface
  ring stands with clamps
  light source

Discussion

What characteristics of a pendulum determine its period, the time taken for one oscillation? Galileo timed the swinging of a chandelier in the cathedral at Pisa using his pulse as a clock. He discovered that the time it took to oscillate back and forth was the same regardless of its amplitude, or the size of its swing. In modern terminology, the period of a pendulum is independent of its amplitude (for small amplitudes, angles less than 10°).

Another quantity that does not affect the period of a pendulum is its mass. In Activity 64, "Tick-Tock," you discovered that a pendulum's period depends only on its length. In this experiment, you will try to determine exactly *how* the length and period of a pendulum are related. Replacing Galileo's pulse, you will use a stopwatch or a computer.

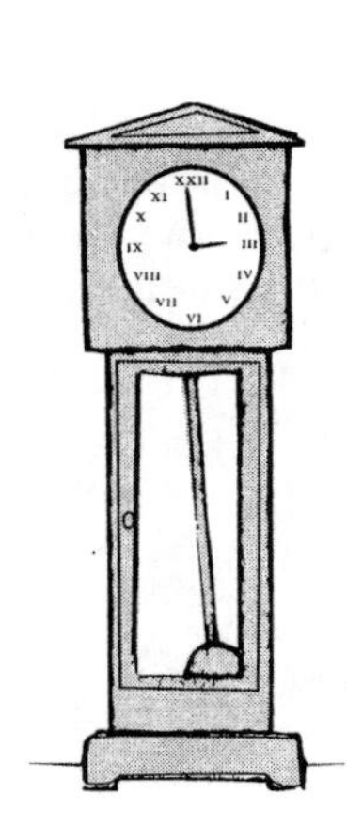

Procedure

Step 1: Set up a pendulum using the listed equipment. Make its length 65 cm. Measure its period three times by timing the oscillations with the stopwatch or the computer system. (If you are using a computer to measure the period of your pendulum, set up the light probe. Use clamps and ring stands to position a light source (such as a flashlight) so that the pendulum bob eclipses the light probe somewhere in its path.) Record the average period in Data Table A.

Set up pendulum.

Step 2: Shorten the pendulum length by 5 cm. Measure its period as in Step 1, and record the average period in Data Table A.

Shorten pendulum and repeat.

LENGTH (cm)	PERIOD (s)
65	
60	
55	
50	
45	
40	
35	
30	
25	
20	
15	
10	

Data Table A

To see the curve of the graph more clearly have your students add the point (0,0) to their graphs.

Step 3: Complete Data Table A for the remaining pendulum lengths indicated there. Measure periods as you did in Step 1.

Step 4: Make a graph of the period (vertical axis) vs. the length of the pendulum (horizontal axis).

1. Describe your graph. Is it a straight line that shows that the period is directly proportional to the length? Or is it a curve that shows that the relationship between period and length is not a direct proportion?

The graph should be a curve with a decreasing slope.

Step 5: Often data points lie on a curve that is not a straight line. It is very difficult to determine the relationship between two variables from such a curve. It is virtually impossible to extrapolate accurately from a curve. Experimenters instead try to produce a straight-line graph by plotting appropriate functions (squares, cubes, logarithms, etc.) of the variables originally used on the horizontal and vertical axes. When they succeed in producing a straight line, they can more easily determine the relationship between variables.

The simplest way to try to "straighten out" a curve is to see if one of the variables is proportional to a power of the other variable. If your graph of period vs. length curves upward (has increasing slope), perhaps period is proportional to the square or the cube of the length. If your graph curves downward (has decreasing slope), perhaps it is the other way around; perhaps the square or the cube of the period is proportional to the length. Try plotting simple powers of one variable against the other to see if you can produce a straight line from your data. If you have a computer with data plotting software, use it to discover the mathematical relationship between the length and the period. A straight-line plot shows that the relationship between the variables chosen is a direct proportion.

2. What plot of powers of length and/or period results in a straight line?

A plot of the square of the period vs. length results in a straight line.

Predict pendulum length.

The pendulum will be nearly a meter long (99.3 cm).

Step 6: From your graph, predict what length of pendulum has a period of exactly two seconds.

predicted length = ___________

Measure pendulum length.

Step 7: Measure the length of the pendulum that gives a period of 2.0 s.

measured length = ___________

3. Compute the percentage difference between the measured length and the predicted length.

Answers will vary, but using the computer students typically come within a few percent of 99.3 cm.

66 Catch a Wave

Purpose

To observe important wave properties.

Setup: <1
Lab Time: >1
Learning Cycle: exploratory
Conceptual Level: moderate
Mathematical Level: moderate

Required Equipment/Supplies

Slinky™ spring toy, long form
tuning fork resonators
portable stereo with two detachable speakers
Doppler ball (piezoelectric speaker with 9-volt battery mounted inside 4"-diameter foam ball)

Discussion

When two waves overlap, the composite wave is the sum of the two waves at each instant of time and at each point in space. The two waves are still there and are separate and independent, not affecting each other in any way. If the waves overlap only for some duration and in some region, they continue on their way afterward exactly as they were, each one uninfluenced by the other. This property is characteristic of waves and is not observable for particles the size of billiard balls. Obviously, you've never seen a cue ball collide with another billiard ball and re-appear on the other side! However, sound and water waves, if they are weak enough, behave this way. That is, two sound waves can overlap so that *no* sound results! Impressive!

You may want to point out to your students that this property of waves *does* apply to particles and *is* observable for particles the size of protons, electrons, or neutrons. This is the crux of wave-particle duality.

Activity

Procedure

Part A: Longitudinal vs. Transverse Waves

(Optional) The *Good Stuff* programs "Waves on a String," "Longitudinal Waves," and "Sum of Two Waves" complement this part of the lab.

Light is a *transverse* wave, meaning that the wave vibrates back and forth perpendicular to the line of propagation. Sound is a *longitudinal* wave, meaning that the wave vibrates along the line of propagation.

Observe transverse waves.

Step 1: Have your partner hold one end of a long Slinky and carefully—so as not to get it tangled or kinked—stretch it out on a smooth (non-carpeted) floor or a long counter top. Give the Slinky a rapid jerk by shaking it to one side then back so that you create a wave pulse that travels to your partner. Be sure your partner holds the other end fixed. Repeat several times, observing what happens to the wave pulse as it travels to your partner and back.

1. How does the Slinky move with respect to the wave pulse?

The slinky moves perpendicular to the line of propagation.

2. Is the pulse transverse or longitudinal?

The pulse is transverse.

3. Why does the amplitude of the pulse decrease as it travels from its source?

Because the friction between the Slinky and the surface decreases the energy of the wave.

4. How does the reflected pulse differ from the original pulse?

It undergoes a 180° phase change; that is, a "positive pulse" is reflected as a "negative pulse."

5. What happens to the wavelength of the pulse as it travels to and from your partner?

The wavelength of the pulse remains unchanged.

Observe reflection of waves.

Step 2: Attach a piece of string about 50 cm long to one end of the Slinky. Have your partner hold the free end of the string. Send a pulse along the Slinky.

6. How does the reflected pulse differ from the original pulse?

It does not undergo a phase change; the pulse is reflected back on the same side.

Observe change in wave speed.

Step 3: Increase the tension in the Slinky by stretching it. Send a pulse along the Slinky. How does the tension affect the speed of the pulse? Does the tension in the Slinky affect other wave properties as well?

The pulse moves faster, but otherwise behaves the same.

Observe longitudinal waves.

Step 4: Repeat steps 1 through 3 but this time jerk the Slinky back and forth along the direction of the outstretched Slinky. What kind of wave is produced? Record your observations.

The pulses are longitudinal and move along the Slinky parallel to the direction of propagation. Phase changes do occur, but are difficult to observe. The speed increases as the tension in the Slinky is increased.

Observe the sum of two waves.

Step 5: Now for some real fun! This time you and your partner will create equal-sized pulses at the same time from opposite ends of the Slinky. It may require some practice to get your timing synchronized. Try it both ways—that is, with the pulses on the same side and then with the pulses on opposite sides of the line of propagation. Pay special attention to what happens as the pulses overlap. Record your observations.

When the pulses are on the same side they constructively interfere to form a larger wave of approximately twice the amplitude. This is relatively easy to see, albeit briefly. When pulses on the opposite side of the Slinky overlap they destructively interfere, briefly canceling each other. This is tricky to observe. It looks like the pulses are flipping when in fact they are just traveling right through one another and reappearing on the other side.

Observe interference of sound waves.

Step 6: Detach two speakers from a portable stereo with both speakers in phase (that is, with the plus and minus connections to each speaker the same). Play in a mono mode so the signals of each speaker are identical. Note the fullness of the sound. Now reverse the polarity of one of the speakers by interchanging the plus and minus wires. Note that the sound is different—it lacks fullness. Some of the waves from one speaker are arriving at your ear out of phase with waves from the other speaker.

Now place the pair of speakers facing each other at arm's length apart. The long waves interfere destructively, detracting from the fullness of sound. Gradually bring the speakers closer to each other. What happens to the volume and fullness of the sound? Bring them face to face against each other. What happens to the volume now?

As you bring the speakers closer together, the fullness of the sound decreases.

Part B: Standing Waves—Resonance

Have your partner hold one end of the Slinky. Shake the Slinky slowly back and forth until you get a wave that is the combination of a transmitted wave and its reflection—a standing wave—like the *fundamental* (or the first harmonic) shown in Figure A.

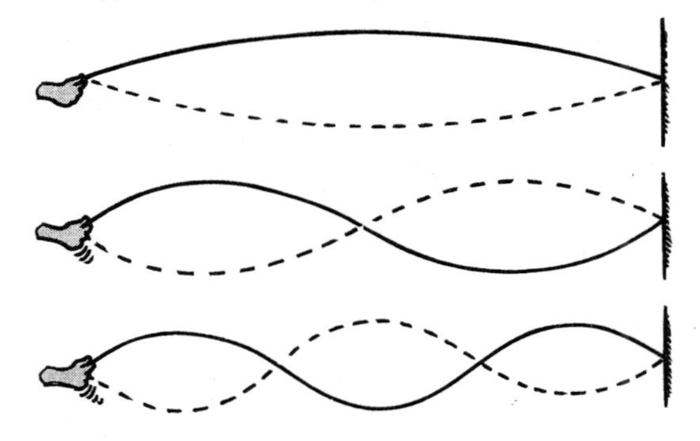

Fig. A

7. How long is the wave compared to the length of the Slinky?

The length of the wave is twice the length of the Slinky.

Observe standing waves.

Step 7: Now shake the Slinky at a higher frequency so that a second-harmonic standing wave is created.

8. How long is the wave compared to the length of the Slinky?

The length of the wave is the length of the Slinky.

Observe harmonics.

Step 8: See how many other harmonics you can create. Record your observations.

Several harmonics can be created by shaking the Slinky at higher frequencies.

(Optional) The *Good Stuff* program "Doppler Effect" complements this part of the lab.

Part C: The Doppler Effect

You are now going to investigate what happens when either the source or the receiver of waves is moving. For example, when a car whizzes by on the road, the pitch of its engine is higher when approaching and lower when receding. This change of frequency due to motion is called the Doppler effect.

Similarly, the whine of an airplane engine changes its pitch as it passes overhead. If the plane moves faster than the speed of sound, the Doppler shift is replaced by a shock wave, which produces a "sonic boom." It is interesting to note that any object traveling at supersonic speed creates its own shock wave—whether or not it is a sound emitter.

The Doppler effect occurs with *any* wave phenomenon (including light and other electromagnetic radiation) whenever there is relative motion between the source and receiver. For example, the decrease of frequency for the light of a receding star gives it a *red shift*. An approaching star is seen to be *blue shifted*. The Doppler effect is far reaching.

Observe Doppler effect.

Step 9: Play catch with your partner using a Doppler ball. How does the movement of the ball affect the pitch of the sound?

You hear a higher pitch than when it is stationary when the ball is tossed towards you and a lower one when tossed away from you.

Name Period Date

67 Ripple While You Work

Purpose

To observe wave phenomena in a ripple tank.

Setup: 1
Lab Time: >1
Learning Cycle: concept development
Conceptual Level: moderate
Mathematical Level: easy

Required Equipment/Supplies

ripple tank with light source and bottom screen
3/4-inch dowel
paraffin blocks
medicine dropper
length of large-diameter rubber hose
wave generator
large glass plate
cut rubber stoppers

Activity

Discussion

Water waves have simple properties when they have small amplitude, and they are familiar to everyone. Observing the behavior of water waves in a ripple tank will introduce you to the analysis of wave motion.

Procedure 

Step 1: The ripple tank is set up for you. Turn on its light. Observe the screen at the base of the tank as you produce a pulse by touching your finger or pencil tip once to the water surface.

Observe pulse.

1. What is the shape of the pulse?

The shape of the pulse is an expanding circle.

2. Does the speed of the pulse seem to be the same in all directions?

Yes, it seems to be the same in all directions.

Step 2: Place the dowel in the water. Produce a straight wave front by rolling the dowel forward 1 cm, with the flat of your hand.

Generate straight wave front.

3. What is the shape of the pulse?

It is a moving straight line with curves at the ends.

Observe reflections.

Step 3: Place a paraffin block in the tank. With the dowel generate a pulse that strikes the barrier straight on.

4. What does the pulse do when it reaches the barrier?

The pulse reflects, or bounces back.

5. After the pulse strikes the barrier, what is the new direction of the pulse?

The new direction of the pulse is straight back.

Step 4: Move the paraffin block to change the angle at which the pulse strikes it.

6. What is the shape of the reflected pulse?

The shape of the reflected pulse is a straight line.

Generate circular wave pulses.

Step 5: Produce circular wave pulses with water drops from the medicine dropper.

7. How do the pulses reflect from the paraffin block?

They reflect as circular arcs.

8. From what point do the reflected pulses appear to be originating?

A reflected circular pulse appears to have come from a point source as far behind the barrier as the actual source is in front of it.

Observe wave pulses reflected by a parabola.

Step 6: Bend a length of large-diameter rubber hose into the approximate shape of a parabola. Place it in the tank.

9. What do you observe when you use this tubing as a reflector for straight pulses?

Straight waves are reflected to a focus.

Step 7: Find the *point* at which the straight pulses reflected by the hose meet and mark it on the screen with your finger. This is the *focus* of the parabola. Generate a circular pulse with the dropper held straight above the focus of the parabola.

Observe wave pulses originating at the focus.

10. What is the shape of the reflected pulse?

The reflected pulse is a straight line.

11. Do any other points give the same pulse shape?

No other points give the same pulse shape.

Step 8: Start the wave generator to produce a straight wave. The distance between bright bars in the wave is the wavelength. Adjust the frequency of the wave generator.

Observe frequency change.

12. What effect does increasing the frequency have on the wavelength?

Increasing the frequency makes the wavelength shorter.

Step 9: Place a paraffin barrier halfway across the middle of the tank. Observe the part of the straight wave that strikes the barrier as well as the part that passes by it. Adjust the frequency of the wave generator so that the combination of the incoming and reflected wave appears to stand still. The combination then forms a standing wave.

Observe standing wave.

13. How does the wavelength of the standing wave compare with the wavelength of the wave traveling past the barrier?

The wavelengths are the same.

Step 10: Support a piece of rectangular slab of glass with rubber stoppers so that it is 1.5 cm from the bottom of the tank and its top is *just* covered with water. Arrange the glass so that incoming wave fronts are parallel to one edge of the glass.

14. What happens as waves pass from deep to shallow water?

They slow down and the wave fronts bunch closer together, so the wavelength decreases.

Compare wave speeds.

Step 11: Now turn the glass so that its edge is no longer parallel to the incoming wave fronts.

15. Are the wave fronts straight both outside and over the glass?

Yes, they are straight in both places.

16. How do the speeds of the waves compare?

The speed is slower over the glass.

Observe wave spreading.

Step 12: Place paraffin blocks across the tank until they reach from side to side with a small opening in the middle. Generate straight waves with the wave generator.

17. How does the straight wave pattern change as it passes through the opening?

The wave fronts leaving the hole are circular or nearly circular.

Step 13: Using a piece of paraffin about 4 cm long, modify your paraffin barrier so that it has two openings about 4 cm apart near the center. Generate a straight wave and allow it to pass through the pair of openings.

18. What wave pattern do you observe?

The pattern should look like the ones in Fig. 31.12 of the text. This pattern is formed when two circular waves intersect. The striped regions are antinodes (maximum displacement) and the gray "spokes" are nodes (minimum displacement).

Step 14: Put two point sources about 4 cm apart on the bar of the wave generator. Turn on the wave generator to produce overlapping circular waves.

19. What pattern do you now observe?

If the spacing is about the same as the two openings and the wave length is the same, then the same pattern as in Step 13 will be observed.

68 Chalk Talk

Purpose

To explore the relationship between sound and the vibrations in a material.

Setup: <1
Lab Time: <1
Learning Cycle: exploratory
Conceptual Level: easy
Mathematical Level: easy

Adapted from PRISMS.

Required Equipment/Supplies

new piece of chalk
chalkboard
masking tape
meterstick

Discussion

Every sound has its source in a vibrating object. Some very unusual sounds are sometimes produced by unusual vibrating objects!

Procedure

Tape chalk to meterstick.

Step 1: Tape the chalk tightly to the thin edge of the meterstick. The end of the chalk should extend beyond the end of the meterstick by 1 cm.

Push or pull on meterstick.

Step 2: Position the chalk perpendicular to the chalkboard's surface. Hold the meterstick firmly with one hand 15 cm from the chalk; very lightly support the far end with the other hand. Pull the meterstick at a constant speed while maintaining firm contact with the board.

Step 3: Modify the procedure of Step 2 by changing the angle, the contact force, or the distance the chalk protrudes beyond the meterstick. Also try making wavy lines. With the proper technique, the sounds of a trumpet, oboe, flute, train whistle, yelping dog, and a sick Tarzan yell can be produced.

Conclude the activity when your instructor can't stand it any longer!

Activity

Analysis

1. What is producing the sound?

The chalk appears to have a constant speed while in contact with the board. However, it is actually slipping and sticking (recall Experiment 29, "Slip-Stick") on the board's surface. This occurs at a rapid rate. The meterstick helps to amplify the sound through forced vibration. The vibration is carried to the observer's ears by sound waves that travel through the air.

2. The end of the meterstick that is loosely supported by the hand should vibrate in different ways for a high and low pitch. Are high or low pitches easier to feel? Explain.

Low-frequency pitches are easier to feel at the end of the meterstick because the slipping and sticking on the board's surface for a low frequency results in a larger amplitude of vibration at the other end of the meterstick.

3. What combinations of angle and pressure tend to produce high pitches?

A high pitch is best produced with the chalk almost perpendicular to the board and with a rather high applied pressure.

4. What combinations of angle and pressure tend to produce low pitches?

An angle greater than 90° between the chalk and direction of motion, together with a very light pressure, will produce the lowest note. At a low enough frequency, the slip-stick nature of the chalk can be readily observed as a series of well-defined, uniform dots.

Name Period Date

69 Mach One

Experiment

Purpose

To determine the speed of sound using the concept of resonance.

Setup: <1
Lab Time: 1
Learning Cycle: concept development
Conceptual Level: moderate
Mathematical Level: moderate

Required Equipment/Supplies

resonance tube approximately 50 cm long or golf club tube cut in half
1-L plastic graduated cylinder
meterstick
1 or 2 different tuning forks of 256 Hz or more
rubber band
Alka-Seltzer® antacid tablet (optional)

Discussion

You are familiar with many applications of *resonance.* You may have heard a vase across the room rattle when a particular note on a piano was played. The frequency of that note was the same as the natural vibration frequency of the vase. Your textbook has other examples of resonating objects.

Gases can resonate as well. A vibrating tuning fork held over an open tube can cause the air column to vibrate at a natural frequency that matches the frequency of the tuning fork. This is resonance. The length of the air column can be shortened by adding water to the tube. The sound is loudest when the natural vibration frequency of the air column is the same as (resonates with) the frequency of the tuning fork. For a tube open at one end and closed at the other, the lowest frequency of natural vibration is one for which the length of the air column is one fourth the wavelength of the sound wave.

In this experiment, you'll use the concept of resonance to determine the wavelength of a sound wave of known frequency. You can then compute the speed of sound by multiplying the frequency by the wavelength.

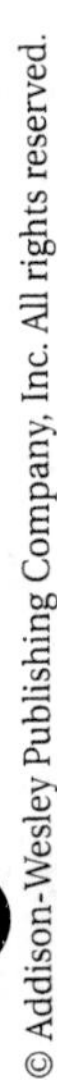

Procedure

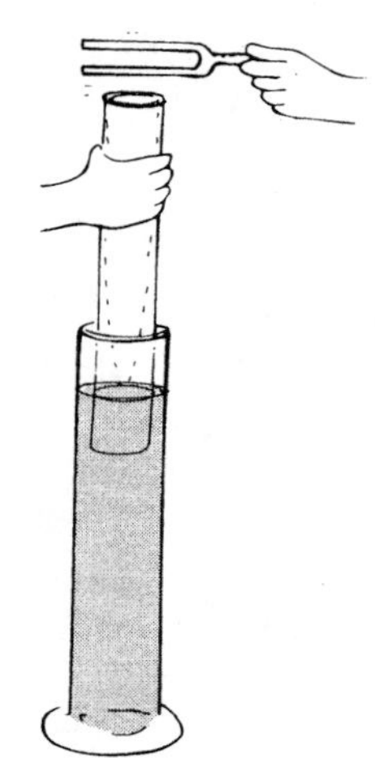

Step 1: Fill the cylinder with water to about two thirds of its capacity. Place the resonance tube in the cylinder. You can vary the length of the air column in the tube by moving the tube up or down.

Step 2: Select a tuning fork, and record the frequency that is imprinted on it.

frequency = ________ Hz

Strike the tuning fork on the heel of your shoe (NOT on the cylinder). Hold the tuning fork about 1 cm above the open end of the tube, horizontally, with its tines one above the other. Move both the fork and the tube up and down to find the air column length that gives the loudest sound. (There are several loud spots.) Mark the water level on the tube for this loudest sound with a rubber band stretched around the cylinder.

Measure air column length.

Step 3: Measure the distance from the top of the resonance tube to the water-level mark.

length of air column = _______ m

Measure the diameter.

Step 4: Measure the diameter of the resonance tube.

diameter of the resonance tube = _______ m

Compute corrected length.

Step 5: Make a corrected length by adding 0.4 times the diameter of the tube to the measured length of the air column. This corrected length accounts for the air just above the tube that also vibrates.

corrected length = _________ m

Compute the wavelength.

Step 6: The corrected length is one fourth of the wavelength of the sound vibrating in the air column. Compute the wavelength of that sound.

wavelength = _________ m

Compute the speed of sound.

Step 7: Using the frequency and the wavelength of the sound, compute the speed of sound in air. Show your work.

speed of sound in air = ______ m/s

Repeat using different tuning fork.

Step 8: If time permits, repeat Steps 2 to 7 using a different tuning fork.

frequency = _______ Hz   length of air column = ________ m

corrected length = ______ m   wavelength = ______ m

speed of sound in air = ________ m/s

Analysis

1. The accepted value for the speed of sound in dry air is 331 m/s at 0°C. This speed increases by 0.6 m/s for each additional degree Celsius above zero. Compute the accepted value for the speed of sound in dry air at the temperature of your room.

   At 20°C the speed will be higher by (0.6 m/s•°C) × (20°C), or 12 m/s. The speed at 20°C is thus (332 m/s) + (12 m/s), or 344 m/s. (Moisture in the air will add only about 1 m/s, an effect that can be neglected.)

2. How does your computed speed of sound compare with the accepted value? Compute the percentage difference.

   % difference = $(|v_{measured} - v_{calculated}|)/v_{calculated}) \times 100\%$

Name Period Date

70 Shady Business

Purpose

To investigate the nature and formation of shadows.

Setup: <1
Lab Time: 1
Learning Cycle: exploratory
Conceptual Level: easy
Mathematical Level: none

Required Equipment/Supplies

bright light source
book or other opaque object
screen or wall
meterstick

Your follow-up discussion of this lab is an excellent time to introduce the terms *umbra* and *penumbra* in relation to eclipses.

Discussion

If you look carefully at a shadow, you will notice that it has a dark central region and a fuzzy and less dark band around the edge of the central region. Why do shadows have two different regions?

Activity

Procedure  

Step 1: Arrange a small light source so that a solid object such as a book casts a shadow on a screen or wall. Sketch the shadow formed, noting both its regions. Sketch the relative positions of the book, light source, and shadow.

Sketch shadow.

Step 2: Move the light source away from the object while keeping the position of the object fixed. Note and sketch any changes in the shadow. Sketch the new relative position of the light source.

Move light source.

1. What happens to the size of the fuzzy region around the edge of the central region when the light source is moved farther away?

The fuzzy region around the edge becomes narrower.

Move object.

Step 3: Move the object closer to the light source while keeping the screen and the light source in the same position. Note and sketch any changes in the shadow. Sketch the new relative position of the object.

2. What happens to the size of the fuzzy region around the edge when the object is moved closer to the light source?

The fuzzy region around the edge becomes wider.

Analysis

3. Which relative positions of the object, light source, and screen result in a sharp distinct shadow with little or no fuzzy region around its edge?

When the distance from the light source to the object is much greater than the distance from the object to the screen, the result is a sharp distinct shadow with little or no fuzzy region around its edge.

4. Which relative positions of the object, light source, and screen result in a large fuzzy region around a small or nonexistent central dark region?

When the distance from the light source to the object is much less than the distance from the object to the screen, the result is a large fuzzy region around a small or nonexistent central dark region.

5. What causes the fuzzy region around the edge of a shadow?

The fuzzy region exists where the light source is *partially* blocked by the object. Its size depends on the relative size of the light source and the object, as well as the relative distances.

71 Absolutely Relative

Experiment

Purpose

To investigate how light intensity varies with distance from a light source.

Setup: <1
Lab Time: 1
Learning Cycle: concept development
Conceptual Level: moderate
Mathematical Level: difficult

Required Equipment/Supplies

computer
light probe with interface
data plotting software
3-socket outlet extender
3 night-lights with clear 7-watt bulbs
2 ring stands
2 clamps
meterstick

Discussion

The light from your desk lamp may seem almost as bright as the sun. If your desk were only a meter away from the sun, your lamp would not seem bright at all. The *brightness* of the sun is far greater than that of the lamp, but the *intensity* of the lamp is almost as great as that of the sun.

The intensity of light decreases with distance from the source. In this experiment, you will use a light probe to measure the intensity of light at various distances to see exactly how the intensity varies with the distance.

Procedure   

Step 1: Install three night-lights in an outlet extender. Use a clamp to secure the extender on a ring stand at least 20 cm above the table. This height is necessary to minimize the effect of the light reflected from the table surface.

Set up light source and light probe.

Clamp the light probe to another ring stand so that it is at the same height as the night lights and is directly facing them 30 cm away. Connect the light probe to the interfacing box. The setup is shown in Figure A.

For best results, instruct students to reduce ambient light to a minimum. Turn off room lights and cover the widows, if possible.

Step 2: Set up the light probe so that the computer automatically measures the intensity of light falling on the light probe and displays the intensity.

Fig. A

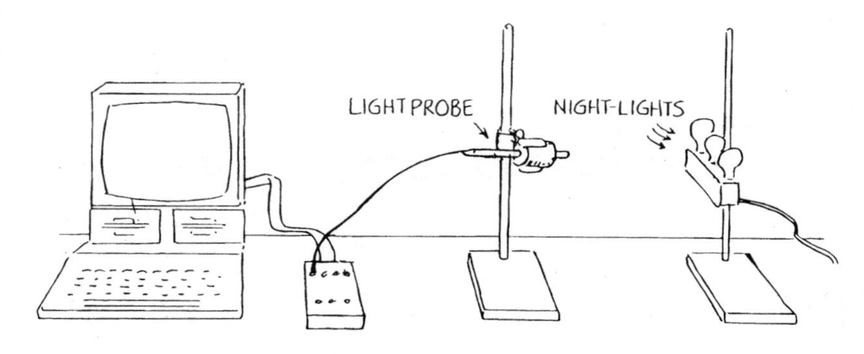

BULB	RELATIVE INTENSITY
A	
B	
C	
A + B	
A + C	
B + C	
A + B + C	

Data Table A

BULB A	
DISTANCE	INTENSITY
30 cm	

Data Table B

Turn on the center bulb (Bulb A). Calibrate the light probe so that the computer automatically compares all other light intensities to the light intensity of Bulb A at this distance. All future intensity readings are, therefore, expressed relative to Bulb A.

Now turn off Bulb A and turn on one of the side bulbs (Bulb B). Leave the light probe in the same position. Record the intensity reading in Data Table A. Turn off Bulb B and turn on the third bulb (Bulb C). Record the intensity reading.

Step 3: Predict what intensity the light probe will read when you turn on Bulbs A and B.

predicted intensity reading = ________

Try it and record your data.

Step 4: Now try all other combinations of the bulbs to complete Data Table A.

1. How do the intensity readings for the different bulb combinations compare with their individual intensity readings?

The intensity reading for each bulb combination should equal the sum of the individual bulb intensities.

2. What might account for any differences in the readings of the bulbs separately compared with when they are combined?

Differences in the readings may be caused by: (1) background light; (2) bulbs not of equal brightness; (3) side bulbs farther from light probe than center bulb is; (4) bulbs not point sources; (5) the light probe blocking some light entering from side bulbs.

Step 5: Take an intensity reading with the light probe 30 cm away from Bulb A. Take nine more readings with the light probe, moving it 5 cm farther away from the bulb each time. Record your data in Data Table B.

3. What happens to the light intensity as the probe gets farther away from the source?

The light intensity decreases.

Step 6: Plot the light intensity (vertical axis) vs. distance (horizontal axis) for Bulb A. If available, use data plotting software to plot the graph.

Plot data.

4. What does the graph look like? What relationship does your graph suggest between the light intensity and the distance?

The graph is a curve that falls toward the horizontal axis (zero intensity) as the distance increases. As the distance increases, the light intensity decreases, rapidly at first, then less rapidly.

Step 7: Use data plotting software to vary the power of the *x*-*y* values so that the plot of intensity vs. distance is a straight line.

Vary the power of x-y values.

5. What does the graph look like?

The graph is a straight line with positive slope.

6. What relationship does your graph suggest between the light intensity and the distance?

The light intensity varies in proportion to the inverse square of the distance.

Analysis

7. Imagine a point source of light, such as a small 1.5-volt flashlight bulb, at the center of a balloon of radius *r*. All the light leaving the bulb strikes the inside surface of the balloon. Also imagine the same light source inside a balloon of twice that radius, 2*r*. All the light leaving the bulb again strikes the inside surface of the balloon. Does the brightness of the bulb increase, decrease, or remain the same?

The brightness of the bulb remains constant. The output of a source of light does not depend upon how far you are from the source.

8. How many times greater is the inside surface area of the larger balloon than that of the smaller balloon?

The inside area of the larger balloon is *four* times as great as the smaller balloon (since the area of a sphere is proportional to r^2).

9. If all the light spreads out evenly onto the surface of the larger balloon, what is the intensity of the light at the inside surface of the larger balloon relative to that for the smaller balloon?

Because the same amount of light from the bulb spreads out and covers four times as much area, the relative intensity is only one fourth as much.

Name Period Date

72 Shades

Purpose

To investigate the effects of polarized light.

Setup: <1
Lab Time: 1
Learning Cycle: exploratory
Conceptual Level: moderate
Mathematical Level: none

Required Equipment/Supplies

3 small polarizing filters
light source
small plane mirror

Thanks to Bruce Ratcliffe for his ideas and suggestions for this lab.

Discussion

The vibrations of light waves reaching your eyes are mostly randomly oriented; they vibrate in many planes at once. In polarized light, the light waves vibrate in one plane only. Polarized light can be made by blocking all the waves except those in one plane with polarizing filters. The filters can also be used to detect polarized light.

Activity

Procedure  

Step 1: Position one polarizing filter between your eyes and a light source. Slowly rotate the filter 360°. Observe the intensity of the light as seen through the filter. Note any intensity changes as you rotate the filter.

1. What happens to the intensity of the light as you rotate the filter?

The intensity does not change.

Step 2: Arrange one filter in a fixed position in front of the light source. Slowly rotate a second filter held between your eyes and the fixed filter. Note any intensity changes of the light as you rotate the filter 360°.

Rotate second filter.

2. What happens to the intensity of the light as you rotate the filter?

During one complete rotation, the intensity changes through two complete cycles from maximum to minimum intensity and back.

Rotate other filter.

Step 3: Hold the filter at your eye in a fixed position while your partner slowly rotates the other filter next to the light source 360°. Note any intensity changes of the light as the filter as rotated.

3. What happens to the intensity of the light as the filter as rotated?

The same effect is seen as in Step 2.

Rotate both filters.

Step 4: Rotate both of the filters through one complete rotation in the same direction at the same time. Note any intensity changes.

4. What happens to the intensity of the light as you rotate both filters together?

The intensity does not change.

Rotate both filters in opposite directions.

Step 5: Rotate both of the filters through one complete rotation at the same time, but in opposite directions. Note any intensity changes.

5. What happens to the intensity of the light as you rotate both filters in opposite directions?

The intensity changes through four complete cycles.

Rotate single filter for light reflected off a mirror.

Step 6: Repeat Step 1, except arrange the light source and a mirror so that you observe only the light coming from the mirror surface. Note any intensity changes of the light as you rotate the filter.

6. What happens to the intensity of the light as you rotate the filter?

The intensity does not change (if the reflecting surface of the mirror is metallic).

7. Is the light reflected off a mirror polarized?

No, light reflecting off a mirror (and other metallic surfaces) is not polarized.

View sky through filter.

Step 7: View different regions of the sky on a sunny day through a filter. Rotate the filter 360° while viewing each region.

CAUTION: *Do not look at the sun!*

Name Period Date

8. What happens to the intensity of the light as you rotate the filter?

The intensity changes through two complete cycles, but does not go all the way to minimum intensity except in certain regions.

9. Is the light of the sky polarized? If so, where is the region of maximum polarization in relation to the position of the sun?

Yes, the light of the sky is partly polarized. The region of maximum polarization is at 90° to the sun's direct rays.

Step 8: View a liquid crystal display (LCD) on a wristwatch or calculator using a filter. Rotate the filter 360°, and note any intensity changes.

View LCD with filter.

10. What happens to intensity of the light as you rotate the filter?

The intensity changes through two complete cycles, as for light that passes through one rotating and one fixed filter.

11. Is the light coming from a liquid crystal display polarized?

Yes, light reflected from an LCD is polarized.

Analysis

12. Why do polarized lenses make good sunglasses?

First, they block one half the unpolarized sunlight. Second, reflected light (glare) from horizontal surfaces is strongly horizontally polarized, so vertically oriented polarized sunglasses block almost 100% of the glare.

13. Explain why the effects seen in Steps 1 to 3 occur.

The effect of a polarizing filter on unpolarized light is to reduce the intensity, but the orientation of the filter does not affect the intensity. However, when the light passes through one filter, it is polarized. When it encounters a second filter, it will not be filtered at all if the axes are in the same direction. If they are crossed but not at right angles, some light will be filtered out. If the axes are crossed at right angles, a minimum of light will get through the second filter.

Going Further

Step 9: Position a pair of filters so that a minimum of light from a light source gets through. Place a third filter between the light source and the pair.

14. Does any light get through?

No more light gets through if the first pair of filters are crossed at 90°.

Step 10: Place the third filter beyond the pair.

15. Does any light get through?

Same as Step 9.

Step 11: This time, sandwich the third filter between the other two filters at a 45° angle.

16. Does any light get through?

If the first pair of filters are crossed, sandwiching a third filter between them passes light through. The amount of light passing through is maximum when that third filter is at 45° to the crossed pair. (See Figure 27.19 of the text.)

73 Flaming Out

Purpose

To observe the spectra of some metal atoms.

Setup: <1

Lab Time: >1

Learning Cycle: concept development

Conceptual Level: easy

Mathematical Level: none

Experiment

Required Equipment/Supplies

safety goggles
spectroscope
incandescent lamp
Bunsen burner with matches or igniter
small beaker
7 20-cm pieces of platinum or nichrome wire with a small loop at one end
6 labeled bottles containing salts of lithium, sodium, potassium, calcium, strontium, and copper

Discussion

Auguste Compte, a famous French philosopher of the nineteenth century stated that humans would never know what elements made up the distant stars. Soon thereafter, the understanding of spectra made this knowledge easy to obtain. An element emits light of certain frequencies when heated to high enough temperatures. Different frequencies are seen as different colors, and each element emits (and absorbs) its own pattern of colors. This allows us to identify these elements, whether they are nearby or far away.

Procedure     

Adjust spectroscope.

Step 1: Practice using the spectroscope by looking at an incandescent lamp. The spectrum will appear as a rainbow of colors. Adjust the spectroscope until the spectrum is horizontal and clear.

Step 2: Put on safety goggles. Ignite the Bunsen burner. Adjust it until a blue, nearly invisible flame is obtained.

Step 3: Dip a loop of wire into the bottle of solid sodium chloride. Hold the loop in the flame until the sodium chloride melts and vaporizes. Observe the spectrum of the sodium atoms through the spectroscope. You may also see a long, broad band image due to the glowing wire. Ignore this image. In Data Table A, sketch any major lines you see, in the appropriate position, for the color of the lines.

Sketch spectral lines.

Step 4: Test each of the other salts as in Step 3. Each wire is to stay in its proper bottle. Mixing the test wires contaminates them. If you should make a mistake and place a wire in the wrong bottle, burn off all the salt until the wire glows red hot, then return it to its proper bottle. In Data Table A, sketch the major lines emitted by each salt.

Data Table A

SALT OF:	MAJOR SPECTRAL LINES				
	RED	YELLOW	GREEN	BLUE	VIOLET
LITHIUM					
SODIUM					
POTASSIUM					
CALCIUM					
STRONTIUM					
COPPER					

Observe spectrum of mixture.

Step 5: Mix small and equal amounts of the copper and lithium salts in the small beaker. Using the extra test wire, hold a loopful of the mixture in the flame, and observe the resulting spectrum through the spectroscope.

1. Does the spectrum of the mixture of copper and lithium salts contain a combination of the copper and lithium lines?

Yes, students should see the spectral lines of both copper and lithium.

Step 6: Mix a trace of lithium salt with a larger amount of copper salt. Observe the resulting spectrum through the spectroscope, using the same wire.

2. Did you see the characteristic colors of both copper and lithium?

Students will probably see the colors of both.

Analysis

3. How do you know that the bright yellow lines you observed when looking at sodium chloride are due to the sodium and not to the chlorine in the compound?

Several salts are chlorides, but only sodium chloride has the yellow line (which is actually two closely spaced lines—a doublet).

4. From your observations in Steps 5 and 6, draw conclusions about the relative amounts of metal elements present in each mixture.

The elements present in greater amounts have brighter lines.

5. In steel mills, large amounts of scrap steel of unknown composition are melted to make new steel. Explain how the laboratory technicians in the steel mills can tell exactly what is in any given batch of steel in order to adjust its composition.

The molten steel sample is vaporized so that the characteristic spectral lines can be observed.

74 Satellite TV

Purpose

To investigate a model design for a satellite TV dish.

Setup: <1
Lab Time: 1
Learning Cycle: application
Conceptual Level: easy
Mathematical Level: none

Required Equipment/Supplies

small transistor radio
large umbrella
aluminum foil

Adapted from PRISMS.

Discussion

When you can barely hear something, you may cup your hand over your ear to capture more sound energy. Does this really work? The answer is yes. In this activity, you will use an umbrella similarly to capture a signal.

Remind your students that this is an application of Step 6 of Activity 67, "Ripple While You Work."

Activity

Procedure

Step 1: Line the inside surface of an open umbrella with aluminum foil.

Step 2: Stand near one of the windows of your classroom. Turn on the radio and locate a weak station. Tune the radio until you get the best reception for that weak station.

Step 3: Have a classmate hold the umbrella horizontally, with the handle pointed at, and just touching, the radio.

Step 4: Move the radio back slowly along the line of the handle until you find the most improved reception for the weak station.

Step 5: Repeat the process in other locations and for other stations.

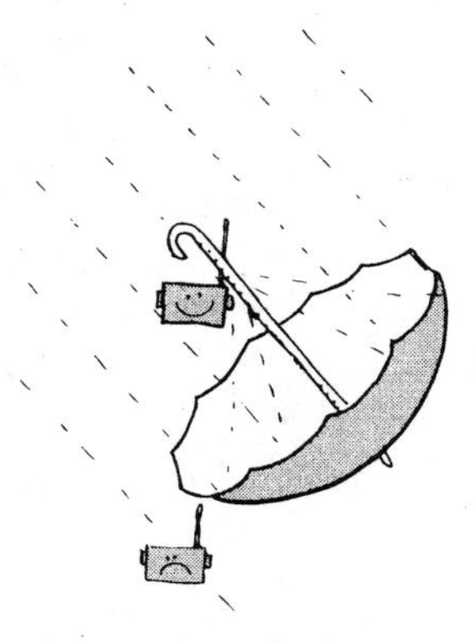

Repeat with other stations.

Analysis

1. Why did the umbrella improve the reception of the radio?

It reflected radio waves and thus increased the signal.

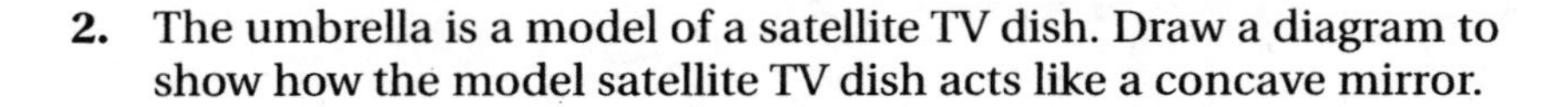

2. The umbrella is a model of a satellite TV dish. Draw a diagram to show how the model satellite TV dish acts like a concave mirror.

Diagram should show rays approaching the inside of the umbrella parallel to the handle, and bouncing to a common focus at a point along the handle.

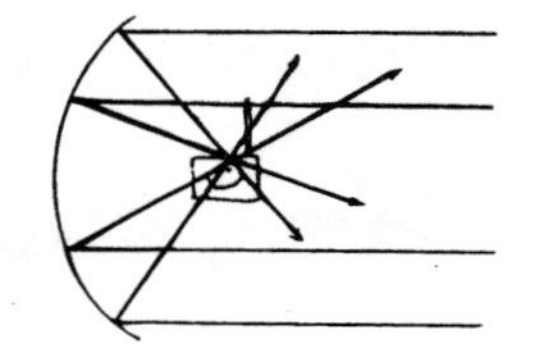

Name Period Date

75 Images

Purpose

To formulate ideas about how reflected light travels to your eyes.

Setup: <1
Lab Time: 1
Learning Cycle: exploratory
Conceptual Level: moderate
Mathematical Level: none

Adapted from PRISMS.

Required Equipment/Supplies

2 small plane mirrors
supports for the mirrors
2 single-hole rubber stoppers
2 pencils
2 sheets of paper
transparent tape

Discussion

Reflections are interesting. Reflections of reflections are fascinating. Reflections of reflections of reflections are . . . you will see for yourself in this activity.

Procedure 

Step 1: Place the pencils in the rubber stoppers. Set one plane mirror upright in the middle of a sheet of paper, as shown in Figure A. Stand one pencil vertically in front of the mirror. Hold your eye steady at the height of the mirror. Locate the image of the pencil formed by the mirror. Place the second pencil where the image of the first appears to be. If you have located the image correctly, the image of the first pencil and the second pencil itself will remain "together" as you move your head from side to side.

Fig. A

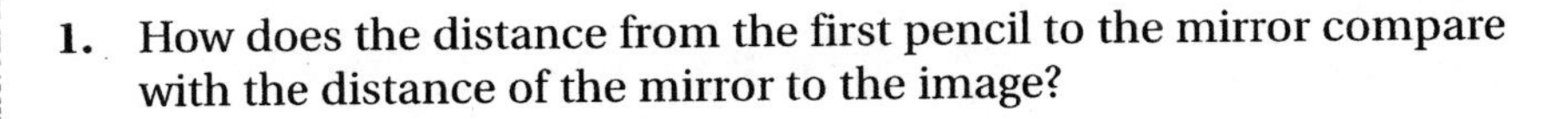

1. How does the distance from the first pencil to the mirror compare with the distance of the mirror to the image?

The distances are the same.

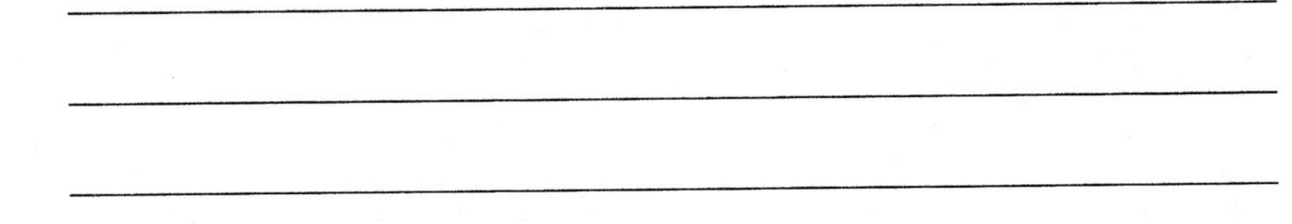

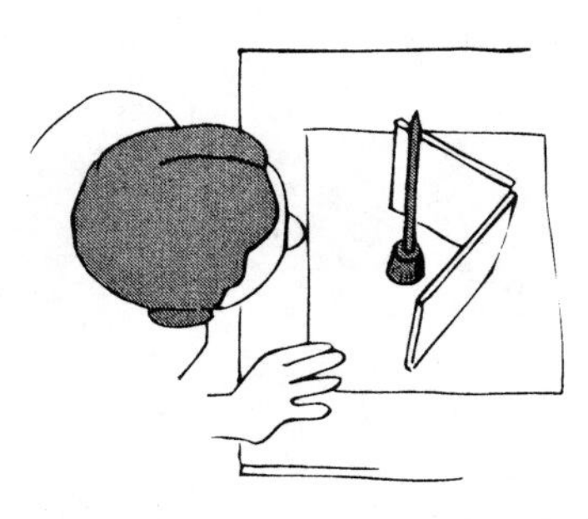

Fig. B

Step 2: On the sheet of paper, draw the path you think the light takes from the first pencil to your eye as you observe the image. Draw a dotted line to where the image appears to be located as seen by the observing eye.

Step 3: Hinge two mirrors together with transparent tape. Set the mirrors upright and at right angles to each other in the middle of a second sheet of paper. Place a pencil in its stopper between the mirrors, as in Figure B.

2. How many images do you see?

   There are three images: one in each mirror and one in the "crack."

Draw ray diagram.

Step 4: On the paper, show where the images are located. Draw the paths you think the light takes as it goes from the pencil to your eye.

Decrease angle between mirrors.

Step 5: Decrease the angle between the two mirrors.

3. What happens to the number of images you get when you decrease the angle between the two mirrors?

   The number of images increases.

76 Pepper's Ghost

Purpose

To explore the formation of mirror images by a plate of glass.

Setup: <1
Lab Time: 1
Learning Cycle: concept development
Conceptual Level: moderate
Mathematical Level: none

Required Equipment/Supplies

safety goggles
two candles of equal size, in holders
1 thick plate of glass, approximately 30 cm × 30 cm × 1 cm
2 supports for the glass plate
matches

Thanks to Dave Wall for his ideas and suggestions for this lab.

Discussion

John Henry Pepper (1821–1900), a professor in London, used his knowledge of image formation to perform as an illusionist. One of his illusions was based on the fact that glass both reflects and transmits light.

Activity

Procedure

Step 1: Put on safety goggles. Light one candle, and place it about 6 cm in front of a vertical thick glass plate. Place a similar, but unlighted, candle at the position on the other side of the glass plate at the point where the flame of the lighted candle appears to be on the unlighted candle.

1. How does the distance from the lighted candle to the glass plate compare with the distance from the glass plate to the unlighted candle?

   The distances are the same.

Step 2: Look carefully, and you should see a double image of the candle flame.

Find double image.

2. How would you explain the double image?

   One image is caused by reflection from the front surface of the thick glass. The second is caused by reflection from the back surface.

Find multiple images.

Step 3: Look at the glass plate such that your line of vision makes a small angle with the surface of the glass. You will see three or more "ghost" images of the candle flame.

3. Explain these "ghost" images.

These "ghost" images are from multiple reflections inside the glass.

Name Period Date

77 The Kaleidoscope

Purpose

To apply the concept of reflection to a mirror system with multiple reflections.

Setup: <1
Lab Time: 1
Learning Cycle: application
Conceptual Level: easy
Mathematical Level: easy

Required Equipment/Supplies

2 plane mirrors, 4 in. × 5 in.
transparent tape
clay
viewing object
protractor
toy kaleidoscope (optional)

Discussion

Have you ever held a mirror in front of you and another mirror in back of you in order to see the back of your head? Did what you saw surprise you?

Procedure 

Step 1: Hinge the two mirrors together with transparent tape to allow them to open at various angles. Use clay and a protractor to hold the two mirrors at an angle of 72°. Place the object to be observed inside the angled mirrors. Count the number of images resulting from this system and record in Data Table A.

Step 2: Reduce the angle of the mirrors by 5 degrees at a time, and count the number of images at each angle. Record your findings in Data Table A.

Step 3: Study and observe the operation of a toy kaleidoscope, if one is available.

ANGLE	NUMBER OF IMAGES
72°	
67°	
62°	
57°	
52°	
47°	
42°	
37°	
32°	
27°	

Data Table A

Activity

Analysis

1. Explain the reason for the multiple images you have observed.

Many images of an object can be seen in a two-mirror system due to multiple reflections, as the light bounces back and is reflected by another mirror.

2. What effect does the angle between the mirrors have on the number of images?

The number of images increases as the angle decreases.

3. Using the information you have gained, explain the construction and operation of a toy kaleidoscope.

A simple kaleidoscope contains three mirrors with their reflecting surfaces facing one another 60° apart along the length of a tube. Colored chips in a compartment are placed between the mirrors. The result is an infinite pattern made of six symmetrical images.

Name Period Date

78 Funland

Experiment

Purpose

To investigate the nature, position, and size of images formed by a concave mirror.

Setup: <1

Lab Time: >1

Learning Cycle: concept development

Conceptual Level: moderate

Mathematical Level: moderate

Required Equipment/Supplies

concave spherical mirror
cardboard
night-light with clear 7-watt bulb
small amount of modeling clay
meterstick

Thanks to Michael Zender for his ideas and suggestions for this lab.

Optional Equipment/Supplies

computer
data plotting software

This lab is a mirror image conceptually of Experiment 82, "Bifocals." If time is short, have your students do one or the other.

Discussion

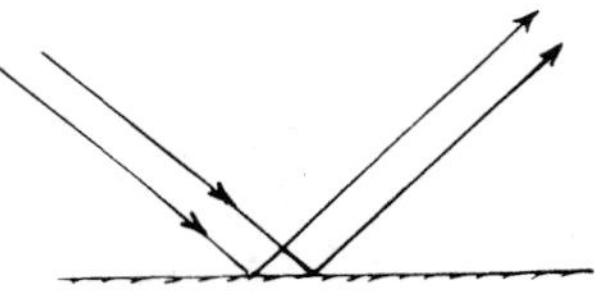

The law of reflection states that the angle of reflection equals the angle of incidence. Parallel light rays that strike a plane mirror head on bounce directly backward and are still parallel. If the parallel rays strike that mirror at another angle, each bounces off at the same angle and the rays are again parallel. A plane mirror cannot bring light rays to a focus because the reflected light rays are still parallel and do not converge. Images observed in a plane mirror are always *virtual* images because *real* images are made only by converging light.

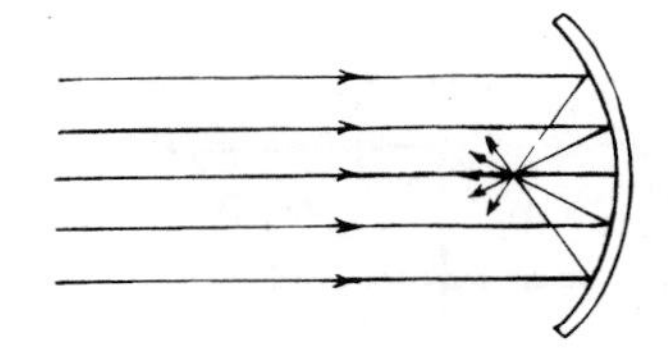

A parabolic mirror is able to focus parallel rays of light to a single point (the *focal point*) because of its variable curvature. A small spherical mirror has a curvature that deviates only a little from that of a parabolic curve and is cheaper and easier to make. Spherical mirrors can, therefore, be used to make real images, as you will do in this experiment.

Procedure  

Measure focal length of mirror.

Step 1: The distance from the center of a spherical mirror surface to the focal point is called the *focal length.* Measure the focal length of your mirror by having it convert a parallel beam of light into a converging beam that comes to a point on a screen. Use the filament of a lit, clear 7-watt bulb as a source of approximately parallel light and a piece of cardboard as a screen. Record your measurement below to the nearest 0.1 cm. Also, record the number of your mirror.

focal length = ________ cm

mirror number = ______

Step 2: The rays of light striking your mirror from the bulb may not be exactly parallel. What effect, if any, would this have on your measured value for the focal length? What effect would moving the light source farther away from the mirror have? Increase the distance between the mirror and the light source to see if the focal length changes. (If a better source of parallel light is available, use it to find the focal point of your mirror). Record your measurement to the nearest 0.1 cm.

focal length f = _______ cm

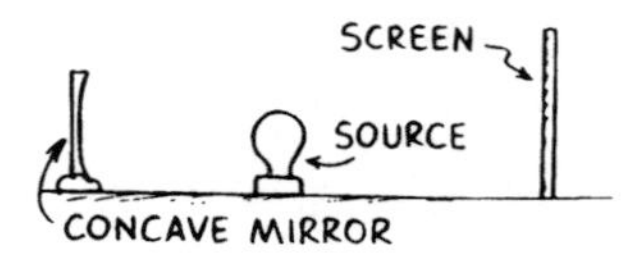

Fig. A

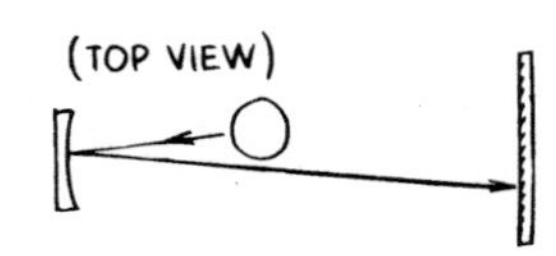

Fig. B

Find a real image with mirror.

Step 3: Use a small amount of modeling clay at the bottom of the mirror to act as a mirror holder. Arrange a screen and a light source as shown in Figure A. Position your screen so that the image is slightly off to one side of the object, as shown in Figure B. Move the mirror so that it is farther than one focal length *f* from the nightlight. Move the screen to form a sharp image of the filament on the screen. Can a real image be formed on the screen? Is it magnified or reduced, compared with the object? Is the image erect (right-side up) or inverted (upside down)? Record your findings in Data Table A.

Data Table A

POSITION OF OBJECT	NATURE OF IMAGE		
	REAL OR VIRTUAL?	MAGNIFIED?	INVERTED OR ERECT ?
BEYOND f			
AT f			
WITHIN f			

Step 4: Where, in relation to one focal length from the mirror, is the object when the image appears right-side up (erect)? What is the relative size of the image (magnified or reduced) compared with the object? Is the image real or virtual? Record your findings in Data Table A.

Step 5: Is there a spacing between object and mirror for which no image appears at all? Where is the object in relation to the focal length? Record this position in Data Table A.

d_o (cm)	d_i (cm)

Data Table B

Step 6: Position the mirror two focal lengths away from the light source. The mirror will then form an image of the filament on a screen placed slightly to one side of the light source. The distance between the *focal point* and the object is the object distance d_o and the distance between the focal point and the image is the image distance d_i. Record the distances d_o and d_i in Data Table B. Move the mirror 5 cm farther away from the light source, and reposition the screen until the image comes back into focus. Progressively extend d_o by repeating these 5-cm movements five more times. Record d_o and d_i each time.

Step 7: Plot d_i (vertical axis) vs. d_o (horizontal axis), then different powers of each to discover the mathematical relation between d_i and d_o. Is there any combination that makes a linear graph through the origin and thus a direct proportion? If available, use data plotting software to plot your data.

1. What mathematical relationship exists between d_i and d_o?

The image distance, d_i, is directly proportional to the inverse of the object image ($1/d_o$) so the image distance is inversely proportional to the object distance, d_o.

Step 8: You can locate the position of the image of the object in Figure C using the ray-diagram method. Draw the path of the light rays that leaves the tip of the arrow parallel to the principal axis.

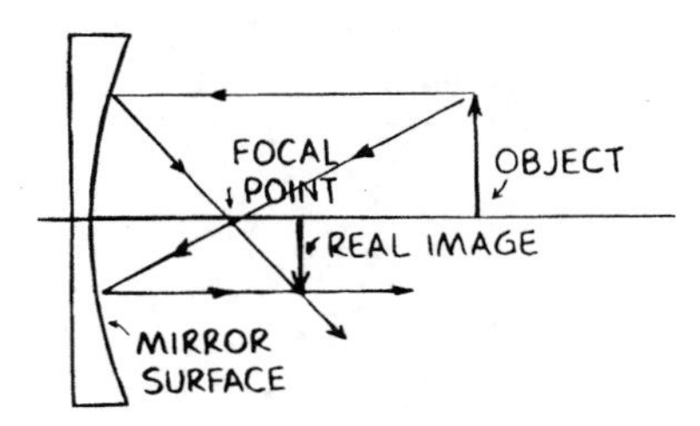

Fig. C

2. Where does this ray go after it is reflected?

The ray parallel to the principal axis is reflected through the focal point.

Draw the light ray that leaves the tip of the arrow and passes through the focal point.

3. Where does this light ray go after it is reflected?

The ray that passes through the focal point is reflected parallel to the principal axis.

Now draw the paths of these two light rays after they are reflected. At the point where they cross, an image of the tip of the arrow is formed.

Step 9: Use the ray-diagram method to locate the image of the object in Figure D. Draw the path of the ray that leaves the tip of the arrow parallel to the principal axis and is reflected by the mirror. Trace another ray that heads toward the mirror in the same direction as if it *originated* from the focal point and is reflected by the mirror.

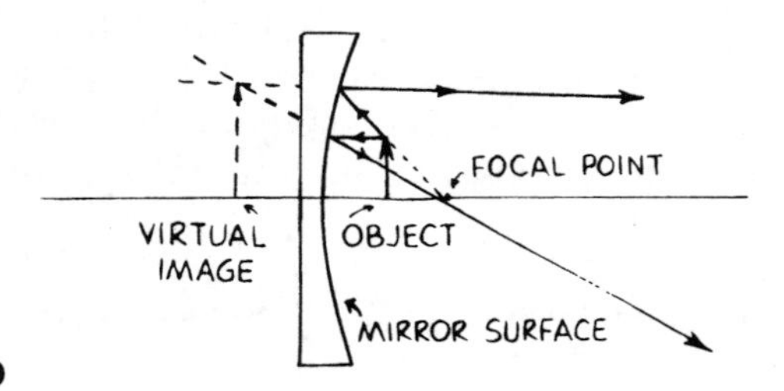

Fig. D

4. Where do the two reflected rays *appear* to cross?

The two reflected rays appear to cross behind the mirror.

5. Could the image be projected onto a screen? Explain.

No; it is a virtual image—no light actually converges at the position where the image appears to be. Only real images can be projected onto a screen.

79 Camera Obscura

Purpose

To observe images formed by a pinhole camera and to compare images formed with and without a lens.

Setup: >1
Lab Time: 1
Learning Cycle: exploratory
Conceptual Level: easy
Mathematical Level: none

Required Equipment/Supplies

covered shoe box with a pinhole and a converging lens set in one end, an open end opposite, and glassine paper inside the box (as in Figure A)
piece of masking tape

Thanks to Wayne Rosenberger for his ideas and suggestions for this lab.

See Teacher Notes for references on how to build actual pinhole cameras.

Discussion

The first camera, known as a *camera obscura,* used a pinhole opening to let light in. The light that passes through the pinhole forms an image on the inner back wall of the camera. Because the opening is small, a long time is required to expose the film sufficiently. A lens allows more light to pass through and still focuses the light onto the film. Cameras with lenses require much less time for exposure, and the pictures have come to be called "snapshots."

Activity

Procedure 

Step 1: Use a pinhole camera constructed as in Figure A. Tape the foil flap down over the lens so that only the pinhole is exposed. Hold the camera with the pinhole toward a brightly illuminated scene, such as the scene through a window during the daytime. Light enters the pinhole and falls on the glassine paper. Observe the image of the scene on the glassine paper.

Observe images on screen of camera with pinhole.

1. Is the image on the screen upside down (inverted)?

   Yes, the image is upside down.

2. Is the image on the screen reversed left to right?

   Yes, left and right are reversed.

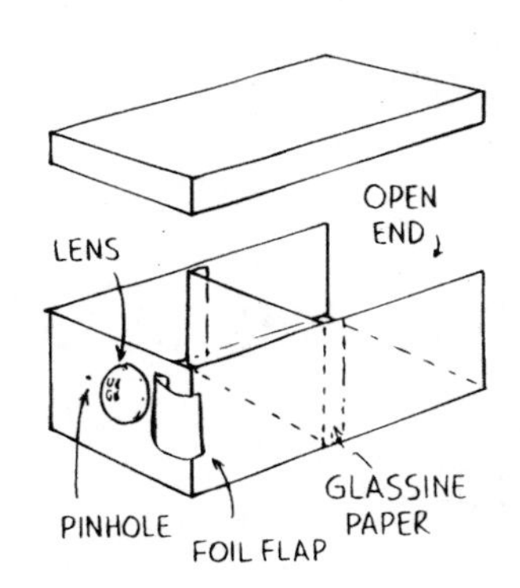

Fig. A

Step 2: Now seal off the pinhole with a piece of masking tape, and open the foil door to allow light through the lens. Move the camera around. You can watch people or cars moving by.

Observe images with lens.

3. Is the image on the screen upside down (inverted)?

Yes, the image is upside down.

4. Is the image on the screen reversed left to right?

Yes, left and right are reversed.

Step 3: A pinhole camera focuses equally well on objects at all distances. Point the camera lens at a nearby object to determine whether the lens focuses on nearby objects.

5. Does the lens focus on nearby objects?

A lens that is adjusted to sharply focus distant objects does not focus as sharply objects that are close.

Draw ray diagram.

Extend this activity by having your students shoot pinhole photographs. A student already familiar with photographic processing can develop the film. Contact Wayne Rosenberger at Crawford High School, San Diego, CA 92115, for more information on techniques.

Step 4: Draw a ray diagram as follows. Draw a ray for light that passes from the top of a distant object through a pinhole and onto a screen. Draw another ray for light that passes from the bottom of the object through the pinhole and onto the screen. Show the image created on the screen by the pinhole.

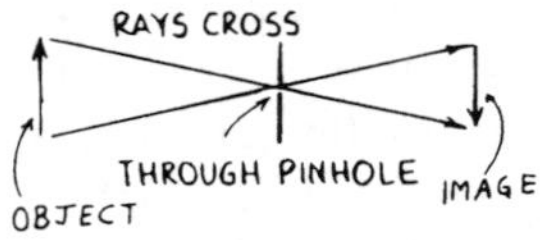

Analysis

6. Why is the image created by the pinhole dimmer than the one created by the lens?

The pinhole admits much less light.

7. How is a pinhole camera similar to your eye? Do you think that the images formed on the retina of your eye are upside down?

The lens of your eye forms an image on your retina that *is* upside down. Your brain, however, interprets the upside-down image as right-side up, even though your eye does *not* have a right-side-up image! In very bright light, the size of your pupil becomes so small that your eye resembles a pinhole camera. Then, if you normally wear glasses or contact lenses, you may see clearly without them because a pinhole produces focused images without a lens.

Name Period Date

80 Thin Lens

This lab requires the use of *Good Stuff* software and an Apple II series computer.

Purpose

To acquire a qualitative understanding of concave and convex lenses.

Setup: <1
Lab Time: 1
Learning Cycle: exploratory
Conceptual Level: easy
Mathematical Level: none

Required Equipment/Supplies

convex lens
concave lens
Good Stuff software
computer

Thanks to Robert H. Good for developing the software that makes this lab possible.

Discussion

Lenses are not to read about, but to experiment with. Before studying Chapter 30 in the text, some hands-on experience is important for understanding lenses. This activity should help guide you to discover some of their interesting properties.

It is best if students have a lens in their hand as they investigate the formation of images on the computer.

Activity

Procedure 

Move an object to different distances from a convex lens and observe the image formed. Select the "Thin Lens" program from *Good Stuff* and the option for a converging lens. A ray diagram consisting of an object arrow and its image as formed by the rays will appear. The *focal length, f,* of the lens is the distance from its center to the point where light parallel to the lens axis (principal axis) converges to a *focus.*

Initially, the object and image are located at a distance $2f$ from the lens. Use the arrow keys to move the object along the principal axis. Is the image larger or smaller than the object? Is the image *erect* (right-side up) or *inverted* (upside down)? Can the image be projected (a *real* image) or not (a *virtual* image)? Is there any position of the object for which no image is formed? Record your observations as to the nature of the image you observe on the computer in Data Table A and Data Table B for both kinds of lenses. Check to see how the image on the screen corresponds to images of objects formed by real lenses in your hand!

Analysis

1. When the image appears right-side up (erect) using a converging lens, how many focal lengths is the object from the lens?

The object is located less than one focal length away from the lens.

2. Under what circumstances is the image formed by a converging lens magnified? Under what circumstances is it reduced? When is it real? When is it virtual?

The image is virtual and magnified when the object is less than one focal length distant from the lens. It is real and magnified when the object is between one and two focal lengths from the lens. It is real and reduced when the object is more than two focal lengths from the lens.

3. Can an object be located in a position where a converging lens forms no real image?

Yes, whenever the object distance ≤ one focal length.

4. For a diverging lens, is the virtual image enlarged or reduced?

Reduced.

5. Can you form a real image with a diverging lens?

No.

6. When the object is moved, does the image formed by a converging lens always move in the same direction? What about the image formed by a diverging lens?

For both lenses, the image always moves in the same direction as the object (except that for a converging lens, the image jumps in the opposite direction from infinite distance on one side to infinite distance on the other side when the object distance passes through the focal distance).

Data Table A

NATURE OF IMAGE - CONVERGING LENS			
POSITION OF OBJECT	REAL OR VIRTUAL	MAGNIFIED OR REDUCED	INVERTED OR ERECT
BEYOND 2f	REAL	REDUCED	INVERTED
AT 2f	REAL	SAME SIZE	INVERTED
AT f	NO IMAGE	NO IMAGE	NO IMAGE
WITHIN f	VIRTUAL	MAGNIFIED	ERECT

Data Table B

NATURE OF IMAGE - DIVERGING LENS			
POSITION OF OBJECT	REAL OR VIRTUAL	MAGNIFIED OR REDUCED	INVERTED OR ERECT
BEYOND 2f	VIRTUAL	REDUCED	ERECT
AT 2f	VIRTUAL	REDUCED	ERECT
AT f	VIRTUAL	REDUCED	ERECT
WITHIN f	VIRTUAL	REDUCED	ERECT

Name Period Date

81 Lensless Lens

Purpose

To investigate the operation of a pinhole "lens."

Setup: <1
Lab Time: <1
Learning Cycle: concept development
Conceptual Level: moderate
Mathematical Level: none

Required Equipment/Supplies

3" × 5" card
straight pin
meterstick

Thanks to Lonnie Grimes for his ideas and suggestions for this lab.

Discussion

The image formed through a pinhole is in focus no matter where the object is located. In this activity, you will use a pinhole to enable you to see nearby objects more clearly than you can without it.

Activity

Procedure 

Step 1: Bring this printed page closer and closer to your eye until you cannot focus on it any longer. Even though your pupil is relatively small, your eye does not function as a pinhole camera because it does not focus well on nearby objects.

Look at print close up.

Step 2: With a straight pin, poke a pinhole about 1 cm from the edge of a 3" × 5" card. Hold the card in front of your eye and read these instructions through the pinhole. Bright light is needed. Bring the page closer and closer to your eye until it is a few centimeters away. You should be able to read clearly. Quickly take the pinhole away and see if you can still read the words.

1. Did the print appear magnified when observed through the pinhole?

The print does appear magnified.

Analysis

2. Did the pinhole actually magnify the print?

Not in the ordinary sense. The pinhole focuses images on your retina no matter how close the object is to your eye. The object *seems* to be magnified because it is closer, making the image larger than it would be if the page were at the normal reading distance. Without the pinhole, the image is the same size but blurry.

3. Why was the page of instructions dimmer when seen through the pinhole than when seen using your eye alone?

The pinhole admits much less light than the pupil (opening) of the eye.

4. A nearsighted person cannot see distant objects clearly without corrective lenses. Yet, such a person can see distant objects clearly through a pinhole. Explain how this is possible. (And if you are nearsighted yourself, try it!)

The pinhole focuses an image on the retina, which the uncorrected eye is unable to do.

82 Bifocals

Purpose

To investigate the nature, position, and size of images formed by a converging lens.

Setup: <1

Lab Time: >1

Learning Cycle: concept development

Conceptual Level: moderate

Mathematical Level: moderate

Required Equipment/Supplies

converging lens
small amount of modeling clay
cardboard
meterstick
night-light with clear, 7-watt bulb

Thanks to Floyd Judd for his ideas and suggestions for this lab.

Optional Equipment/Supplies

data plotting software
computer

This lab is a mirror image conceptually of Experiment 78, "Funland." If time is short, have your students do one or the other.

Discussion

The use of lenses to aid vision may have occurred as early as the tenth century in China. Eyeglasses came into more common use in Europe in the fifteenth century. Have you ever wondered how they work? In Experiment 78, "Funland," you learned that the size and location of an image formed by a concave mirror is determined by the size and location of the object. In this experiment, you will investigate these relationships for a converging glass lens.

Procedure 

Step 1: A converging lens focuses parallel light rays to a *focal point.* The distance from the center of a lens to the focal point is called the *focal length, f.* Measure the focal length of the lens by having it convert a parallel beam of light into a converging beam that comes to a small spot on a screen. Use the filament of a lit, clear, 7-watt bulb as a source of approximately parallel light and a piece of cardboard as a small screen. Record your measurement below to the nearest 0.1 cm. Also, record the number of your lens.

Measure focal length.

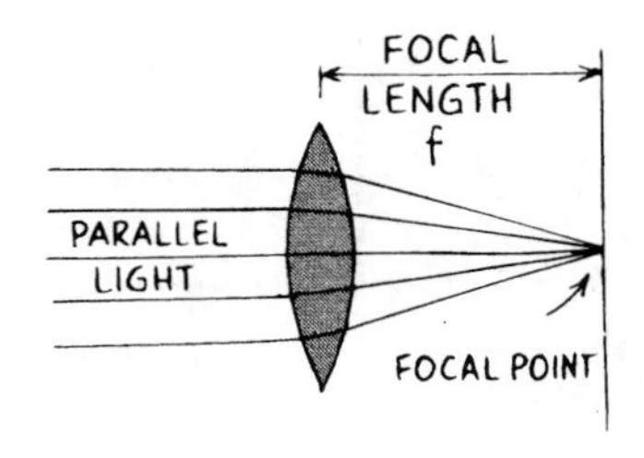

focal length = __________ cm

lens number = __________

Step 2: The rays of light striking the lens may not be parallel. What effect, if any, would this have on your measured value for the focal length? What effect would moving the light source farther away have? Move it farther away and record the focal length to the nearest 0.1 cm. (If a better source of parallel light is available, use it to find the focal length of your lens.)

focal length f= _______ cm

Find inverted image with lens.

7-W BULB LENS SCREEN

Fig. A

Step 3: Use a small amount of modeling clay at the bottom of the lens as a lens holder. Arrange a screen and a light source as shown in Figure A. Observe the image of the filament on the screen, and move the screen until the image of the filament is as sharp as possible. Where, in relation to one focal length from the lens, is the object when the image appears upside down (inverted)? What is the relative size of the image (magnified or reduced) and the object (the filament)? Is the image real or virtual? Record your findings in Data Table A.

Data Table A

POSITION OF OBJECT	NATURE OF IMAGE		
	REAL OR VIRTUAL?	MAGNIFIED?	INVERTED OR ERECT ?
BEYOND f			
AT f			
WITHIN f			

Find erect image with lens.

Step 4: Where, in relation to one focal length from the lens, is the object when the image appears right-side up (erect)? What is the relative size of the image compared with the object? Is the image real or virtual? Record your findings in Data Table A.

Step 5: Is there a distance of the object from the lens for which no image appears at all? If so, what is this distance relative to the focal length? Record this position in Data Table A.

Measure d_i and d_o.

Step 6: Position the lens two focal lengths away from the light source to form an image on the screen on the other side of the lens as in Figure A. The distance between the object and the *focal point* closest to it is the distance d_o, and the distance between the other focal point and the image is the image distance d_i. Record the distances d_o and d_i in Data Table B. Move the lens 5 cm farther away from the light source, and reposition the screen until the image comes back into focus. Repeat these 5-cm movements five more times, recording d_o and d_i each time.

d_o (cm)	d_i (cm)

Data Table B

Step 7: Plot d_i (vertical axis) vs. d_o (horizontal axis), then different powers of each, to discover the mathematical relation between d_i and d_o. Does any combination give a linear graph through the origin and, thus, a direct proportion? If available, use data plotting software to plot your data.

1. What mathematical relationship exists between d_i and d_o?

Image distance d_i, is inversely proportional to the object distance, d_o.

Name Period Date

Step 8: You can locate the position of the image in Figure B using the ray-diagram method. Draw the path of the light ray that leaves the tip of the arrow parallel to the principal axis.

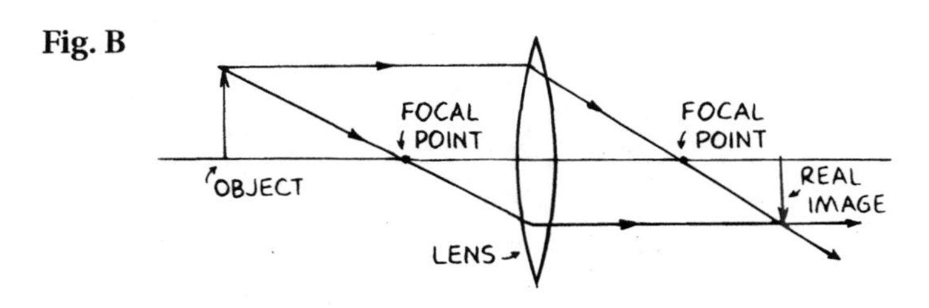

2. Where does this ray go after it is refracted?

The refracted ray goes through the focal point.

Draw the light ray that leaves the tip of the arrow and passes through the focal point.

3. Where does this light ray go after it is refracted?

The refracted ray is parallel to the principal axis.

Now, draw the paths of these two light rays after they are refracted. At the point where they cross, an image of the tip of the arrow is formed.

Step 9: Use the ray-diagram method to locate the image of the object in Figure C. Draw the path of the ray that leaves the tip of the arrow parallel to the principal axis and is refracted by the lens. Trace another ray that

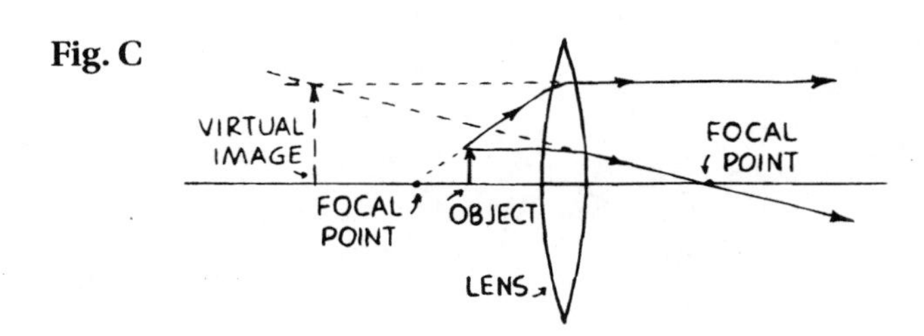

heads toward the lens in the same direction as if it *originated* from the focal point and is refracted by the lens.

4. Do the refracted rays *actually* cross?

No, they do not cross.

5. Where do they *appear* to cross?

They appear to cross on the same side of the lens as the image.

6. Could the image be projected onto a screen?

No, it is a virtual image.

Going Further

Step 10: Use two converging lenses to see if you can create a magnified image of a distant object. Sketch your arrangement of lenses with their relative positions and focal lengths. (Had you been the *first* person to have discovered, 400 years ago, that two converging lenses can make a telescope, your name would be the answer to questions in science classes today!)

If f_o is the focal length of the objective (the lens closer to the object) and f_e is the focal length of the eyepiece (the lens closer to the eye), then the two lenses should be separated by a distance a little less than $f_o + f_e$ (the sum of the focal lengths). (The objective creates a real image at its focal length. The eyepiece serves as a magnifying glass to look at the real image. The eyepiece must be a little closer to the image than its focal length.)

83 Where's the Point?

Purpose

To measure the focal length of a diverging lens.

Experiment

Setup: <1

Lab Time: >1

Learning Cycle: concept development

Conceptual Level: moderate

Mathematical Level: moderate

Required Equipment/Supplies

low-powered helium-neon laser
diverging lens
sheet of white paper
meterstick
graph paper (if computer is not used)

Thanks to Herb Gottlieb for his ideas and suggestions for this lab.

Optional Equipment/Supplies

computer
data plotting software
printer

Discussion

Parallel light rays are brought to a focus at the focal point of a converging lens. Ray diagrams are useful for understanding this. Parallel light rays are *not* brought to a focus by a diverging lens. Ray diagrams or other techniques are essential for understanding this. In this experiment, you will use a laser to simulate a ray diagram for a diverging lens. Lasers are fun!

Procedure  

Step 1: Carefully (lasers are delicate!) place the laser on a table. Do not point it at any mirrors, windows, or persons.

Set up laser.

Step 2: Place a diverging lens in front of the laser. Turn on the laser. Use the sheet of white paper to observe that the beam is very narrow as it comes out of the laser, but after going through the lens, it spreads out into an ever-widening cone.

Observe diverging laser beam.

Step 3: Place the sheet of paper against a book. Hold the paper 5 cm beyond the lens that is in the laser beam. The red spot made by the beam should be in the center of the paper. With a pen or pencil, trace the outline of the red spot. Label the outline "distance 5 cm."

DISTANCE FROM LENS (cm)	DIAMETER OF LASER SPOT (cm)
5	
10	
15	
20	

Data Table A

Step 4: Move the paper 5 cm farther away from the lens, and again trace the outline of the spot that is produced. Label the outline with the new distance. Repeat this procedure, increasing the distance between the lens and the paper by 5-cm intervals, until the spot completely fills the paper.

Step 5: Measure the diameters of the traces of the laser beam on the paper for each position of the paper. Record these diameters and distances from the lens in Data Table A.

Step 6: Plot a graph of the beam diameter (vertical axis) vs. the distance (horizontal axis) between the lens and the paper. Allow room on your graph for negative distances (to the left of the vertical axis.) If available, use data plotting software to plot your data.

Step 7: To find the distance from the lens at which the beam diameter would be zero, extend your line until it intersects the horizontal axis. The negative distance along the horizontal axis is the focal length of the diverging lens. If you are using data plotting software, include a printout of the graph with your lab report.

focal length of lens = __________ cm

Analysis

The focal length of a convex lens is the distance from the lens where parallel light rays are brought to a focus. Why is it impossible to find the focal length of a diverging lens in this manner?

The focal length of a diverging lens cannot be found in the same manner as that of a convex lens, because the parallel light rays never converge to a focus. The focal length of a diverging lens is the distance to the lens from a point *behind* the lens from which diverging light rays *appear* to have originated.

Name Period Date

84 Air Lens

Purpose

To apply your knowledge of light behavior and glass lenses to a different type of lens system.

Setup: 1
Lab Time: <1
Learning Cycle: application
Conceptual Level: easy
Mathematical Level: none

Adapted from PRISMS.

Required Equipment/Supplies

2 depression microscope slides
light source
screen

Discussion

Ordinary lenses are made of glass. A glass lens that is thicker at the center than at the edge is convex in shape, converges light, and is called a converging lens. A glass lens that is thinner at the middle than at the edge is concave in shape, diverges light, and is called a diverging lens.

Suppose you had an air space that was thicker at the center than at the edges and was surrounded by glass. This would comprise a sort of "convex air lens." What would it do to light? This activity will let you find out.

Activity

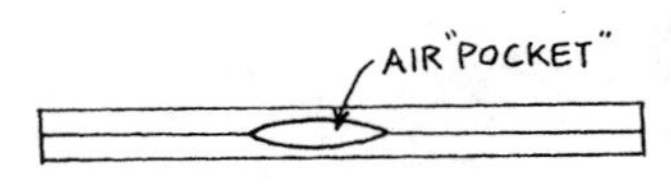

Fig. A

Procedure

Step 1: A convex air lens encased in glass can be produced by placing two depression microscope slides together, as shown in Figure A.

Construct convex air lens.

1. Predict whether this arrangement makes a diverging or converging lens. Explain your prediction.

Some students may realize that, just as light bends toward the normal as it goes from air to glass, it bends away from the normal as it goes from glass into air. These students should accurately predict that the convex air lens acts the opposite from the convex glass lens and diverges light (see Concept-Development Practice Page 29-3).

Step 2: Use your lens with a light source and screen to check your prediction.

2. What do you discover?

Students will not be able to form an image of a distant light source on the screen. They should infer that the convex air lens behaves as a diverging lens.

Analysis

3. Why is the statement "The shape of a lens determines whether it is a converging or diverging lens" not always true?

You need to know more than the shape. You also need to know whether light travels faster or slower in the lens than in the surrounding medium.

4. Draw ray diagrams for both a *convex* and a *concave* air lens encased in glass to show what these lenses do to light rays passing through them.

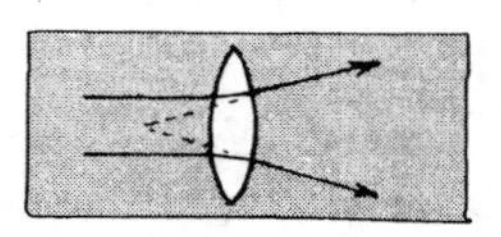

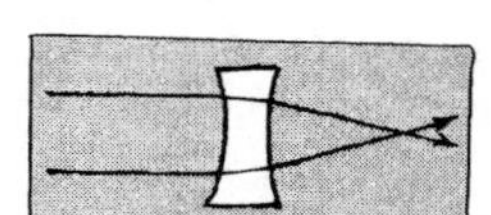

85 Rainbows Without Rain

Purpose

To observe and develop a hypothesis about a phenomenon of light interference.

Setup: 1
Lab Time: 1
Learning Cycle: exploratory
Conceptual Level: moderate
Mathematical Level: none

Adapted from PRISMS.

Required Equipment/Supplies

soap-bubble solution
wire frame for soap films
large, flat, rimmed pan (such as a cookie sheet)
oil
2 microscope slides or glass plates
2 rubber bands

Discussion

A rainbow is produced by the refraction and reflection of light from drops of water in the sky. Rainbow colors, however, can be produced in a variety of ways. Some of these ways will be explored in this lab activity.

Procedure   

Step 1: Pour some of the bubble solution into the flat pan. Place the loop of the wire frame into the solution. Hold the frame in a vertical position. Look at the soap film with the room lights behind you, reflecting off the film.

1. List as many observations as you can of what you saw in the soap film.

   Students should note many different colors. If the film is held vertically for a long enough time, the top will become dark.

2. How would you explain these observations?

   Answers will vary widely. See Teacher Notes for this activity in the front of the teacher's edition for a full explanation.

Step 2: During or after a rainfall, you may have noticed brilliant colors on a wet driveway or parking lot where oil has dripped from a car engine. To reproduce this situation, cover the bottom of the pan with a thin layer of water. Place a drop of oil on the water, and look at the oil slick with various angles of incident light.

3. List as many observations as you can of the oil on the water.

Students should see different colors in the oil film, depending on the angle at which they look at the film.

4. How would you explain these observations?

Answers will vary widely. See Teacher Notes for a full explanation.

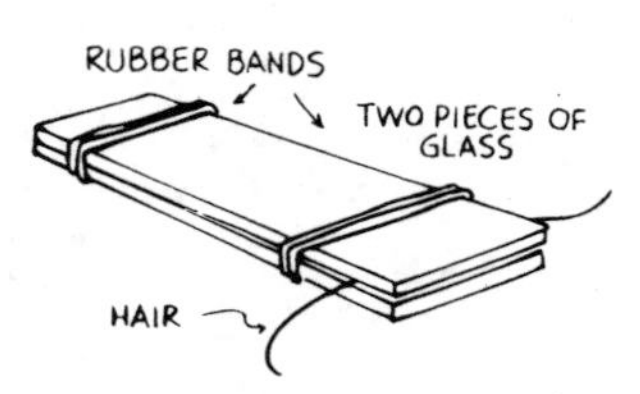

Step 3: Make a very thin wedge of air between two glass plates or microscope slides. You can do this by placing a hair across one end between the two. Fasten both ends together with rubber bands. Try to observe small repeating colored bands in the air wedge.

5. List your observations.

With optically flat plates of glass, the interference bands will be fairly straight across the plates. If you use microscope slides or longer strips of glass, the interference bands will still be there but not in straight patterns because of slight irregularities in the surfaces.

6. How would you explain these observations?

Answers will vary widely. See Teacher Notes for a full explanation.

Analysis

7. Summarize any patterns in your observations.

In all three situations, most or all colors of light are selectively blocked out. All three situations involve interference of reflected waves.

Name Period Date

86 Static Cling

Purpose

To observe some of the effects of static electricity.

Setup: <1
Lab Time: 1
Learning Cycle: concept development
Conceptual Level: moderate
Mathematical Level: none

Required Equipment/Supplies

electroscope
hard rubber rod and fur *or* glass rod and silk
plastic golf-club tube
foam rubber
Styrofoam (plastic foam) "peanuts" or packing material
coin with insulated connected string
empty can from soup or soda pop, with insulated connecting string

Activity

Discussion

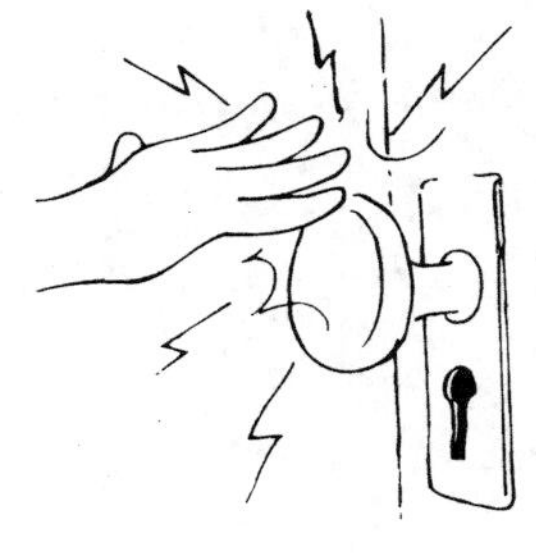

Have you ever been shocked after walking on a carpet and reaching for a doorknob? Have you ever found your sock hiding inside one of your shirts just after it came out of the clothes dryer? Have you ever seen a lightning bolt from closer range than you might like? All of these situations arise due to *static electricity*. After this activity, you should understand its behavior a bit better.

Procedure 

Step 1: Make sure the electroscope is discharged (neutral) by touching the probe with your finger. The leaves will drop down as far as possible.

CAUTION: *Do not open the electroscope in an effort to adjust the position of the leaves. NEVER touch the leaves.*

Step 2: Rub a hard rubber rod with fur, or a glass rod with silk. The rubber rod will become negatively charged, while the glass rod will become positively charged. Touch the probe of the electroscope with the charged rod.

1. What happens to the leaves of the electroscope?

The leaves spread apart.

2. What kind of charge is on the leaves?

The charge on the leaves will be the same as on the charged rod (negative if a rubber rod is used, positive if a glass rod is used).

Charge plastic tube.

Step 3: Discharge the electroscope by touching the probe with your finger. Charge a plastic tube by rubbing it with a piece of foam rubber. Observe what happens when you bring the charged tube close to (but not touching) the electroscope, and then move the tube away.

3. Record what happens.

The leaves spread apart when the tube is close, but collapse when the tube is moved away.

Charge electroscope by induction.

Step 4: Devise a way to leave a charge on the electroscope, using the charged plastic tube but without touching the tube to the probe.

4. Record the method you used to charge the electroscope by induction.

Bring the charged tube up to (but not touching) the electroscope probe, and touch the probe with your finger. Without moving the tube away, lift your finger off the probe. Then move the tube away.

Test charge on electroscope.

Step 5: Test whether the charge on the electroscope is positive or negative by bringing a charged glass or rubber rod close to (but not touching) the probe.

5. Is the charge on the electroscope positive or negative? Explain how you can tell.

Positive. A negatively charged rubber rod will cause the leaves to collapse, while a positively charged glass rod will make the leaves spread farther apart.

Explore Styrofoam "peanuts."

Step 6: Charge the plastic tube. Now put some Styrofoam packing "peanuts" out on the table. Try to pick them up with the charged tube, or pour some over the charged tube. See how many different ways you can make the "peanuts" interact with the charged tube.

6. Describe and explain the behavior of the "peanuts."

The "peanuts" jump and hop and are attracted to the tube.

Explore charge on tube.

Step 7: Charge the plastic tube by rubbing it with different materials. Each time, charge the electroscope by induction, as in Step 4, and test whether the charge on the electroscope is positive or negative.

87 Sparky, the Electrician

Purpose

To study various arrangements of a battery and bulbs and the effects of those arrangements on bulb brightness.

Setup: <1
Lab Time: >1
Learning Cycle: exploratory
Conceptual Level: moderate
Mathematical Level: moderate

Adapted from PRISMS.

Required Equipment/Supplies

size-D dry cell (battery)
6 pieces of bare copper wire
3 flashlight bulbs
3 bulb holders
second size-D dry cell (optional)

Activity

Discussion

A dry cell (commonly called a battery) is a source of electric energy. Many arrangements are possible to get this energy from dry cells to flashlight bulbs. In this activity, you will test these arrangements to see which makes the bulbs brightest.

Procedure 

Step 1: Arrange one bulb (without a holder), one battery, and wire in as many ways as you can to make the bulb emit light. Sketch each of your arrangements, including failures as well as successes. Label the sketches of the successes.

1. Describe the similarities among your successful trials.

There must be a direct connection (by wire or contact) between the metal bottom of the bulb and one end of the battery. There must also be a direct connection between the metal side of the bulb and the opposite end of the battery. There can be no direct connection between the two ends of the battery or between the metal side and bottom of the bulb.

Step 2: Use a bulb in a bulb holder (instead of a bare bulb), one battery, and wire. Arrange these in as many ways as you can to make the bulb light.

2. What two parts of the bulb does the holder make contact with?

The holder makes contact with the metal side and metal bottom of the bulb.

Step 3: Using one battery, light as many bulbs in holders as you can. Sketch each of your arrangements, and note the ones that work.

3. Compare your results with those of other students. What arrangement(s), using only one battery, made the most bulbs glow?

If students discover them, parallel arrangements of the bulbs will be most successful in lighting the most bulbs with one battery.

Step 4: Diagrams for electric circuits use symbols like the ones in Figure A.

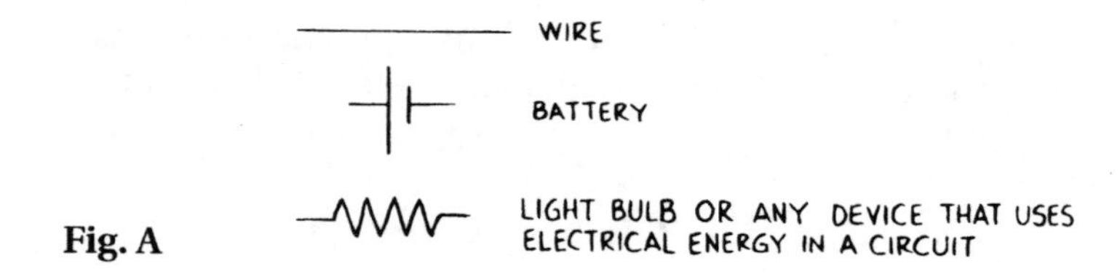

Fig. A

Connect the bulbs in holders, one battery, and wire as shown in each circuit diagram of Figure B. Circuits like these are examples of *series circuits.*

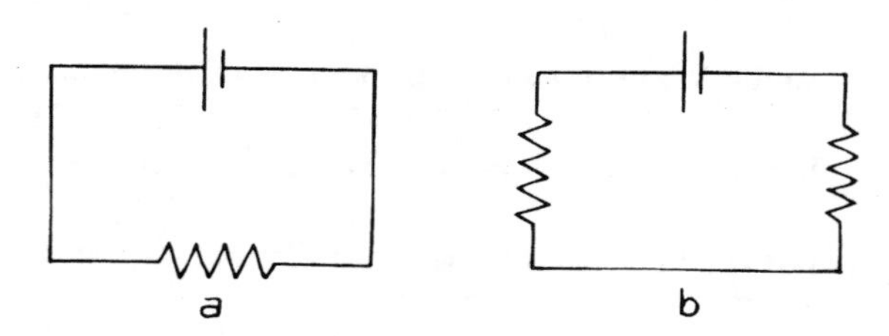

Fig. B

Name Period Date

4. Do the bulbs light in each of these series circuits?

The bulbs should light if the circuit is connected properly, but the two bulbs in series will be less bright than the single bulb.

Step 5: In the circuit with two bulbs, unscrew one of the bulbs.

5. What happens to the other bulb?

The other bulb goes out.

Step 6: Set up the circuit shown in the circuit diagram of Figure C. A circuit like this is called a *parallel circuit.*

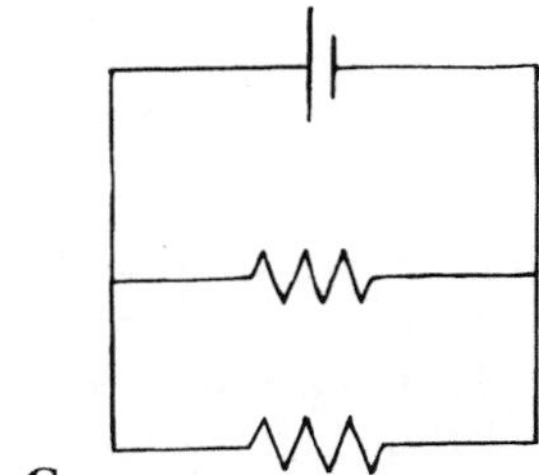

Fig. C

6. Do both bulbs light in this parallel circuit?

The bulbs should light if the circuit is connected properly.

Step 7: Unscrew one of the bulbs in the parallel circuit.

7. What happens to the other bulb?

The other bulb remains lit.

8. In your own words, describe the differences between series and parallel circuits.

Answers may emphasize the single vs. separate paths or the dependence vs. independence of circuit elements.

Going Further

Step 8: Using two batteries, light as many bulbs as you can. Sketch each of your arrangements, and note the ones that work.

9. What arrangement(s), using two batteries, lit the most bulbs?

The most effective way to use the two batteries is to connect the end of one to the opposite end of the other (put the batteries in series).

Step 9: Using three bulbs and two batteries, discover the arrangements that give different degrees of bulb brightness. Sketch each of your arrangements, and note the bulb brightness on the sketches.

When one of two bulbs in a parallel circuit is unscrewed, the other bulb normally not only remains lit but maintains the same brightness. With flashlight bulbs, however, the resistance of the connecting wire is appreciable compared with that of the remaining bulb, so its brightness may increase slightly.

10. How many different degrees of brightness could you obtain using three bulbs and two batteries? Did other students use different arrangements?

There are quite a number of degrees of brightness. The weakest setup is to put the batteries in parallel and the bulbs in series; the strongest is to put the batteries in series and the bulbs in parallel.

Name Period Date

88 Brown Out

Purpose

To investigate charging and discharging a capacitor through a bulb.

Setup: <1
Lab Time: <1
Learning Cycle: exploratory
Conceptual Level: moderate
Mathematical Level: none

Required Equipment/Supplies

CASTLE Kit (available from PASCO)
or
1 25,000 μF capacitor (20 volts nonpolar)
2 #14 lightbulbs (round) (no substitutions allowed!)
2 #48 lightbulbs (long) (no substitutions allowed!)
4 lightbulb sockets
1 packet of 12 alligator leads
1 D-cell battery holder and 3 D-cells

It's important to use a voltage source between 3.6 and 4.5 volts (3 D-cells) for the specified bulbs. A greater voltage is likely to burn out the bulbs; a smaller voltage is not sufficient.

Activity

Discussion

When you switch on a flashlight, the maximum brightness of the bulb occurs immediately. If a capacitor is in the circuit, however, there is a noticeable delay before maximum brightness occurs. When the circuit contains a capacitor, the flow of charge through the circuit may take a noticeable time. How much time depends upon the resistance of the resistor and the charge capacity of the capacitor. In this activity, we will place a resistor (lightbulb) between the battery and the capacitor to be charged. By using bulbs of different resistances, the charging and discharging times are easily observed.

Procedure 

Assemble the circuit and charge the capacitor.

Step 1: Connect a battery, two long bulbs, and a blue capacitor (25,000 μF) as shown in Figure A. Leave one wire (lead) to the battery disconnected. *Close* the circuit by connecting the lead to the battery and observe how long the bulb remains lit. You are *charging the capacitor.*

Step 2: Disconnect the leads from the battery and remove the battery from the circuit. Connect the two leads that were connected to the battery to each other as shown in Figure B. Observe the length of time the bulbs remain lit. This process is called *discharging the capacitor.*

Step 3: Replace the long (#48) bulbs in the circuit with round (#12) bulbs and charge the capacitor. Observe the length of time the bulbs remain lit. Remove the battery from the circuit as in Step 2 and discharge the capacitor through the round bulbs. Observe the time the bulbs remain lit. Which bulbs remain lit longer as the capacitor charges and discharges—long or round bulbs?

Fig. A

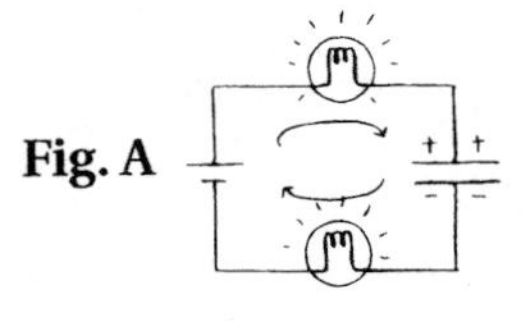

Fig. B

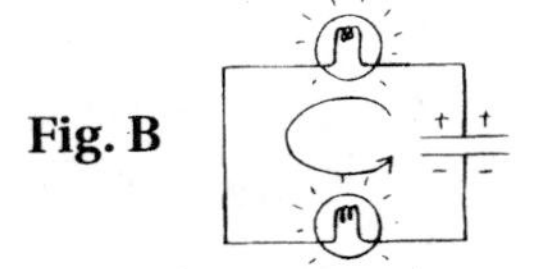

The long bulbs remain lit a longer time than the round bulbs.

To account for the different times the bulbs are lit, we can make the following hypotheses:

- If bulbs affect the amount of charge that passes through them, bulbs that remain lit longer allow more charge to pass through them. The charge would be stored in the capacitor.
- If, however, bulbs affect the rate of charge flow rather than the amount of charge that flows, then bulbs that remain lit longer will increase the time during which charge flows through them.

Charge the capacitor.

Step 4: Charge the capacitor through two long bulbs. Now remove the long bulbs from their sockets and replace them with round bulbs, being careful not to accidentally discharge the capacitor.

1. Suppose the capacitor stores the same amount of charge no matter what type of bulbs are used during charging. Will discharging through round bulbs take more, less, or the same time as in Step 3—when the capacitor was *charged* through round bulbs?

   The discharging should take the same time.

2. If, instead, a longer charging time indicates more charge is stored in the capacitor than occurs with a shorter charging time, will discharging now through round bulbs take more, less, or the same time as it did in Step 4?

   The discharging will take more time.

Discharge the capacitor.

Step 5: Remove the battery from the circuit and discharge the capacitor. Is the time the bulbs remain lit longer, shorter, or the same as in Step 3?

The discharging time through the single bulb is somewhat shorter than through two similar bulbs.

Analysis

3. (a) What is one use of a capacitor?

   To store charge (somewhat like a battery, but with important differences).

   (b) Does the amount of charge stored in a capacitor depend on the type of bulbs through which it was charged? Explain.

   No. Switching the bulbs after the capacitor has been charged does not change the discharge *time*.

4. Is it true that the type of bulb affects the rate charge flows through it? Why or why not?

   True, evidenced by the difference between the round and long bulbs.

Name Period Date

89 Ohm Sweet Ohm

Experiment

Purpose

To investigate how the current in a circuit varies with voltage and resistance.

Setup: 1
Lab Time: 1
Learning Cycle: concept development
Conceptual Level: moderate
Mathematical Level: none

Required Equipment/Supplies

nichrome wire apparatus with bulb
2 1.5-volt batteries
or
2 Genecon® handheld generators

Discussion

Normally, it is desirable for wires in an electric circuit to stay cool. Red-hot wires can melt and cause short circuits. There are notable exceptions, however. Nichrome wire is a high-resistance wire capable of glowing red-hot without melting. It is commonly used as the heating element in toasters, ovens, stoves, hair dryers, and so forth. In this experiment, nichrome wire is used as a variable resistor. Doubling the length of a piece of wire doubles the resistance; tripling the length triples the resistance, and so on.

Tungsten wire is capable of glowing white-hot and is used as filaments in lightbulbs. Light and heat are generated as the current heats the high-resistance tungsten filament. The hotter the filament, the brighter the bulb. For the same voltage, a bright bulb (such as a 100-watt bulb) has a *lower* resistance than a dimmer bulb (such as a 25-watt bulb). Just as water flows with more difficulty through a thinner pipe, electrical resistance is greater for a thinner wire. Manufacturers make bulbs of different wattages by varying the thickness of the filaments, so we find that a 100-W bulb has a lower resistance and a thicker filament than a 25-W bulb.

In this lab, the brightness of the bulb will be used as a current indicator. A bright glow indicates a large current is flowing through the bulb; a dim glow means a small current is flowing.

Procedure

Assemble the battery with four D-cells.

Be *sure* students understand the circuit before attempting to do this lab.

Step 1: Connect four D-cell batteries in series, so that the positive terminal is connected to the negative terminal in a battery holder as shown in Figure A. This arrangement, with Terminal #1 as ground, will provide you with a variable voltage supply as indicated in Table A.

Data Table A

TERMINAL #'s	VOLTAGE
1-2	1.5
1-3	3.0
1-4	4.5
1-5	6.0

Assemble the circuit and draw a circuit diagram.

Step 2: Assemble the circuit as shown in Figure B. Label one binding post of the nichrome wire "A" and the other "B". Attach the ground lead (#1) of the voltage supply to one side of a knife switch. Connect the other side of the switch to binding Post A on the thickest nichrome wire. Connect the 3-volt lead (#3) from the voltage supply to a clip lead of a test bulb. Attach the other clip lead of the test bulb to the other binding Post B on the nichrome wire.

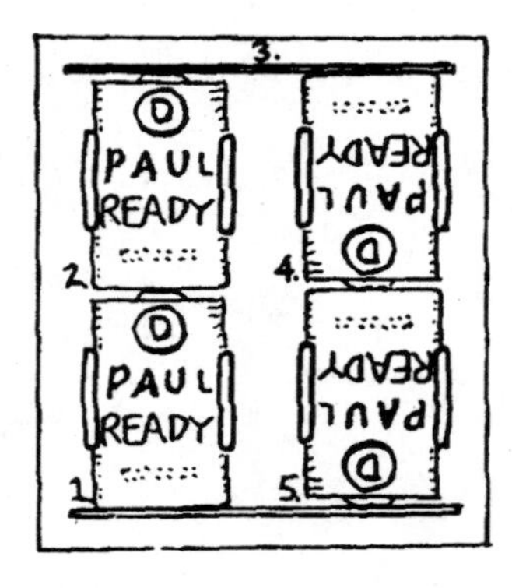

Fig. A

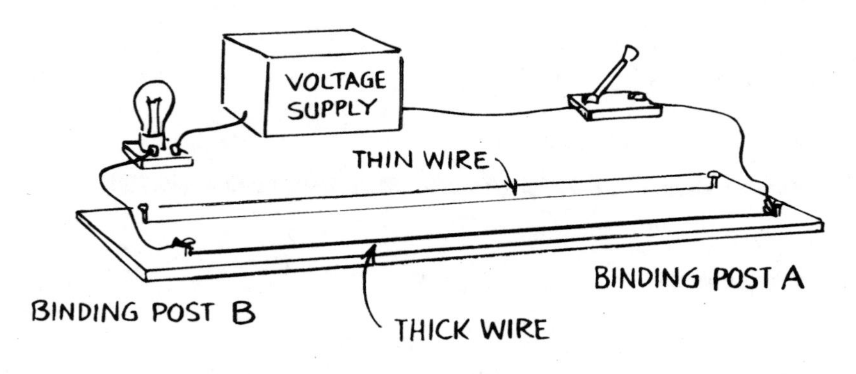

Fig. B

The voltage supply is now connected so that the current passes through two resistances: the bulb and the nichrome wire. You will vary the resistance in the circuit by moving the clip lead of the test bulb from binding Post B to binding Post A. Using the standard symbols for the circuit elements, draw a diagram that represents this *series* circuit.

Note: *Always apply power from battery packs by closing a switch and make your measurements quickly.* Leave the power on just long enough to make your measurements and then open the switch. *Leaving the power on in the circuit for long periods of time will drain your batteries and heat the wire, thereby changing its resistance.*

Observe the brightness of the bulb.

Step 3: After carefully checking all your connections, apply power to the circuit by closing the switch. Observe the intensity of the bulb as you move the test bulb lead from binding Post B toward A.

1. What happens to the brightness of the bulb as you move it from Binding Post B to A?

The brightness of the bulb decreases from B to A.

Repeat with thinner wire.

Step 4: Repeat using the thinner nichrome wire. Observe the relative brightness of the bulb as you move the bulb's lead closer to Binding Post A.

2. How does the brightness of the bulb with the thinner wire compare with the brightness of the bulb when connected to a thicker wire?

The bulb is dimmer when the bulb is connected in the same relative position and dims when the lead is moved from B to A.

3. What effects do the thickness and length of the wire have on its resistance?

Based upon the brightness of the bulb, the resistance apparently increases with length and decreases with diameter (and hence cross-sectional area) of the wire.

4. Does the current of the circuit increase or decrease as you move the lead closer to Binding Post B? As you move the lead from B to A, does the resistance of the circuit increase or decrease?

Based upon the brightness of the bulb, the current in the circuit decreases as the resistance increases.

Step 5: Repeat Steps 1–2 using the 4.5 and 6-volt leads instead of the 3-volt leads.

5. How does the brightness of the test bulb compare for the two nichrome wires using 4.5 volts instead of 3 volts?

The brightness of the bulb is greater with the greater voltage, but dims as the resistance of the circuit increases as you move the lead from B to A, just as in the case with 3 volts.

6. Combining your results from Questions 4 and 5, how does the current in the circuit depend upon voltage and resistance?

The current increases as voltage increases and decreases as resistance increases.

Going Further

Step 6: Now insert an ammeter into the circuit as illustrated in Figure C. With the thicker piece of nichrome wire in the circuit, place the ammeter in series with the voltage supply between Terminal #1 of the voltage supply and the switch. The ammeter will read total current in the circuit. Measure the current in the circuit as you move the test bulb lead from B to A. Be sure to apply power *only* while making the measurements to prevent draining the batteries. Repeat using the thinner wire.

Note: *If you are not using a digital meter, you may have to reverse the polarity of the leads if the needle of the meter goes the wrong way (–) when power is applied.*

Install ammeter in the circuit and measure current.

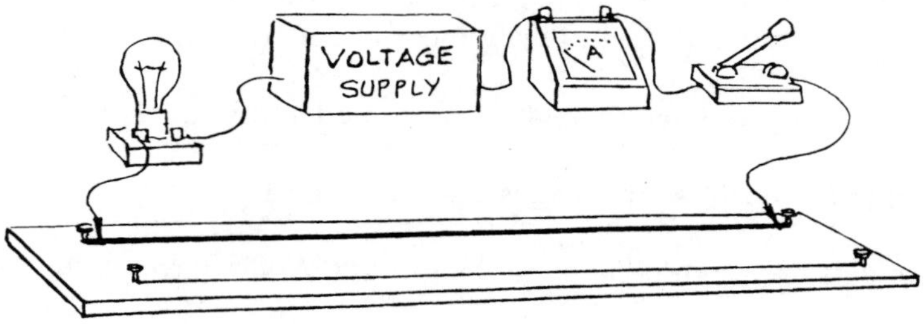

Fig. C

7. Do your results show a decrease in current as the resistance (or length of the wire) is increased?

Yes, the current decreases as the length of the wire increases.

8. Do your results show an increase in current as the voltage is increased?

Yes, the current increases as the voltage increases.

9. How do the currents in the thicker and thinner wires compare when the same voltage is applied to the same lengths of wire?

The current is less in the thinner wire. It has greater resistance.

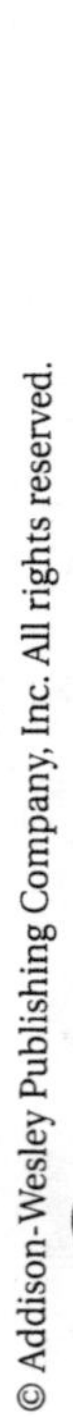

Getting Wired

Purpose

To build a model that illustrates electric current.

Setup: 1
Lab Time: 1
Learning Cycle: exploratory
Conceptual Level: easy
Mathematical Level: none

Required Equipment/Supplies

4 D-cells (1.5 volt)
D-cell battery holder
3 alligator clip leads
2 bulbs and bulb holders

Discussion

While we can float on a raft gliding down the Mississippi or ride in cars moving in traffic, nobody can *see* electric current flow. Even in the case of lightning, we are seeing the flash of hot glowing gases produced by electric current—not the current itself. However, we can infer the presence of electric current using lightbulbs and magnetic compasses in much the same way as a flag indicates the presence of wind. In this activity, you will build a *model* to study electric current.

Activity

Part A: What Is Happening in the Wires?

Procedure 

Step 1: With the circuit arranged as shown in Figure A, turn the bulbs on and off by connecting and disconnecting one of the wires.

Observe the wires when the circuit is closed.

Encourage students to visualize what is going on *in* the wires.

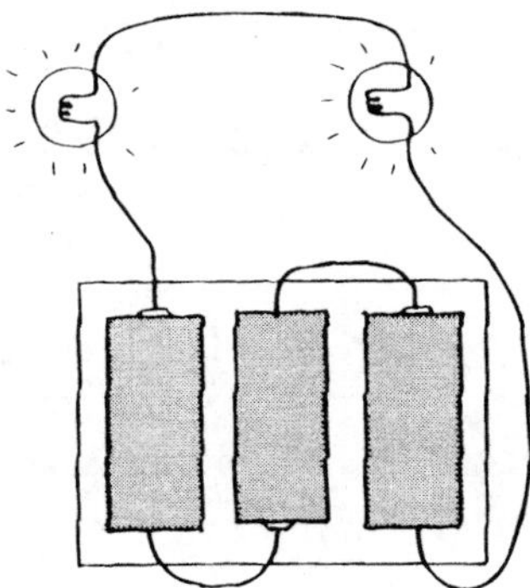

Fig. A

1. Is there any visual evidence that something is moving around the circuit when the bulbs are lit? For example, does one bulb light before the other? Does one bulb go out before the other? Is one bulb brighter than the other?

Nothing visually indicates something is moving in the closed circuit—at least nothing is moving at a speed that is noticeable. As long as the bulbs are the same, they will have the same brightness, and neither bulb will go out before the other when disconnected.

Position a compass underneath the wire.

Step 2: Place a magnetic compass on the table near the circuit with the needle pointing to the "N." With the bulbs unlit, place one of the wires on top of the compass parallel to the needle as in Figure B. Connect and disconnect a lead in the circuit several times while you observe the compass needle. Observe what happens to the needle when the bulb lights. Observe what happens to the needle when the bulbs go out.

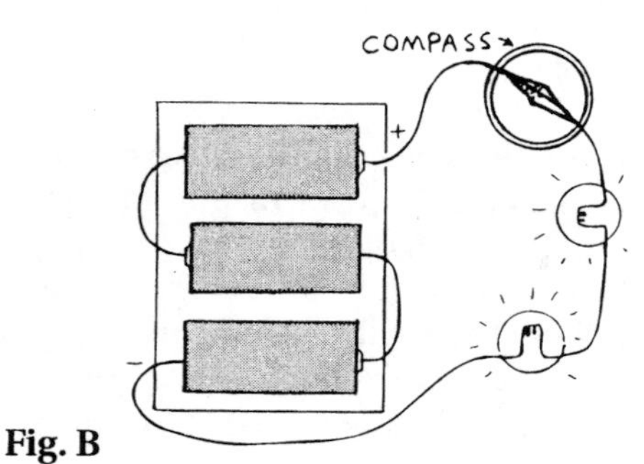

Fig. B

Step 3: Place the compass beneath the wire in different parts of the circuit. Be sure the needle is parallel to the wire when the bulbs are not lit. Observe the needle as you open and close the circuit several times. Look to see if the needle deflects in the same direction as before. Look to see if the amount of the needle's deflection is the same as before. Also, observe whether the bulbs must be lit to deflect the compass needle.

2. What evidence supports the notion that something is happening in the wires while the bulbs are energized?

The deflection of the compass needle indicates something is going on while the bulb is lit.

3. What evidence supports the notion that whatever is happening occurs uniformly in all parts of the circuit?

The size and direction of the deflection of the compass needle is the same everywhere in the circuit.

Name Period Date

Part B: Is There Directionality to What Is Happening in the Circuit?

Step 4: Arrange the circuit as in Figure A. Place one of the wires on top of the compass parallel to the needle. Open and close the circuit while you carefully observe the needle. Note whether the needle deflects clockwise or counterclockwise.

Observe the deflection of the compass.

Step 5: Reverse the leads from the battery without altering the circuit and compass. Do this by simply exchanging the lead connected to the positive terminal of the battery with the lead connected to the negative terminal. Open and close the circuit while you watch the compass needle. Watch the needle deflect and note whether it is clockwise or counterclockwise.

Reverse the leads from the battery and observe the deflection of the compass.

4. Is the direction of the deflection the same as in Step 4, before the leads to the battery were reversed? Is the amount of the needle's deflection the same as before?

The deflection of the compass needle is in the same amount but the opposite direction as in Step 4.

5. Suppose something is flowing in the wires. Do you think the *direction* the needle is deflected is caused by the amount of the flow or the direction of the flow?

The direction of the compass needle's deflection is due to the direction of the flow in the wires.

Step 6: Remove one of the D-cells from the battery holder so that the battery holder only has two cells instead of three. Carefully observe deflection of the needle while you repeat Steps 4 and 5.

Remove one of the cells from the battery.

6. How do the size and direction of the compass needle's deflections compare when you use two cells instead of three?

The deflection of the compass needle is less but in the same direction as with three D-cells.

Step 7: Install two more D-cells in the battery holder so that it has four D-cells. Carefully observe deflection of the needle as you repeat Steps 4 and 5.

Add two cells to the battery.

7. How do the deflections of the compass needle compare with those with two and three D-cells? Do you think the *size* of the needle's deflection is caused by *amount of the flow* or the *direction of the flow*?

The deflection of the compass needle is greater but in the same direction as with three D-cells. The size of the deflection depends on the amount of flow—deflection is more vigorous when bulbs are brightest—i.e., when a stronger battery is used.

Analysis

8. Hypothesize what is happening in the circuit when bulbs are lit.

Some form of electrical energy is moving in the circuit.

9. Hypothesize what is happening in the circuit when the direction of the compass deflection reverses.

The current, or the electricity, reverses direction.

10. Hypothesize what is happening in the wires when the amount of the needle's deflection increases or decreases.

More electrical energy moves in the circuit.

11. What do you think the battery does?

The battery is a source of energy that supplies the current, or electricity that flows through the circuit.

91 Cranking Up

Experiment

Purpose

To observe and compare the work done in a series circuit and the work done in a parallel circuit.

Setup: 1

Lab Time: 1

Learning Cycle: concept development

Conceptual Level: moderate

Mathematical Level: moderate

Required Equipment/Supplies

4 lightbulbs, sockets and clip leads
Genecon
parallel bulb apparatus
voltmeter
ammeter

Discussion

Part A: Qualitative Investigation  

Step 1: Assemble four bulbs in series as shown in Figure A. Screw all the bulbs into their sockets. Connect the sockets with clip leads.

Connect one lead of a Genecon to one end of the string of bulbs and the other lead to the other end of the string. Crank the Genecon so that all the bulbs light up. Now, disconnect one of the bulbs from the string and reconnect the Genecon. Crank the Genecon so that the three remaining bulbs are energized to the same brightness as the four-bulb arrangement. How does the crank feel now? Repeat, removing one bulb at a time and comparing the cranking torque each time.

Assemble the circuit and crank the Genecon as you unscrew bulbs in series.

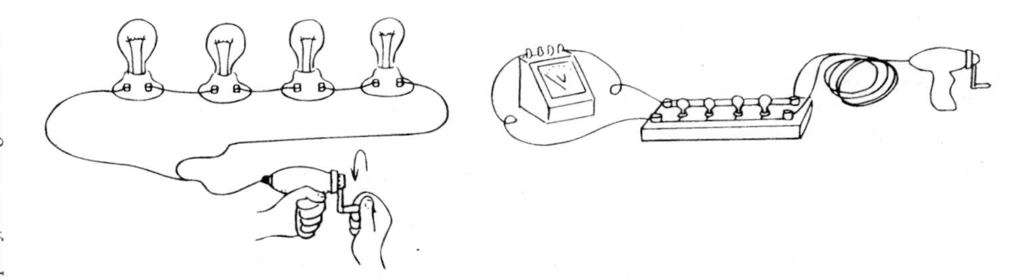

Fig. A

Fig. B

The series arrangements require bulbs of *identical* resistance (or as nearly so as possible) if they are to flow reasonably well with D-cells. This subtlety can be very confusing for students if their bulbs have different resistances.

Step 2: Assemble the circuit with the parallel bulb apparatus as shown in Figure B. Each end of the bulb apparatus has two terminals. Connect the leads of a voltmeter to one pair of terminals on one end of the apparatus. Connect the leads of the Genecon to the terminals on the other end of the apparatus. Crank the Genecon with all the bulbs unscrewed in the sockets so that they don't light. Then, have your partner screw them in one at a time as you crank on the Genecon. Try to keep the bulbs energized at the same brightness as each bulb is screwed into its socket.

Assemble the parallel circuit and crank the Genecon as you screw in the bulbs.

1. What do you notice about the *torque* required to crank the Genecon at a constant speed as more bulbs are added to the circuit?

More torque is required.

2. How would you describe the amount of torque required to crank the Genecon to energize four bulbs in series compared with that required for four bulbs in parallel?

The torque is the same.

3. If all the bulbs in the series and parallel circuits are glowing equally brightly, is the energy expended (the work you are doing to crank the Genecon) the same?

Yes.

Part B: Quantitative Investigation—Resistors in Series

Now repeat Part A in a quantitative fashion using a voltmeter and an ammeter.

Assemble the circuit in series.

Step 3: Assemble four bulbs in a series circuit and connect the meters as shown in Figure C. Connect the voltmeter in parallel with all four bulbs so you can measure the total voltage applied to the circuit. Then you will connect it in parallel with single bulbs to measure the voltage across each bulb. Connect the 3-volt lead from the voltage supply to one terminal of the bulbs and the ground connection to one lead of an ammeter. Connect the other lead of the ammeter to the second terminal of the bulbs. The ammeter will measure the *total* current in the circuit.

To see the quantitative differences between series and parallel circuits, it is best to use batteries. Although the Genecon is great for hands-on experience, it is not a good tool for getting accurate quantitative data.

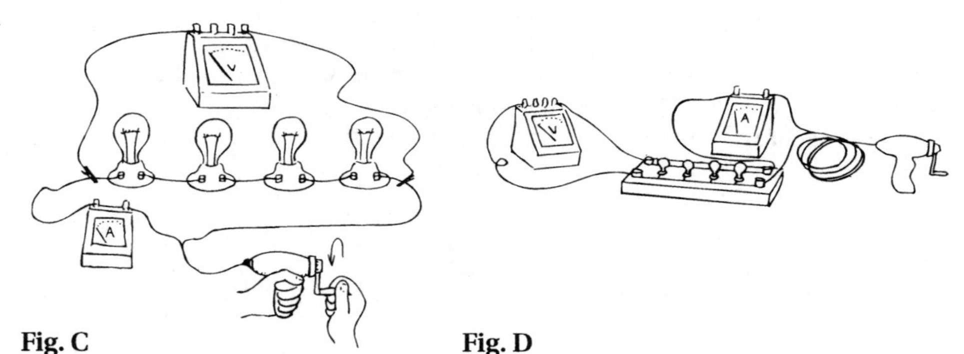

Fig. C

Fig. D

Note: *If you are not using digital meters, you may have to reverse the polarity of the leads if the needle of the meter goes the wrong way (–) when power is applied.*

Close the switch, apply power to the circuit, and measure the current in the circuit, the voltage applied to the circuit, and the voltage across each bulb. Record your results in Data Table A.

Step 4: Now remove one of the bulbs from the string, close the gap in the circuit, and repeat your measurements for three bulbs. Then remove the other bulbs one at a time, closing the gap in the circuit each time, and repeat the measurements. Record your data in Data Table A.

Name Period Date

# BULBS	TOTAL CURRENT (A)	TOTAL VOLTAGE (V)	VOLTAGE ACROSS EACH BULB (V)			
1						
2						
3						
4						

Data Table A

# BULBS	TOTAL CURRENT (A)	TOTAL VOLTAGE (V)	VOLTAGE ACROSS EACH BULB (V)			
1						
2						
3						
4						

Data Table B

Step 5: Repeat using the 4.5-volt terminal of the voltage supply instead of the 3-volt terminal. Record your data in Data Table B.

Repeat using a different voltage.

4. Is there any change in brightness as the number of bulbs changes?

   Yes, the brightness decreases as more bulbs are added to the circuit.

5. Does the voltage applied in the circuit change as you add more bulbs?

   No, the voltage remains unchanged regardless of the number of bulbs.

6. How are the voltages across each bulb related to the applied voltage?

   The voltage drop across all bulbs is about the same. The sum of the voltage drops across all the bulbs equals the applied voltage.

7. How does the current supplied by the battery change when more bulbs are added?

   Total current in the circuit decreases as the number of bulbs increases.

8. Did any of the rules you discovered relating voltages and currents change when you applied 4.5 volts instead of 3 volts?

   No.

Part C: Quantitative Investigation—Resistors in Parallel

Step 6: Assemble the circuit and connect the meters as shown in Figure D. Connect the voltmeter in parallel with the bulbs by connecting the voltmeter to two terminals on one end of the parallel bulb apparatus. Connect the 3-volt lead from the voltage supply to one terminal of the parallel bulb apparatus. Connect the ground lead from the voltage supply to one lead of an ammeter; connect the other lead of the ammeter to the second terminal of the parallel bulb apparatus. The ammeter will measure the total current in the circuit.

Assemble the circuit in parallel.

Make sure the bulbs are not loose in their sockets. Close the switch and apply power to the circuit. Observe the brightness of the bulbs, then unscrew the bulbs one at a time.

Record your measurements of current and voltage.

Step 7: Screw the bulbs back in, one at a time, each time measuring the current in the circuit, the voltage applied to the circuit, and the voltage drop across each bulb. Record your data in Data Table C.

# BULBS	TOTAL CURRENT (A)	TOTAL VOLTAGE (V)	VOLTAGE ACROSS EACH BULB (V)			
1						
2						
3						
4						

Data Table C

# BULBS	TOTAL CURRENT (A)	TOTAL VOLTAGE (V)	VOLTAGE ACROSS EACH BULB (V)			
1						
2						
3						
4						

Data Table D

Repeat using a different voltage.

Step 8: Repeat Steps 3 and 4 using the 4.5-volt terminal of the voltage supply instead of the 3-volt terminal. Record your data in Data Table D.

9. Is there any change in brightness as the number of bulbs changes?

No, this time the bulbs remain the same brightness.

10. Does the voltage across each bulb change as more bulbs are added to or subtracted from the circuit?

No, not appreciably.

11. Does the applied voltage to the circuit change as you add more bulbs?

No, not appreciably.

12. How does the current supplied by the battery change as the number of bulbs in the circuit changes?

In direct proportion to the number of bulbs. When 3 volts is applied, the current amounts to about 0.2 amp/bulb. The amount of increase decreases slightly due to the drop in the terminal voltage (due to internal resistance) of the battery as more bulbs are added to the circuit.

13. Did the ratio of voltage and current change when you applied 4.5 volts instead of 3 volts?

No; the ratio of voltage to current equals the resistance.

Name Period Date

92 3-Way Switch

Purpose

To explore ways to turn a lightbulb on or off from either one of two switches.

Setup: <1
Lab Time: 1
Learning Cycle: application
Conceptual Level: moderate
Mathematical Level: none

Required Equipment/Supplies

2.5-V dc lightbulb with socket
connecting wire
2 single-pole double-throw switches
2 1.5-V size-D dry cells connected in series in a holder

Thanks to Manuel DaCosta for his ideas and suggestions for this lab.

Discussion

Frequently, multistory homes have hallways with ceiling lights. It is convenient if you can turn a hallway light on or off from a switch located at either the top or bottom of the staircase. Each switch should be able to turn the light on or off, regardless of the previous setting of either switch. The same arrangement is often adopted in a room with two doors. In this activity, you will see how simple, but tricky, such a common circuit really is!

Procedure 

Step 1: Examine a 3-volt battery (formed from two 1.5-volt dry cells with the positive terminal of one connected to the negative terminal of the other). Connect a wire from the positive terminal of the battery to the center terminal of a single-pole double-throw switch. Connect a wire from the negative terminal of the same battery to one terminal of the lightbulb socket. Connect the other terminal of the lightbulb socket to the center terminal of the other switch.

Step 2: Now interconnect the free terminals of the switches so that the bulb turns on or off from either switch. That is, when both switches are closed in either direction, moving either switch from one side to the other will always turn an unlit bulb on or a lit bulb off.

Devise working circuit.

Step 3: Draw a simple circuit diagram of your successful circuit.

Diagram 3-way switch.

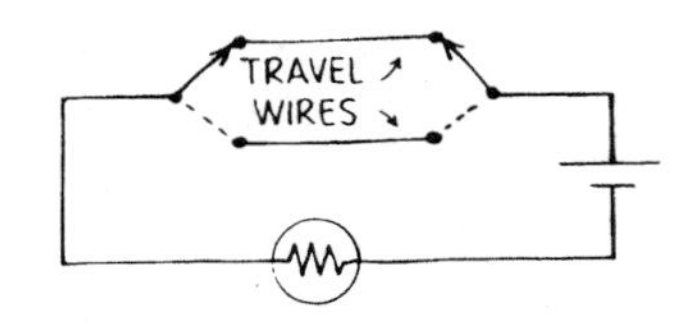

Activity

Reverse polarity of battery.

Step 4: The polarity of a battery can be reversed in a circuit by switching the connections to the positive and negative terminals. Predict whether your successful circuit will work if you reverse the polarity of the battery.

prediction: ____________________

Now reverse the polarity and record the result.

result: **The circuit will work.**

Interchange battery and bulb.

Step 5: Predict whether your successful circuit will work if you reconnect the circuit so that the battery is where the lightbulb is now, and vice versa.

prediction: ____________________

Now try it and record your results.

results: **The circuit will work.**

Analysis

An ordinary switch has an "on" setting, which closes the circuit at that point, and an "off" setting, which opens the circuit at that point. On the switches you used in this activity, what function do the two "closed" settings on each switch have? Can either setting keep the circuit open independently of how the other switch is set?

The two switches are connected to each other by two wires in parallel (called *travel wires*). One setting of each switch closes the connection between the switch and one travel wire. The other setting opens that connection but closes the connection between the switch and the other travel wire. Thus, neither setting can keep the circuit open independently of the setting on the other switch.

Name Period Date

93 3-D Magnetic Field

Purpose

To explore the shape of magnetic fields.

Setup: <1
Lab Time: 1
Learning Cycle: exploratory
Conceptual Level: easy
Mathematical Level: none

Adapted from PRISMS.

Required Equipment/Supplies

2 bar magnets
iron filings
strong horseshoe magnet
sheet of clear plastic
5 to 10 small compasses
jar of iron filings in oil
paper

Discussion

A magnetic field cannot be seen directly, but its overall shape can be seen by the effect it has on iron filings.

Activity

Procedure  

Step 1: Vigorously shake the jar of iron filings. Select the strongest horseshoe magnet available. Place the jar over one of the poles of the magnet and observe carefully. Place the jar at other locations around the magnet to observe how the filings line up.

1. What happened to the iron filings when they were acted upon by the magnetic field of the magnet?

The iron filings align along paths that are closer together near the poles of the magnet. (A torque on the magnetic domains of the iron filings aligns them along the direction of the magnetic field lines.) At the poles, the filings accumulate like spikes radiating outward.

Step 2: From all your observations, draw a sketch showing the direction of the magnetic field all around your magnet, as observed from the side. Also, draw a sketch as viewed from the end of the magnet.

Sketch magnetic field.

Observe orientation of compasses.

Step 3: Obtain two bar magnets and 5 to 10 small compasses. Note which end of each compass points toward the north. As you proceed with the activity, represent each compass as an arrow whose point is the north-pointing end.

Trace magnetic field lines.

Step 4: Trace one of the bar magnets on a piece of paper. Move the compasses around the magnet, and use arrows to draw the directions they point at each location. Link the arrows together by continuous lines to show the magnetic field.

Sketch magnetic field lines.

Step 5: Obtain a small quantity of iron filings and a sheet of clear plastic. Place the plastic on top of one of the bar magnets, and sprinkle a small quantity of iron filings over the plastic. It may be necessary to gently tap or jiggle the plastic sheet. The filings will line themselves up with the magnetic field. In the following space, sketch the pattern that the filings make. Repeat this step using the other bar magnet.

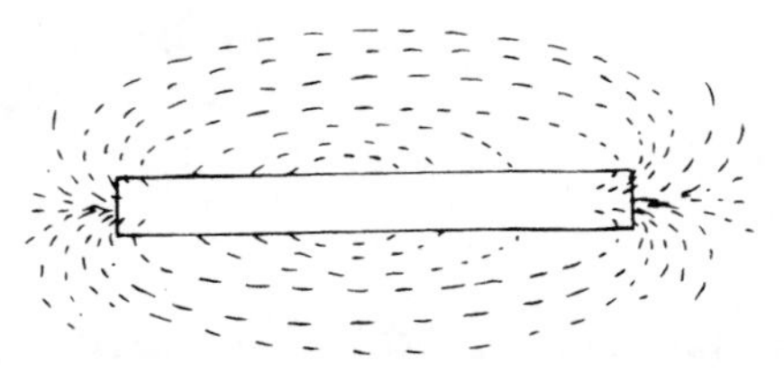

Repeat using two magnets.

Step 6: Repeat Step 5 for two bar magnets with like poles facing each other, such as N and N or S and S, and with unlike poles facing each other. Sketch the pattern of the filings in both situations.

2. Compare the methods of Steps 4 and 5 in terms of their usefulness in obtaining a quick and accurate picture of the magnetic field.

The small compasses indicate the direction of the magnetic field at their locations. The iron filings give a quick indication of the overall magnetic field lines in a plane. Also, where the iron filings are closest together, the field is strongest.

3. Are there any limitations to either method?

The compasses do not align well if the magnetic field is weak. The iron filings clump into uneven piles and make the magnetic field lines difficult to see.

4. What generalizations can you make about magnetic field lines?

The field lines begin and end on a magnet (either the same magnet or one nearby). Also, the field lines never cross. Field lines indicate the direction and strength of the magnetic field.

Name Period Date

94 You're Repulsive

Purpose

To observe the force on an electric charge moving in a magnetic field, and the current induced in a conductor moving in a magnetic field.

Setup: <1
Lab Time: 1
Learning Cycle: exploratory
Conceptual Level: moderate
Mathematical Level: none

Adapted from PRISMS.

Required Equipment/Supplies

cathode ray oscilloscope or computer with monitor
horseshoe magnet
bar magnet
compass
50 cm insulated wire
galvanometer or sensitive milliammeter
masking tape

Activity

Discussion

In this lab activity, you will explore the relationship between the magnetic field of a horseshoe magnet and the force that acts on a beam of electrons that move through the field. You will see that you can deflect the beam with different orientations of the magnet. If you had more control over the strength and orientation of the magnetic field, you could use it to "paint" a picture on the inside of a cathode ray tube with the electron beam. This is what happens in a television set.

A moving magnetic field can do something besides make a television picture. It can induce the electricity at the generating station to power the television set. You will explore this idea, too.

CAUTION: *Do not bring a strong, demonstration magnet near a color TV or monitor.*

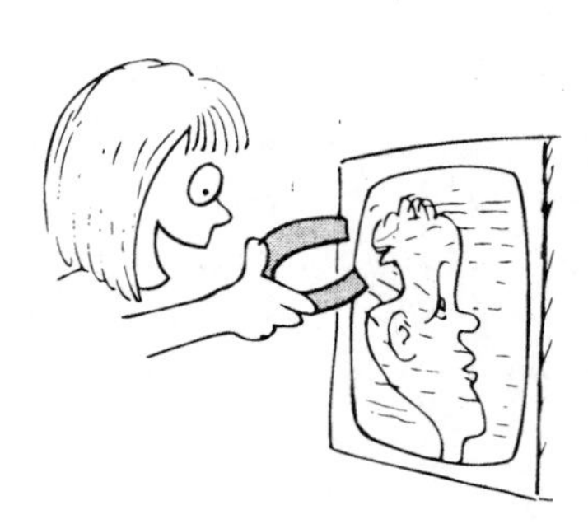

Procedure 

Step 1: If you are using an oscilloscope, adjust it so that only a spot occurs in the middle of the screen. This will occur when there is no horizontal sweep.

If you are using a computer monitor for the cathode ray tube, use a graphics program to create a spot at the center of the screen. Make a white spot on a black background.

Adjust monitor or oscilloscope so dot is centered.

1. The dot on the screen is caused by an electron beam that hits the screen. In what direction are the electrons moving?

The electrons are moving straight toward the screen from the back.

Make sketches of different orientations of the magnet.

Step 2: If the north and south poles of your magnets are not marked, use a compass to determine whether a pole is a north or south pole, and label it with tape.

CAUTION: *Do not use large, demonstration magnets in this activity. Such magnets should not be brought close to any cathode ray tube, as they can cause permanent damage.*

Place the poles of the horseshoe magnet 1 cm from the screen. Try the orientations of the magnet shown in Figure A. Sketch arrows on Figure A to indicate the direction in which the spot moves in each case. Try other orientations of the magnet, and make sketches to show how the spot moves.

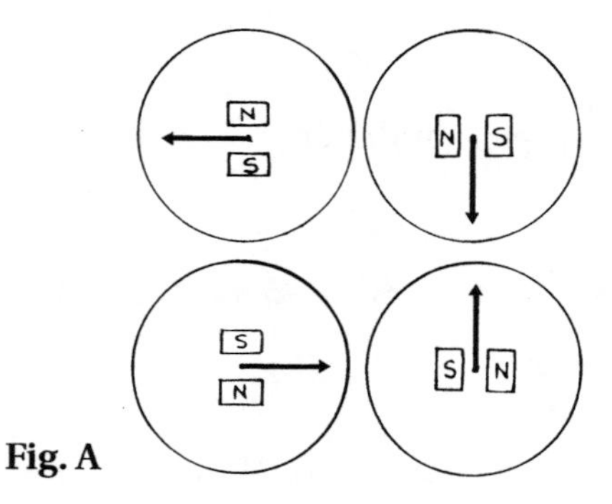

Fig. A

2. Recall that the magnetic field lines outside a magnet run from the north pole to the south pole. The spot moves in the direction of the magnetic force on the beam. How is the direction of the magnetic force on the beam related to the direction of the magnetic field?

When the magnetic field is parallel to the screen, the beam is deflected along the screen perpendicular to the magnetic field. Thus the force on the beam is perpendicular to the magnetic field.

Note effects of bar magnet on spot.

Step 3: Aim one pole of a bar magnet directly toward the spot. Record your observations.

The spot does not move when the magnetic field is parallel to the electron beam.

Name Period Date

Step 4: Change your aim so that the pole of the bar magnet points to one side of the spot. Record your observations.

The spot should move when the magnetic field is at an angle to the electron beam.

3. In general, what are the relative directions of the electron beam, the magnetic field, and the magnetic force on the beam for maximum deflection of the electron beam?

The electron beam, the magnetic field, and the magnetic force on the beam are all perpendicular to the each other.

Explore how moving magnet affects coil of wire.

Step 5: With a long insulated wire, make a three-loop coil with a diameter of approximately 8 cm. Tape the loops together. Connect the ends of the wire to the two terminals of a galvanometer or sensitive milliammeter. Explore the effects of moving a bar magnet into and out of the coil to induce electric current and cause the galvanometer to deflect. Vary the directions, poles, and speeds of the magnet. Also, vary the number of loops in the coil, and try different strengths of magnets.

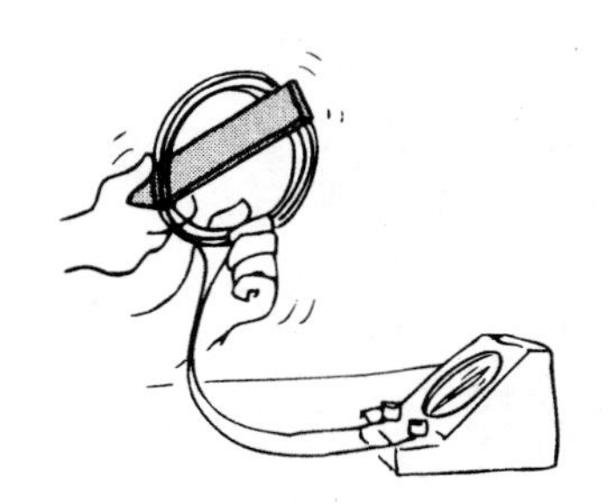

4. Under what conditions can you induce the largest current and get the largest deflection of the galvanometer?

The largest current can be induced by (1) thrusting the magnet through the coil as quickly as possible; (2) using the maximum number of turns of wire on the coil, and (3) using the strongest bar magnet.

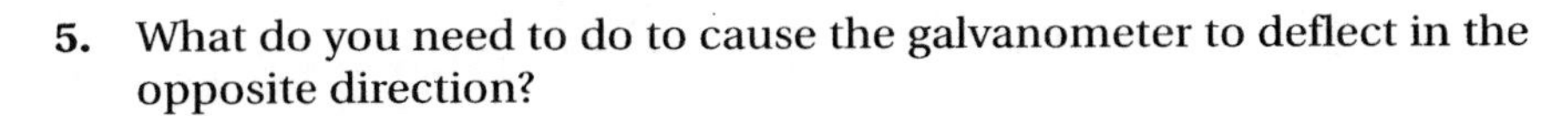

5. What do you need to do to cause the galvanometer to deflect in the opposite direction?

The galvanometer can be deflected in the opposite direction by (1) changing which pole moves through the coil first (while keeping the direction of motion the same); (2) reversing the direction in which the magnet moves (while keeping the orientation of the magnet the same).

Name Period Date

95 Jump Rope Generator

Purpose

To demonstrate the generator effect of a conductor cutting through the earth's magnetic field.

Setup: <1
Lab Time: 1
Learning Cycle: application
Conceptual Level: easy
Mathematical Level: none

Adapted from PRISMS.

Required Equipment/Supplies

50-ft extension cord with ground prong
galvanometer
2 lead wires with alligator clips at one end

Activity

Discussion

When the net magnetic field threading through a loop of wire is changed, a voltage and, hence, a current are induced in the loop. This is what happens when the armature in a generator is rotated, when an iron car drives over a loop of wire embedded in the roadway to activate a traffic light, and when a piece of wire is twirled like a jump rope in the earth's magnetic field.

Since the deflection is small, the more sensitive the galvanometer the better. Digital meters are commonly very sensitive, but it is much more difficult for students to see the sinusoidal variation. The galvanometer must respond rapidly enough so that it doesn't average out the alternating input.

Procedure 

Step 1: Attach an alligator clip to the ground prong of the extension cord (see Figure A). Attach the wire's other end to the galvanometer. Jam the other alligator clip into the ground receptacle on the other end of the extension cord. Attach the other end of this second wire to the other contact on the galvanometer.

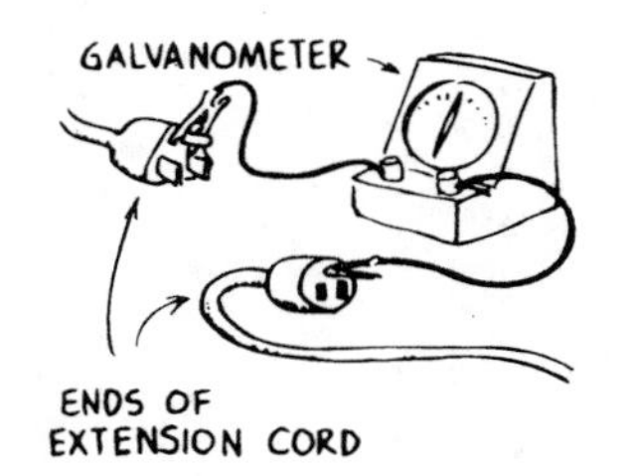

Fig. A

Jump rope with the extension cord.

Step 2: Align the extension cord in the east-west direction. Leaving both ends of the extension cord on the ground, pick up the middle half and twirl it like a jump rope with the help of another person. Observe the galvanometer. Twirl the cord faster, and observe the galvanometer.

1. What effect does the rotational speed of the cord have on the deflection of the galvanometer?

Increasing the rotational speed increases the alternating deflection on the galvanometer.

Change directions.

Step 3: Repeat Step 2, but align the extension cord in the north-south direction. Observe the difference in the deflection on the galvanometer.

2. Is it harder to spin the cord in one direction or the other?

It is slightly harder to spin the extension cord when it is aligned in the east-west direction, but the difference may be imperceptible.

Analysis

3. Describe the conditions in which you had maximum current through the galvanometer.

The maximum current occurs when the rotational speed of the extension cord is at a maximum and the cord is aligned in the east-west direction. (Under these conditions, the greatest number of magnetic field lines are cut per unit time.)

4. Describe the conditions in which you had minimum current through the galvanometer.

The minimum current occurs when the speed is small and the cord is aligned in the north-south direction.

96 Particular Waves

Purpose

To observe the photoelectric effect.

Setup: 1

Lab Time: 1

Learning Cycle: concept development

Conceptual Level: moderate

Mathematical Level: none

Required Equipment/Supplies

electroscope
electrostatic kit, containing strips of white plastic and clear acetate and swatches of wool and silk
zinc plate (approximately 5 cm × 5 cm × 0.1 cm) or magnesium ribbon, 10 cm long
steel wool
lamp with 200-watt bulb
60-watt ultraviolet light source (mercury vapor lamp)

Thanks to Lonnie Grimes for his ideas and suggestions for this lab.

Activity

Discussion

Albert Einstein was best known for his discovery of special and general relativity. Interestingly enough, he was awarded the Nobel Prize for something entirely different—the rules governing the photoelectric effect. In this activity, you will observe the photoelectric effect, which is the ejection of electrons from certain metals (in this case, zinc or magnesium) when exposed to light.

Procedure  

Step 1: Scrub the zinc plate (or magnesium ribbon) with the steel wool to make it shiny; then place the plate on the probe of the electroscope.

Scrub metal until shiny.

Step 2: Rub the white plastic strip with the wool cloth to charge the strip negatively. Touch the strip to the zinc plate on the electroscope. Notice that the leaves separate—evidence that the electroscope is charged.

Observe electroscope.

Step 3: Observe the leaves of the electroscope for one minute.

1. Did the electroscope discharge by itself during that time?

The electroscope is unlikely to discharge by itself in so short a time.

Step 4: Touch the zinc plate on the electroscope with your finger. Note that the electroscope discharges.

2. Why did the electroscope discharge when you touched it?

The excess electrons on the negatively charged electroscope are repelled to the ground via your finger. Electrons can flow onto or off the electroscope only if they have a conductive path to carry them.

Step 5: Rub the clear acetate strip with the silk cloth to charge the strip positively. Touch the strip to the zinc plate on the electroscope. Observe the leaves for one minute. Now touch the zinc plate to see whether the electroscope discharges.

3. Describe what happened to the electroscope from the time you touched the charged acetate strip to the zinc plate.

When the positively charged strip is touched to the zinc plate, the leaves spread apart. They drop when the zinc plate is touched by a finger.

Shine white light from 200-watt bulb on zinc plate.

Step 6: Perhaps electrons can be "blown away" from a negatively charged electroscope by bombarding it with high-intensity light. Charge the electroscope negatively, as in Step 2. Shine white light from a 200-watt lightbulb onto the zinc plate from a distance of 10 cm.

4. How does the electroscope react to the high-intensity white light?

The high-intensity white light has little or no effect on the electroscope. The leaves remain in the "charged" position.

Shine weak ultraviolet light on zinc plate.

Step 7: Recharge the electroscope negatively if its leaves have dropped. Now shine light from a 60-watt ultraviolet light source on the zinc plate.

5. How does the electroscope react to the weak, ultraviolet light?

The electroscope leaves drop, indicating that electrons leave the negatively charged electroscope.

Step 8: Possibly the ultraviolet light is somehow making the air conductive. Charge the strip with a positive charge as in Step 5, and shine the ultraviolet light on the zinc plate again.

6. Does the electroscope discharge?

No. Ultraviolet light does not discharge a positively charged electroscope.

Step 9: In Data Table A, indicate in each box whether or not the electroscope discharged during your tests.

	BRIGHT VISIBLE LIGHT CAUSED THE ELECTROSCOPE TO:	WEAK ULTRAVIOLET LIGHT CAUSED THE ELECTROSCOPE TO:
NEGATIVELY CHARGED ELECTROSCOPE		
POSITIVELY CHARGED ELECTROSCOPE		

Data Table A

Analysis

7. If the *intensity* of the light is responsible for discharging the electroscope, which light source should discharge the electroscope better?

The 200-watt lightbulb should discharge better.

8. If the electroscope could not be discharged by high-intensity white light, but was discharged by weak, ultraviolet light, the *intensity* of the light did not cause it to discharge. How is ultraviolet light different from visible light?

Ultraviolet light has higher frequencies than visible light.

9. Although light is a wave phenomenon, it also behaves like a stream of particles called *photons*. Each photon carries a discrete amount of energy. When a photon is absorbed, its energy is given to whatever absorbs it. According to your data, which type of photons seem to have more energy—those of visible light or those of ultraviolet light?

A photon of ultraviolet light has more energy than a photon of visible light.

10. The energy of a photon is proportional to the frequency of the light. Photons of ultraviolet light possess enough energy to "kick free" electrons trapped in the zinc metal. When the electroscope was positively charged, why did it not discharge when it was exposed to ultraviolet light?

Any electrons "kicked free" by the ultraviolet light only increase the net positive charge of the zinc surface. The freed electrons are drawn back to the positive surface.

11. Which would be better at discharging a negatively charged electroscope, an infrared heat lamp or a dentist's X-ray machine? Why?

The X-ray machine is better because the photons it emits have even more energy than photons from an ultraviolet source. Infrared photons have even less energy than photons of visible light and cannot eject electrons from a metal surface.

97 Nuclear Marbles

This activity may also be performed using the simulation of the same name.

Purpose

To determine the diameter of a marble by indirect measurement.

Setup: <1

Lab Time: 1

Learning Cycle: concept development

Conceptual Level: moderate

Mathematical Level: moderate

Required Equipment/Supplies

7 to 10 marbles
3 metersticks

Adapted from PRISMS.

Discussion

People sometimes have to resort to something besides their sense of sight to determine the shape and size of things, especially for things smaller than the wavelength of light. One way to do this is to fire particles at the object to be investigated, and to study the paths of the particles that are deflected by the object. Physicists do this with particle accelerators. Ernest Rutherford discovered the tiny atomic nucleus in his gold-foil experiment. In this activity, you will try a simpler but similar method with marbles.

You are not allowed to use a ruler or meterstick to measure the marbles directly. Instead, you will roll other marbles at the target "nuclear" marbles and, from the percentage of rolls that lead to collisions, determine their size. This is a little bit like throwing snowballs at a tree trunk while blindfolded. If only a few of your throws result in hits, you can infer that the trunk is small.

First, use a bit of reasoning to arrive at a formula for the diameter of the nuclear marbles (NM). Then, at the end of the experiment, you can measure the marbles directly and compare your results.

When you roll a marble toward a nuclear marble, you have a certain probability of a hit between the rolling marble (RM) and the nuclear marble (NM). One expression of the probability P of a hit is the ratio of the path width required for a hit to the width L of the target area (see Figure A). The path width is equal to two RM radii plus the diameter of the NM, as shown in Figure B. The probability P that a rolling marble will hit a lone nuclear marble in the target area is

$$P = \frac{\text{path width}}{\text{target width}}$$

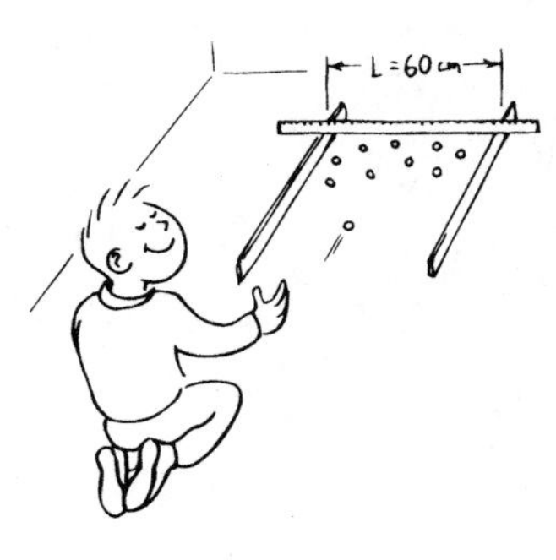

Fig. A

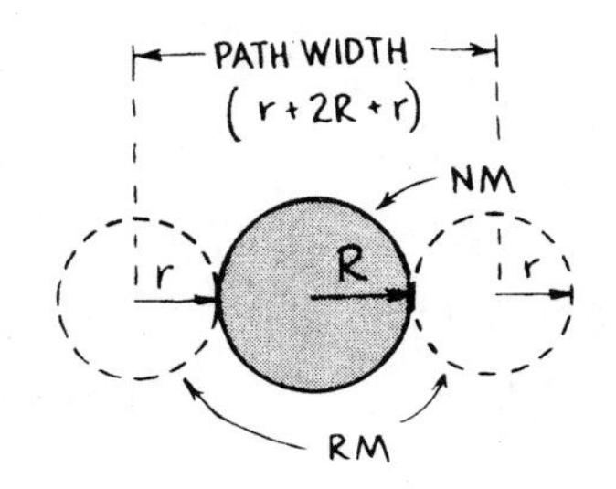

Fig. B

Activity

$$= \frac{2R + 2r}{L} = \frac{2(R + r)}{L}$$

where R = the radius of the NM

r = the radius of the RM

$R + r$ = the distance between the centers of an RM and an NM that are touching

and L = the width of the target area.

If the number of nuclear marbles is increased to N, the probability of a hit is increased by a factor of N (provided N is small enough that the probability of multiple collisions is small). Thus, the probability that the rolling marble will hit one of the N widely spaced nuclear marbles is

$$P = \frac{2N(R + r)}{L}$$

The probability of a hit can also be determined experimentally by the ratio of the number of hits to the number of trials.

$$P = \frac{H}{T}$$

where H = the number of hits

and T = the number of trials.

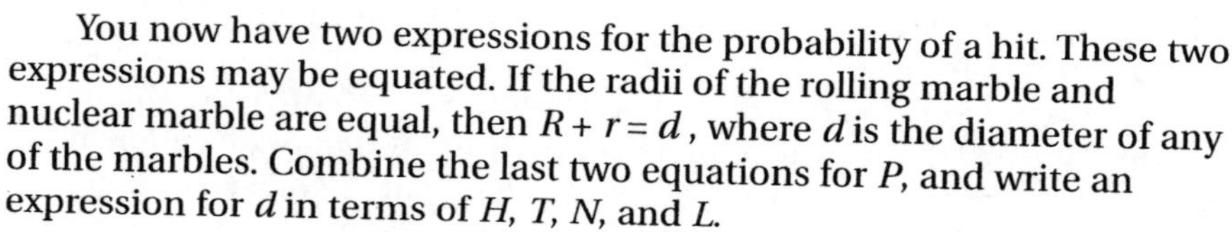

FOR ONE TARGET MARBLE,
P= PATH WIDTH / TARGET WIDTH = 2(R+r) / L
BUT PROBABILITY BY EXPERIMENT EQUALS THE NUMBER OF HITS PER NUMBER OF TRIALS!
P = H / T

You now have two expressions for the probability of a hit. These two expressions may be equated. If the radii of the rolling marble and nuclear marble are equal, then $R + r = d$, where d is the diameter of any of the marbles. Combine the last two equations for P, and write an expression for d in terms of H, T, N, and L.

marble diameter d = $\frac{HL}{2TN}$

This is the formula you are now going to test.

Procedure

Set up nuclear targets.

Step 1: Place 6 to 9 marbles in an area 60 cm wide (L = 60 cm), as in Figure A. Roll additional marbles randomly, one at a time, toward the whole target area from the release point. If a rolling marble hits two nuclear marbles, count just one hit. If a rolling marble goes outside the 60-cm-wide area, do not count that trial. A significant number of trials—more than 200—need to be made before the results become statistically significant. Record your total number of hits H and total number of trials T.

H = ________

T = ________

Name Period Date

Step 2: Use your formula from the Discussion to find the diameter of the marble. Show your work.

computed diameter = __________

Step 3: Measure the diameter of one marble.

measured diameter = __________

Analysis

1. Compare your results for the diameter determined indirectly in the collision experiment and directly by measurement. What is the percentage difference in these two ways of measuring the diameter?

   The collision experiment should yield a diameter that is within 20% of the directly measured diameter. Students should be impressed that they got as close as they did to the measured value for the diameter of the marble just by rolling marbles into a target area.

2. State a conclusion you can draw from this experiment.

   It is possible to infer the size of something indirectly in a collision experiment.

98 Half-Life

Purpose

To develop an understanding of half-life and radioactive decay.

Setup: <1

Lab Time: 1

Learning Cycle: concept development

Conceptual Level: easy

Mathematical Level: easy

Adapted from PRISMS.

Required Equipment/Supplies

shoe box and lid
200 or more pennies
graph paper

Optional Equipment/Supplies

jar of 200 brass fasteners
computer
data plotting software

Discussion

Many things grow at what is called an exponential rate: population, money bearing interest in the bank, and the thickness of paper that is repeatedly folded over onto itself. Many other things decrease exponentially: the value of money stuffed under a mattress, the amount of vacant area in a place where population is growing, and the remaining amount of a material that is undergoing radioactive decay. A useful way to describe the rate of decrease is in terms of *half-life*—the time it takes for the quantity to be reduced to half its initial value. For *exponential* decrease, the half-life stays the same. This means that the time to go from 100% of the quantity to 50% is the same as the time to go from 50% to 25% or from 25% to 12.5%, or from 4% to 2%.

Radioactive materials are characterized by their rates of decay and are rated in terms of their half-lives. You will explore this idea in this activity.

Procedure

Step 1: Place the pennies in a shoe box, and place the lid on the box. Shake the box for several seconds. Open the box and remove all the pennies with the head-side up. Count these and record the number in Data Table A. Do *not* put the removed pennies back in the box.

Remove all pennies with head-side up.

Step 2: Repeat Step 1 over and over until one or no pennies remain. Record the number of pennies removed each time in Data Table A.

Repeat Step 1.

Data Table A

TOTAL PENNIES:					
SHAKE NUMBER	NUMBER OF PENNIES REMOVED	NUMBER OF PENNIES REMAINING	SHAKE NUMBER	NUMBER OF PENNIES REMOVED	NUMBER OF PENNIES REMAINING
1			6		
2			7		
3			8		
4			9		
5			10		

Compute number of pennies remaining each time.

Step 3: Add the numbers of pennies removed to find the total number of pennies. Now, find the number of pennies remaining after each shake by subtracting the number of pennies removed each time from the previous number remaining, and record in Data Table A.

Plot your data.

Step 4: Graph the number of pennies remaining (vertical axis) vs. the number of shakes (horizontal axis). Draw a smooth line that best fits the points.

Analysis

1. What is the meaning of the graph you obtained?

   The graph is a "decay curve" for the pennies. After each shake, the number of pennies remaining is reduced by about the same percentage (in this case 50%—see answer to Question 2).

2. Approximately what percent of the remaining pennies were removed on each shake? Why?

   About 50% of the pennies are removed after each shake, because each penny has a 50% chance of landing with its head side up.

3. Each shake represents a half-life for the pennies. What is meant by a half-life?

   A half-life is the time it takes for the amount of something to be reduced to half its original value, or from any value to half that value.

Going Further

Step 5: Shake a jar of brass paper fasteners and pour them onto the table. Remove the fasteners that are standing on their heads, as you did the pennies with the head side up. Repeat until all the fasteners are gone.

Step 6: Plot your data on the computer, using data plotting software. Try to discover what changes in the value of your vertical or horizontal axis will make your graph come out a straight line. Ask your teacher for assistance if you need help.

Name Period Date

99 Chain Reaction

Purpose

To simulate a simple chain reaction.

Setup: <1
Lab Time: <1
Learning Cycle: concept development
Conceptual Level: easy
Mathematical Level: none

Adapted from PRISMS.

Required Equipment/Supplies

100 dominoes
large table or floor space
stopwatch

Discussion

You could give your cold to two people; they in turn could give it to two others, each of whom in turn could give it to two others. Before you knew it, everyone in school would be sneezing. You would have set off a chain reaction. Similarly, electrons in a photomultiplier tube in an electronic instrument multiply in a chain reaction so that a tiny input produces a huge output. Another example of a chain reaction occurs when one neutron triggers the release of two or more neutrons in a piece of uranium, and each of these neutrons triggers the release of more neutrons (along with the release of nuclear energy). The results of this kind of chain reaction can be devastating.

In this activity, you will explore chain reactions using dominoes.

Activity

Procedure

Step 1: Set up a string of dominoes about half a domino length apart in a straight line, as you did back in Activity 3, "The Domino Effect." Push the first domino and measure how long it takes for the entire string to fall over. Also notice whether the number of dominoes being knocked over per second increases, decreases, or remains about the same as the pulse runs down the row of dominoes.

Set up dominoes in a straight line.

Step 2: Set up the dominoes in an arrangement similar to the one in Figure A. When one domino falls, another one or two will be knocked over. When you finish setting up all your dominoes, push the first domino over, and time how long it takes for all or most of the dominoes to fall over. Also, notice whether the number of dominoes being knocked over per unit time increases, decreases, or remains about the same.

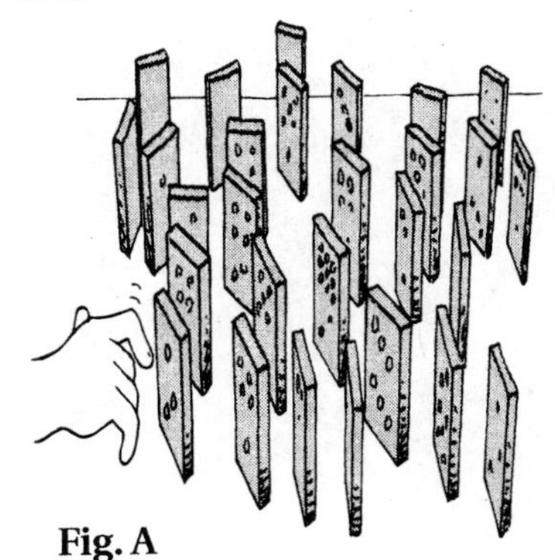

Fig. A

Analysis

1. Which approach results in the shorter time to knock over all the dominoes, the one with a line of dominoes or the one with randomly arranged dominoes?

   The procedure of Step 2 will knock over the dominoes in a shorter time, since the reaction "grows"—involves more dominoes per unit time—as it proceeds.

2. How did the number of dominoes being knocked over per unit time change in each procedure?

   In the procedure of Step 1, the number of dominoes knocked over per unit time remains the same. In the procedure of Step 2, it increases.

3. What made each sequence of falling dominoes?

   Each stops when the potential energy represented by the upright dominoes is exhausted.

4. Imagine that the dominoes represent the neutrons released by uranium atoms when they fission (split apart). Neutrons from the nucleus of each fissioning uranium atom hit other uranium nuclei and cause them to fission. In a big enough piece of uranium, this chain reaction continues to grow if there are no controls. An atomic explosion would then result in a split second. How is the domino reaction in Step 2 similar to the atomic fission process?

   The domino reaction of Step 2 is similar to the fission process in that it is a chain reaction in which the number of atoms (dominoes) involved increases over time. This increase continues until there is no more "fuel" (in the form of domino potential energy) to allow the process to continue.

5. How is the domino reaction in Step 2 dissimilar to the atomic fission process?

   The domino reaction in Step 2 is dissimilar to the fission process in that it is slower, the "multiplication factor" (number of dominoes tipped over by one domino) is less than the neutron multiplication factor, and it has a directional property (dominoes tend to fall forward) that the fission process does not have.

Appendix: Significant Figures and Uncertainty in Measurement

Units of Measurement

All measurements consist of a unit that tells what was measured and a number that tells how many units were measured. Both are necessary. If you say that a friend is going to give you 10, you are telling only *how many*. You also need to tell *what:* 10 fingers, 10 cents, 10 dollars, or 10 corny jokes. If your teacher asks you to measure the length of a piece of wood, saying that the answer is 36 is not correct. She or he needs to know whether the length is 36 centimeters, feet, or meters. All measurements must be expressed using a number and an appropriate unit. Units of measurement are more fully covered in Appendix A of the *Conceptual Physics* text.

Numbers

Two kinds of numbers are used in science—those that are counted or defined and those that are measured. There is a great difference between a counted or defined number and a measured number. The exact value of a counted or defined number can be stated, but the exact value of a measured number cannot be known.

For example, you can count the number of chairs in your classroom, the number of fingers on your hand, or the number of nickels in your pocket with absolute certainty. Counted numbers are not subject to error (unless the number counted is so large that you can't be sure of the count!).

Defined numbers are about exact relations, defined to be true. The exact number of seconds in an hour and the exact number of sides on a square are examples. Defined numbers also are not subject to error.

Every measured number, no matter how carefully measured, has some degree of uncertainty. What is the width of your desk? Is it 98.5 centimeters, 98.52 centimeters, 98.520 centimeters, or 98.5201 centimeters? You cannot state its exact measurement with absolute certainty.

Uncertainty in Measurement

The uncertainty in a measurement depends on the precision of the measuring device and the skill of the person who uses it. There are nearly always some human limitations in a measurement. In addition, uncertainties contributed by limited precision of your measuring instruments cannot be avoided.

Uncertainty in a measurement can be illustrated by the two different metersticks in Figure A. The measurements are of the length of a tabletop. Assuming that the zero end of the meterstick has been carefully and accurately positioned at the left end of the table, how long is the table?

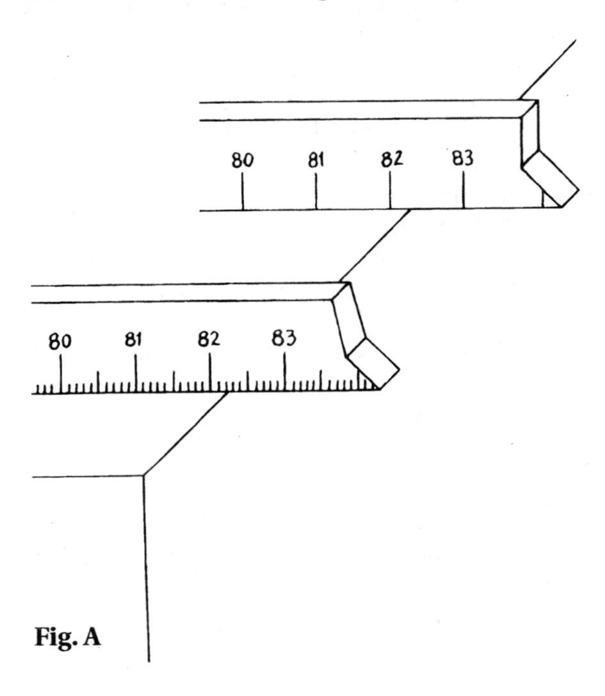

Fig. A

The upper scale in the figure is marked off in centimeter intervals. Using this scale, you can say with certainty that the length is between 82 and 83 centimeters. You can say further that it is closer to 82 centimeters than to 83 centimeters; you can estimate it to be 82.2 centimeters.

The lower scale has more subdivisions and has a greater precision because it is marked off in millimeters. With this meterstick, you can say that the length is definitely between 82.2 and 82.3 centimeters, and you can estimate it to be 82.25 centimeters.

Note how both readings contain some digits that are exactly known, and one digit (the last one) that is estimated. Note also that the uncertainty in the reading of the lower meterstick is less than that of the top meterstick. The lower meterstick can give a reading to the hundredths place, and the top meterstick to the tenths place. The lower meterstick is more precise than the top one.

No measurements are exact. A reported measurement conveys two kinds of information: (1) the magnitude of the measurement and (2) the precision of the measurement. The location of the decimal point and the number value gives the magnitude. The precision is indicated by the number of significant figures recorded.

Significant Figures

Significant figures are the digits in any measurement that are known with certainty plus one digit that is uncertain. The measurement 82.2 centimeters (made with the top meterstick in Figure A) has three significant figures, and the measurement 82.25 centimeters (made with the lower meterstick) has four significant figures. The right-most digit is always an estimated digit. Only one estimated digit is ever recorded as part of a measurement. It would be incorrect to report that, in Figure A, the length of the table as measured with the lower meterstick is 82.253 centimeters. This five-significant-figure value would have two estimated digits (the 5 and 3) and would be incorrect because it indicates a precision greater than the meterstick can provide.

Standard rules have been developed for writing and using significant figures, both in measurements and in values calculated from measurements.

Rule 1: In numbers that do not contain zeros, all the digits are significant.

Examples:

3.1428	five significant figures
3.14	three significant figures
469	three significant figures

Rule 2: All zeros between significant digits are significant.

Examples:

7.053	four significant figures
7053	four significant figures
302	three significant figures

Rule 3: Zeros to the left of the first nonzero digit serve only to fix the position of the decimal point and are not significant.

Examples:

0.0056	two significant figures
0.0789	three significant figures
0.000001	one significant figure

Rule 4: In a number with digits to the right of a decimal point, zeros to the right of the last nonzero digit are significant.

Examples:

43	two significant figures
43.0	three significant figures
43.00	four significant figures
0.00200	three significant figures
0.40050	five significant figures

Rule 5: In a number that has no decimal point, and that ends in one or more zeros (such as 3600), the zeros that end the number may or may not be significant. The number is ambiguous in terms of significant figures. Before the number of significant figures can be specified, further information is needed about how the number was obtained. If it is a measured number, the zeros are probably not significant. If the number is a defined or counted number, all the digits are significant (assuming perfect counting!).

Confusion is avoided when numbers are expressed in scientific notation. All digits are taken to be significant when expressed this way.

Examples:

3.6×10^5	two significant figures
3.60×10^5	three significant figures
3.600×10^5	four significant figures
2×10^{-5}	one significant figure
2.0×10^{-5}	two significant figures
2.00×10^{-5}	three significant figures

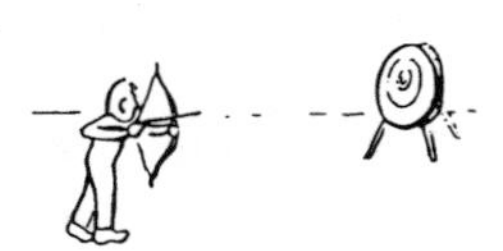

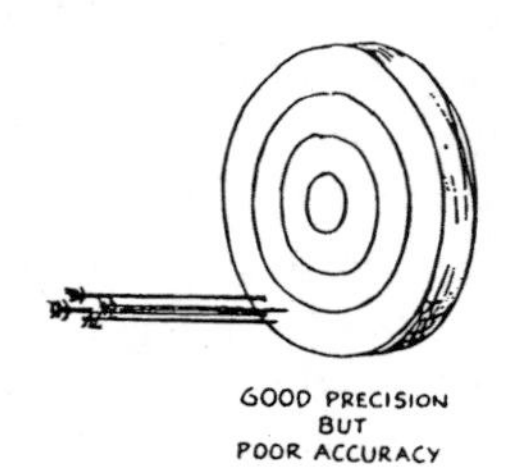

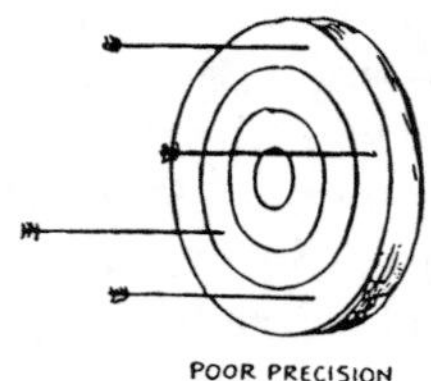

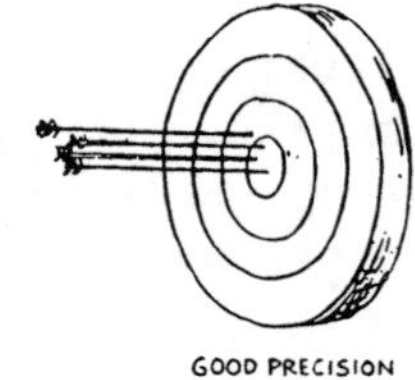

Rounding Off

A calculator displays eight or more digits. How do you round off such a display of digits to, say, three significant figures? Three simple rules govern the process of deleting unwanted (nonsignificant) digits from a calculator number.

Rule 1: If the first digit to be dropped is less than 5, that digit and all the digits that follow it are simply dropped.

Example:

54.234 rounded off to three significant figures becomes 54.2.

Rule 2: If the first digit to be dropped is a digit greater than 5, or if it is a 5 followed by digits other than zero, the excess digits are all dropped and the last retained digit is increased in value by one unit.

Example:

54.36, 54.359, and 54.3598 rounded off to three significant figures all become 54.4.

Rule 3: If the first digit to be dropped is a 5 not followed by any other digit, or if it is a 5 followed only by zeros, an odd-even rule is applied. That is, if the last retained digit is even, its value is not changed, and the 5 and any zeros that follow are dropped. But if the last digit is odd, its value is increased by one. The intention of this odd-even rule is to average the effects of rounding off.

Examples:

54.2500 to three significant figures becomes 54.2.

54.3500 to three significant figures becomes 54.4.

Significant Figures and Calculated Quantities

Suppose that you measure the mass of a small wooden block to be 2 grams on a balance, and you find that its volume is 3 cubic centimeters by poking it beneath the surface of water in a graduated cylinder. The density of the piece of wood is its mass divided by its volume. If you divide 2 by 3 on your calculator, the reading on the display is 0.6666666. It would be incorrect to report that the density of the block of wood is 0.6666666 gram per cubic centimeter. To do so would be claiming a degree of precision that is not warranted. Your answer should be rounded off to a sensible number of significant figures.

The number of significant figures allowable in a calculated result depends on the number of significant figures in the data used to obtain the result, and on the type of mathematical operation(s) used to obtain the result. There are separate rules for multiplication and division, and for addition and subtraction.

Multiplication and Division For multiplication and division, an answer should have the number of significant figures found in the number with the fewest significant figures. For the density example above, the answer would be rounded off to one significant figure, 0.7 gram per cubic centimeter. If the mass were measured to be 2.0 grams, and if the volume were still taken to be 3 cubic centimeters, the answer would still be rounded to one significant figure, 0.7 gram per cubic centimeter. If the mass were measured to be 2.0 grams and the volume to be 3.0 or 3.00 cubic centimeters, the answer would be be rounded off to two significant figures: 0.67 gram per cubic centimeter.

Study the following examples. Assume that the numbers being multiplied or divided are measured numbers.

Example A:

$8.536 \times 0.47 = 4.01192$ (calculator answer)

The input with the fewest significant figures is 0.47, which has two significant figures. Therefore, the calculator answer 4.01192 must be rounded off to 4.0.

Example B:

$3840 \div 285.3 = 13.459516$ (calculator answer)

The input with the fewest significant figures is 3840, which has three significant figures. Therefore, the calculator answer 13.459516 must be rounded off to 13.5.

Example C:

$360.0 \div 3.000 = 12$ (calculator answer)

Both inputs contain four significant figures. Therefore, the correct answer must also contain four significant figures, and the calculator answer 12 must be written as 12.00. In this case, the calculator gave too few significant figures.

Addition and Subtraction For addition or subtraction, the answer should not have digits beyond the last digit position common to all the numbers being added and subtracted. Study the following examples:

Example A:

$$\begin{array}{r} 34.6 \\ 17.8 \\ +\ 15 \\ \hline 67.4 \end{array}$$ (calculator answer)

The last digit position common to all numbers is the units place. Therefore, the calculator answer of 67.4 must be rounded off to the units place to become 67.

Example B:

20.02
20.002
+ 20.0002
60.0222 (calculator answer)

The last digit position common to all numbers is the hundredths place. Therefore, the calculator answer of 60.0222 must be rounded off to the hundredths place, 60.02.

Example C:

345.56 − 245.5 = 100.06 (calculator answer)

The last digit position common to both numbers in this subtraction operation is the tenths place. Therefore, the answer should be rounded off to 100.1.

Percentage Uncertainty

If your aunt told you that she had made $100 in the stock market, you would be more impressed if this gain were on a $100 investment than if it were on a $10 000 investment. In the first case, she would have doubled her investment and made a 100% gain. In the second case, she would have made a 1% gain.

In laboratory measurements, the *percentage* uncertainty is usually more important than the *size* of the uncertainty. Measuring something to within 1 centimeter may be good or poor, depending on the length of the object you are measuring. Measuring the length of a 10-centimeter pencil to ±1 centimeter is quite a bit different from measuring the length of a 100-meter track to the same ±1 centimeter. The measurement of the pencil shows a relative uncertainty of 10%. The track measurement is uncertain by only 1 part in 10 000, or 0.01%.

Percentage Error

Uncertainty and error can easily be confused. *Uncertainty* gives the range within which the actual value is *likely* to lie relative to the measured value. It is used when the actual value is not known for sure, but is only inferred from the measurements. Uncertainty is reflected in the number of significant figures used in reporting a measurement.

The *error* of a measurement is the amount by which the measurement differs from a known, accepted value as determined by skilled observers using high-precision equipment. It is a measure of the accuracy of the method of measurement as well as the skill of the person making the measurement. The *percentage* error, which is usually more important than the actual error, is found by dividing the difference between the measured value and the accepted value of a quantity by the accepted value, and then multiplying this quotient by 100%.

$$\% \text{ error} = \frac{|\text{accepted value} - \text{measured value}|}{\text{accepted value}} \times 100\%$$

For example, suppose that the measured value of the acceleration of gravity is found to be 9.44 m/s^2. The accepted value is 9.80 m/s^2. The difference between these two values is (9.80 m/s^2) − (9.44 m/s^2), or 0.36 m/s^2.

$$\% \text{ error} = \frac{0.36 \text{ m/s}^2}{9.80 \text{ m/s}^2} \times 100\%$$

$$= 3.7\%$$